SCHAUM'S OUTLINE

THEORY AND PROBLEMS

OF

REINFORCED CONCRETE DESIGN

Second Edition

•

BY

NOEL J. EVERARD, MSCE, Ph.D.

Professor of Civil Engineering
The University of Texas at Arlington
and Consulting Engineer

AND

JOHN L. TANNER III, MSCE

Technical Consultant
Texas Industries, Inc.

•

SCHAUM'S OUTLINE SERIES

McGRAW-HILL BOOK COMPANY

New York St. Louis San Francisco Auckland Bogotá Hamburg London
Madrid Mexico Milan Montreal New Delhi Panama Paris
São Paulo Singapore Sydney Tokyo Toronto

Schaum's Outline of Theory and Problems of
REINFORCED CONCRETE DESIGN

1 2 3 4 5 6 7 8 9 10 11 12 13 14 15 16 17 18 19 20 SHP SHP 8 9 8 7

ISBN 0-07-019771-7

Sponsoring Editor, David Beckwith
Production Manager, Nick Monti
Editing Supervisor, Marthe Grice

Library of Congress Cataloging-in-Publication Data

Everard, Noel
   Schaum's outline of theory and problems of
reinforced concrete design.

   (Schaum's outline series)
   Includes index.
1. Reinforced concrete construction—Outlines,
syllabi, etc.   I. Tanner, John.   II. Title.
III. Title: Theory and problems of reinforced concrete
design.
TA683.2.E94   1987   624.1'8341   86-21363
ISBN 0-07-019771-7

Preface

Since the first edition of this book was published, in 1966, many changes have taken place in the state of the art concerning the design of reinforced concrete structures. Two versions of the *Building Code Requirements for Reinforced Concrete* have come and gone. Those versions of the Code were published in 1971 and 1977, with interim changes provided in 1974 and 1980. It is the practice of American Concrete Institute Committee 318, the committee that develops the Code, to provide interim changes three years after a Code version has been officially adopted, and to publish a completely revised Code every six years. This second edition conforms entirely with the 1983 ACI Code, which has been published in two versions: ACI-318-83, written entirely in U.S. customary units, and ACI-318M-83, written entirely in metric (SI) units. The senior author, Noel J. Everard, has been a member of ACI Committee 318 since 1964; he has therefore been aware of proposed Code changes well before they were officially adopted.

Reinforced concrete structures are now being designed almost exclusively by the *strength design method* (formerly called the *ultimate strength design method*). The *alternate design method* (formerly called the *working stress design method*) is now used primarily for establishing the serviceability of structures under the influence of service-level loads. Emphasis is now placed on structural integrity by designing for *strength* and insuring proper performance of the structure at service-level loads by controlling deflections and cracking of the concrete.

The manuscript for this second edition was revised entirely by the senior author, and most of the supplementary problems were solved using either the IBM 4341 mainframe computer located at the University of Texas at Arlington, via his terminal, or using his own IBM-XT-PC, with 640K memory and a ten-megabyte fixed disk, in his office at home. He again wishes to express sincere appreciation to his wife of thirty-five years, Courtney, for her love, understanding and encouragement (to the point of being a catalyst and even a driving force) while this revised manuscript was in preparation.

NOEL J. EVERARD
JOHN L. TANNER III

Contents

CONTENTS

CONTENTS

CONTENTS

Chapter 1

Materials and Components
for Reinforced Concrete Construction

DEFINITIONS

Concrete is a non-homogeneous manufactured stone composed of graded, granular inert materials which are held together by the action of cement and water. The inert materials usually consist of gravel or large particles of crushed stone, and sand or pulverized stone. Manufactured lightweight materials are also used. The inert materials are called *aggregates*. The large particles are called *coarse aggregates* and the small particles are called *fine aggregates*.

Concrete behaves very well when subjected to compressive forces, but ruptures suddenly when small tension forces are applied. Therefore in order to utilize this material effectively, steel reinforcement is placed in the areas subjected to tension.

Reinforced concrete is a composite material which utilizes the concrete in resisting *compression forces*, and some other material, usually steel bars or wires, to resist the *tension forces*. Steel is also often used to assist the concrete in resisting compression forces. Concrete is always assumed to be incapable of resisting tension, even though it actually can resist a small amount of tension.

A number of definitions are presented in the following list taken from the *Building Code Requirements for Reinforced Concrete*, ACI 318-83, of the American Concrete Institute. (This code is usually referred to as the ACI Code.) For conversion factors between U.S. customary units and metric (SI) units, refer to Appendix A-1. **Note**: In this book the word "density" shall be understood as **weight density (unit weight)** in the context of U.S. customary units, and as **mass density** in the context of SI units.

Admixture—A material other than Portland Cement, aggregate or water added to concrete to modify its properties.

Aggregate—Inert material which is mixed with Portland Cement and water to produce concrete.

Aggregate, lightweight—Aggregate having a dry, loose density of $70 \, \text{lb/ft}^3$ ($1100 \, \text{kg/m}^3$) or less.

Building official—City Engineer, Plan Examiner, etc.

Column—An upright compression member the length of which exceeds three times its least lateral dimension.

Combination column—A column in which a structural steel member, designed to carry the principal part of the load, is encased in concrete of such quality and in such a manner that an additional load may be placed thereon.

Composite column—A column in which a steel or cast-iron structural member is completely encased in concrete containing spiral and other longitudinal reinforcement.

Composite concrete flexural construction—A precast concrete member and cast-in-place reinforced concrete so interconnected that the component elements act together as a flexural unit.

Compressive strength of concrete (f'_c)—Specified compressive strength of concrete in pounds per square inch, psi (megapascals, MPa). Compressive strength is determined by tests of standard cylinders made and tested in accordance with ASTM (American Society for Testing and Materials) specifications at 28 days or such earlier age as the concrete is to receive its full service load or maximum stress. The height of a standard cylinder is required to be twice the diameter; 4×8 in. (100×200 mm) and 6×12 in. (150×300 mm) are common dimensions. In many European countries, cubes are used instead of cylinders. In the case of modern, very strong concrete—up to 12,000 psi (83 MPa)—the 45 day strength of the cylinders or cubes may be tested. The assumption is that live load will not be introduced onto the structure before 45 days

1

after the concrete has been emplaced. If materials for outside walls and studs, dry wall materials or door frames are piled on slabs that were cast only a few days before, then these factors must be taken into account in determining strength and serviceability.

Concrete—A mixture of Portland Cement, fine aggregate, coarse aggregate and water.

Concrete, structural lightweight (low-density)—A concrete containing lightweight aggregate weighing not more than 115 lb/ft^3 (1900 kg/m^3).

Deformed bar—A reinforcing bar conforming to "Specifications for Minimum Requirements for the Deformations of Deformed Steel Bars for Concrete Reinforcement" (ASTM A-305) or "Specifications for Special Large Size Deformed Billet-Steel Bars for Concrete Reinforcement" (ASTM A-408). Welded wire fabric with welded intersections not farther apart than 12″ (305 mm) in the direction of the principal reinforcement and with cross wires not more than six gage numbers smaller in size than the principal reinforcement may be considered equivalent to a deformed bar when used in slabs.

Effective area of concrete—The area of a section which lies between the centroid of the tension reinforcement and the compression face of the flexural member.

Effective area of reinforcement—The area obtained by multiplying the right cross-sectional area of the reinforcement by the cosine of the angle between the axis of the reinforcement and the direction for which the effectiveness is to be determined.

Pedestal—An upright compression member whose height does not exceed three times its average least lateral dimension.

Plain bar—Reinforcement that does not conform to the definition for a deformed bar.

Plain concrete—Concrete that does not conform to the definition for reinforced concrete.

Precast concrete—A plain or reinforced concrete element cast in other than its final position in the structure.

Prestressed concrete—Reinforced concrete in which there have been introduced internal compressive stresses of such magnitude and distribution that the tension stresses resulting from service loads are counteracted to a desired degree.

Reinforced concrete—Concrete containing reinforcement, designed on the assumption that the two materials act together in resisting forces.

Reinforcement—Steel bars used in concrete to resist tension forces.

Service dead load—The calculated dead weight supported by a member.

Service live load—The live load (specified by the general building code) for which a member must be designed.

Splitting tensile strength—The results of splitting tests of cylinders.

Stress—Force per unit area.

Surface water—Water carried by an aggregate, excluding that water held by absorption within the aggregate particles themselves.

Yield strength or yield point (f_y)—Specified minimum yield strength or yield point of reinforcement in pounds per square inch (megapascals). Yield strength or yield point shall be determined in tension according to applicable ASTM specifications.

MATERIALS FOR CONCRETE

Cement as used in plain or reinforced concrete has the ability to form a paste when mixed with water. The paste hardens with passage of time, holding all of the larger inert particles together in a common bond. Cement may be obtained from nature (natural cement) or it may be manufactured. When manufactured, cement usually conforms to certain specifications of the ASTM. When the material is so manufactured, it is classified as *Portland Cement*, and concrete made using this material is called *Portland Cement Concrete*, or simply, *concrete*.

In general, Portland Cement is manufactured using definite proportions of various calcareous materials, which are burned to form clinkers. The clinkers are pulverized to a powder-like form, which then becomes cement.

Portland Cement is generally available in a number of different types:

(1) *Normal Portland Cement*—used for general purposes when specific properties are not required.

(2) *Modified Portland Cement*—for use when low heat of hydration is desired, such as in mass concrete, huge piers, heavy abutments and heavy retaining walls, particularly when the weather is hot. (Type 1 may be more desirable in cold weather.)

(3) *High-early-strength Portland Cement*—for use when very high strength is desired at an early age.

(4) *Low-heat-of-hydration Portland Cement*—for use in large masses such as dams. Low heat of hydration is desirable to reduce cracking and shrinkage.

(5) *Sulfate-resistant Portland Cement*—for use when the structure will be exposed to soil or water having a high alkali content.

(6) *Air-entrained Portland Cement*—for use when severe frost action is present, or when salt application is used to remove snow or ice from the structure.

(7) *Shrinkage-compensating Portland Cement* (*expansive cement*)—which expands as the concrete cures, thus compensating for some shrinkage.

Table 1.1 shows the variations in strength of identical mixtures using different types of cement.

Table 1.1   Approximate Relative Strengths of Concrete as Affected by Type of Cement

Type of Portland Cement	Compressive strength—percent of strength of normal Portland Cement Concrete		
	3 days	28 days	3 months
1—Normal	100	100	100
2—Modified	80	85	100
3—High-early-strength	190	130	115
4—Low-heat	50	65	90
5—Sulfate-resistant	65	65	85

It should be noted here that other types of cement are also available. Some cements are made using blast furnace slag, while others consist of a mixture known as *Portland-Pozzolan Cement*. Pozzolan cement is generally a natural cement. These types of cement have rather special areas of application which will not be discussed in detail in this book.

AGGREGATES

Aggregates form the bulk of the concrete components. Fine aggregates consist of sand or other fine grained inert material usually less than $\frac{1}{4}''$ (6.4 mm) maximum size. Coarse aggregates consist of gravel or crushed rock usually larger than $\frac{1}{4}''$ (6.4 mm) size and usually less than $3''$ (76 mm) size.

WATER

Water is an important ingredient in the concrete mixture. The water must be clean and free from salts, alkalis or other minerals which react in an undesirable manner with the cement. Thus sea water is not recommended for use in mixing concrete.

WATER-CEMENT RATIO

The *water-cement ratio* is the most important single factor involved in mixing concrete. Using average materials, the strength of the concrete and all of the other desirable properties of concrete are directly related to the water-cement ratio.

ULTIMATE STRENGTH AND DESIGN OF CONCRETE MIXES

The *ultimate strength* of concrete is generally defined as the compressive strength of a molded concrete cylinder or prism after proper curing for 28 days and is designated as f'_c, measured in psi (MPa). Using average materials, the ultimate strength in compression for various water-cement ratios can be predicted with reasonable accuracy using Table 1.2.

Table 1.2 Compressive Strength of Concrete for Various Water-Cement Ratios*

Water-Cement Ratio†		Probable compressive strength at 28 days, psi (MPa)	
By wt.	Gal/sack‡	Non-air-entrained	Air-entrained
0.35	4.0	6000 (42)	4800 (33)
0.44	5.0	5000 (35)	4000 (28)
0.53	6.0	4000 (28)	3200 (22)
0.62	7.0	3200 (22)	2600 (18)
0.71	8.0	2500 (17)	2000 (14)
0.80	9.0	2000 (14)	1800 (12)

* Strengths are based on 6″ × 12″ (150 × 300 mm) cylinders moist cured under standard conditions for 28 days.

† Including free surface moisture on aggregates.

‡ 1 U.S. gallon = 3.785 liters.

Concrete mixtures can be designed to provide a given strength and other desirable properties by properly proportioning all of the materials. Two general methods are currently in use: (1) The arbitrary proportions method and (2) the trial batch method.

The *arbitrary proportions method* is based on the assumption that average materials will be used and that ordinary weight aggregates (i.e. sand and gravel or crushed stone) will be used. When lightweight structural aggregates are used, or if the design strength exceeds 4000 psi (30 MPa), this method may not be used. If the aggregates are properly graded and proportioned the 28 day compressive strength of concrete may be based on the water-cement ratios shown in Table 1.3.

Table 1.3 Arbitrary Proportions Method for Concrete Strength

Specified compressive strength at 28 days, psi (MPa) f'_c	Maximum permissible water-cement ratio*			
	Non-air-entrained concrete		Air-entrained concrete	
	U.S. gal.† per 94-lb (43-kg) sack of cement	Absolute ratio by weight	U.S. gal.† per 94-lb (43-kg) sack of cement	Absolute ratio by weight
2500 (17)	$7\frac{1}{4}$	0.642	$6\frac{1}{4}$	0.554
3000 (21)	$6\frac{1}{2}$	0.576	$5\frac{1}{4}$	0.465
3500 (24)	$5\frac{3}{4}$	0.510	$4\frac{1}{2}$	0.399
4000 (28)	5	0.443	4	0.354

* Including free surface moisture on aggregates.

† 1 U.S. gallon = 3.785 liters.

The *trial batch method* is a scientific method of proportioning concrete. Various proportions of fine and coarse aggregate are used with different predetermined water-cement ratios in order to provide a mixture having the desired *consistency* and a probable strength about 15 percent higher than the proposed design strength. Test specimens are molded for each mix and are tested at 28 days to establish the value of f'_c for each. The most desirable mix is adopted for use. A number of factors which enter into the trial batch method are discussed in the following paragraphs.

FINENESS MODULUS

The *fineness modulus* indicates the relative fineness of the aggregates. The aggregates are sieved using standard screens and the weight of all of the particles larger than a given size is tabulated. The percent retained on sieves No. 4, 8, 16, 30, 50 and 100 is tabulated, as well as that passing the No. 100 sieve. The sum of the weights retained on all of the sieves larger than a given sieve is accumulated and then added together and divided by 100. The result, always larger than 1.0, is the fineness modulus.

Specifications for sieve analysis are usually provided as part of a general specification for concrete on major projects. One example of the sieve analysis specifications is that of the U.S. Dept. of Interior, which is shown in Table 1.4.

Table 1.4   Ranges of Percentages of Aggregate Sizes for Concrete   Construction

Sieve Size	Percent Retained (Cumulative)
No. 4	0 to 5
No. 8	10 to 20
No. 16	20 to 40
No. 30	40 to 70
No. 50	70 to 88
No. 100	92 to 98

AGGREGATE SIZE AND EFFECTS ON CONCRETE

Aggregate gradation enters into the consideration of the strength and workability of concrete. Table 1.5 illustrates the effects of aggregate gradation on the cement requirement for quality concrete.

Table 1.5   Effects of Aggregate Gradation on Cement Requirement

Grading of coarse aggregate (percent by weight)			Optimum* amount of sand	Cement required at percent of sand indicated, sacks per cu. yd.†	
No. 4–$\frac{3}{8}$ in. (9.5 mm)	$\frac{3}{8}$–$\frac{3}{4}$ in. (9.5–19 mm)	$\frac{3}{4}$–1$\frac{1}{2}$ in. (19–38 mm)	Percent	Optimum	|35 percent
35.0	00.0	65.0	40	5.4	5.7
30.0	17.5	52.5	41	5.4	5.8
25.0	30.0	45.0	41	5.4	6.2
20.0	48.0	32.0	41	5.4	6.0
00.0	40.0	60.0	46	5.4	7.0

* Amount giving best workability with aggregate used. Water content 6.3 gal. per sack of cement.

† 1 cu. yd. = 0.765 m^3. One sack @ 94 lb = 42.6 kg.

The average quantities of fine and coarse aggregate required for a unit volume of concrete are functions of the fineness modulus and the maximum size of the coarse aggregate. Values obtained experimentally are shown in Tables 1.6(a) and 1.6(b) for the coarse and fine aggregates respectively.

The type of construction often dictates the desirable maximum size of aggregate. Recommended values are listed in Table 1.7.

Table 1.6(a)*   Volume of Coarse Aggregate per Unit Volume of Concrete†

Maximum Size of Aggregate, in. (mm)	Volume of dry-rodded coarse aggregate per unit volume of concrete for different fineness moduli of sand					
	2.40	2.60	2.80	3.00	3.20	3.40
$\frac{3}{8}$ (9.5)	0.46	0.44	0.42	0.40	0.38	0.36
$\frac{3}{4}$ (19)	0.65	0.63	0.61	0.59	0.57	0.55
1 (25)	0.70	0.68	0.66	0.64	0.62	0.60
$1\frac{1}{2}$ (38)	0.76	0.74	0.72	0.70	0.68	0.66
2 (51)	0.79	0.77	0.75	0.73	0.71	0.69
3 (76)	0.84	0.82	0.80	0.78	0.76	0.74

* From report of A.C.I. Committee 613: Recommended Practice for Selecting Proportions for Concrete.

† For less workable concrete such as required for concrete pavement construction, increase volumes by approximately 10%.

Table 1.6(b)*   Approximate Percentages of Sand for Different Gradings and Maximum Sizes of Coarse Aggregate

Maximum Size of Coarse Aggregate, in. (mm)	Cement Factor, sacks per cu. yd.,† for Rounded Aggregate				Cement Factor, sacks per cu. yd.,† for Angular Aggregate			
	4	5	6	7	4	5	6	7
	Fine Sand—F.M. 2.3 to 2.4							
$\frac{3}{4}$ (19)	39	37	34	32	46	44	42	39
1 (25)	38	36	33	31	45	43	41	38
$1\frac{1}{2}$ (38)	36	34	31	29	42	40	38	36
2 (51)	35	33	30	28	41	39	37	35
3 (76)	33	31	29	27	39	37	35	34
	Medium Sand—F.M. 2.6 to 2.7							
$\frac{3}{4}$ (19)	42	40	37	34	49	47	45	42
1 (25)	40	38	36	33	47	45	43	41
$1\frac{1}{2}$ (38)	38	36	34	31	45	43	41	39
2 (51)	36	34	33	30	43	41	39	37
3 (76)	34	32	31	29	41	39	37	35
	Coarse Sand—F.M. 3.0 to 3.1							
$\frac{3}{4}$ (19)	47	44	41	38	55	52	49	47
1 (25)	45	42	39	37	53	50	47	45
$1\frac{1}{2}$ (38)	42	40	37	35	50	47	45	43
2 (51)	40	38	35	33	47	45	43	41
3 (76)	37	35	33	31	44	42	40	38

* Table from N.R.M.C.A. Publication No. 52, Calculating the Proportions for Concrete.

† 1 cu. yd. = 0.765 m³. One sack @ 94 lb = 42.6 kg.

Table 1.7 Maximum Sizes of Aggregates Recommended for Various Types of Construction

Minimum Dimension of Section, in. (mm)	Maximum Size Aggregates, in. (mm)			
	Reinforced Walls, Beams and Columns	Unreinforced Walls	Heavily Reinforced Slabs	Lightly Reinforced or Unreinforced Slab
$2\frac{1}{2}$–5 (64–127)	$\frac{1}{2}$–$\frac{3}{4}$ (13–19)	$\frac{3}{4}$ (19)	$\frac{3}{4}$–1 (19–25)	$\frac{3}{4}$–$1\frac{1}{2}$ (19–38)
6–11 (152–279)	$\frac{3}{4}$–$1\frac{1}{2}$ (19–38)	$1\frac{1}{2}$ (38)	$1\frac{1}{2}$ (38)	$1\frac{1}{2}$–3 (38–76)
12–29 (305–737)	$1\frac{1}{2}$–3 (38–76)	3 (76)	$1\frac{1}{2}$–3 (25–76)	3 (76)

CONSISTENCY

The consistency of concrete is important since it is necessary to have the concrete flow freely around corners and between the reinforcing bars. The standard method of determining the relative consistency of concrete is the *slump test*. In this test, a standard slump cone is filled in three layers, rodding each layer 25 times. The concrete is smoothed off at the top of the cone. The cone is then lifted vertically, permitting the concrete to slump downward. The distance between the original and final surfaces of the concrete is called the *slump* and is measured in inches (millimeters). Recommended slumps for various types of construction are listed in Table 1.8.

Table 1.8 Recommended Slumps for Various Types of Construction*

Types of Construction	Slump, in. (mm)†	
	Max.	Min.
Reinforced foundation walls and footings	5 (127)	2 (51)
Plain footings and caissons	4 (102)	1 (25)
Slabs, beams and reinforced walls	6 (152)	3 (76)
Building columns	6 (152)	3 (76)
Pavements	3 (76)	2 (51)
Heavy mass construction	3 (76)	1 (25)

* Adapted from Table 4 of the 1940 Joint Committee Report on Recommended Practice and Standard Specifications for Concrete and Reinforced Concrete.
† When high frequency vibration is used, the values given should be reduced by about one-third.

Experience has shown that the slump of concrete is directly related to the water-cement ratio and the sizes, gradation and quantities of the aggregates. Table 1.9 indicates the approximate mixing water requirements for the variables which affect the slump of concrete.

ENTRAINED AIR AND DISPERSING AGENTS

Air is entrapped naturally in concrete during the mixing process. Experience has shown that if air is entrained artificially using *air-entraining agents* the quantity of mixing water required to produce a given consistency will be reduced. The entrained air reduces friction between the particles and lessens the need for water as a lubricant. *Dispersing agents* break the surface tension in water bubbles and accomplish some of the same effects as air-entraining agents.

Among the many products commercially available are PDA-Protex dispersing agent and Protex air-entraining agent. The Protex Company has provided Tables 1.10(*a*), (*b*), (*c*) and (*d*) in order to

Table 1.9   **Approximate Mixing Water Requirements for Slumps and Maximum Sizes of Aggregates**

Slump, in. (mm)	Water, gals. per cu. yd. (L/m³) of concrete						
	$\frac{3}{8}$ (10)	$\frac{3}{4}$ (19)	1 (25)	$1\frac{1}{2}$ (38)	2 (51)	3 (76)	6 (152)
	Unit water requirement without admixture						
1–2 (25–51)	41(205)	37(185)	35(175)	33(165)	31(155)	29(125)	27(135)
3–4 (76–102)	45(225)	40(200)	38(190)	36(180)	34(170)	32(160)	30(150)
5–6 (127–152)	48(240)	42(210)	40(200)	38(190)	36(180)	34(170)	32(160)
Approximate amt. of entrapped air, %	3	2	1.5	1	0.5	0.3	0.2
	Unit water requirements for air-entrained concrete						
1–2 (25–51)	39(195)	35(175)	33(165)	31(155)	29(145)	27(135)	25(125)
3–4 (76–102)	43(215)	38(190)	36(180)	34(170)	32(160)	30(150)	28(140)
5–6 (127–152)	45(225)	40(200)	38(190)	36(180)	34(170)	32(160)	30(150)
Recommended avg. total air, %	8	6	5	4.5	4	3.5	3

Table 1.10(a)

Required compressive strength—200 psi (14 MPa)
Coefficient of variation—15%
Maximum size aggregate—$1\frac{1}{2}''$ (38 mm)

Percent of Tests above Req. Str.	Required Design Strength, psi (MPa)	W/C G/S	3–4″ (76–102 mm) Slump				5–6″ (127–152 mm) Slump			
			Plain		PDA		Plain		PDA	
			Water G/Y	CF S/Y	Water G/Y	CF S/Y	Water G/Y	CF S/Y	Water G/Y	CF S/Y
70	2180 (15)	8.75	36	4.1	32	3.7	38	4.35	34	3.0
80	2300 (16)	8.5	36	4.25	32	3.8	38	4.5	34	4.0
90	2480 (17)	8.0	36	4.5	32	4.0	38	4.75	34	4.25
99	3100 (21)	7.25	36	5.0	32	4.4	38	5.25	3	4.7
			Maximum size aggregate—$\frac{3}{4}''$ (19 mm)							
70	2180 (15)	8.75	40	4.6	36	4.1	42	4.8	38	4.35
80	2300 (16)	8.5	40	4.75	36	4.25	42	4.9	38	4.5
90	2480 (17)	8.0	40	5.0	36	4.5	42	5.25	38	4.75
99	3100 (21)	7.25	40	5.5	36	5.0	42	5.8	38	5.25

show the results of experiments with air-entrained concrete and non-air-entrained concrete. The tables provide data pertaining to the water-cement ratio (W/C), gallons of water per sack of cement (G/S), gallons per cubic yard (G/Y) and the cement factor (CF) in sacks of cement per cubic yard of concrete. (1 gal/sack = 3.785 L/sack; 1 gal/cu. yd. = 4.95 L/m³; 1 sack/cu. yd. = 1.31 sacks/m³)*.

The tables provide the probable *over-design* strength required to insure against *under-design*. The indicated water-cement ratios and compressive strengths are from "ACI Recommended Practice for Measuring, Mixing, Transporting and Placing Concrete" (AC1-304-73). The unit water contents are derived from local job experience.

---

* Based on 94 lb (43 kg) per sack.

Table 1.10(*b*)

			3–4″ (76–102 mm) Slump				5–6″ (127–152 mm) Slump			
Required compressive strength—2500 psi (17 MPa) Coefficient of variation—15% Maximum size aggregate—1½″ (38 mm)										
Percent of Tests above Req. Str.	Required Design Strength, psi (MPa)	W/C G/S*	Plain		PDA		Plain		PDA	
			Water* G/Y	CF* S/Y	Water G/Y	CF S/Y	Water G/Y	CF S/Y	Water G/Y	CF S/Y
70	2700	7.75	36	4.6	32	4.1	38	4.9	34	4.4
80	2850	7.5	36	4.8	32	4.25	38	5.1	34	4.5
90	3100	7.25	36	5.0	32	4.4	38	5.3	34	4.7
99	3850	6.25	36	5.8	32	5.1	38	6.1	34	5.45
Maximum size aggregate—¾″ (19 mm)										
70	2700	7.75	40	5.15	36	4.6	42	5.4	38	4.9
80	2850	7.5	40	5.35	36	4.8	42	5.6	38	5.1
90	3100	7.25	40	5.5	36	5.0	42	5.8	38	5.3
99	3850	6.25	40	6.4	36	5.8	42	6.7	38	6.1

* 1 gal/sack = 3.785 L/sack; 1 gal/cu. yd. = 4.95 L/m³; 1 sack/cu. yd. = 1.31 sacks/m³. Based on 94 lb (43 kg) per sack.

Table 1.10(*c*)

			3–4″ (76–102 mm) Slump				5–6″ (127–152 mm) Slump			
Required compressive strength—3000 psi (21 MPa) Coefficient of variation—15% Maximum size aggregate—1½″ (38 mm)										
Percent of Tests above Req. Str.	Required Design Strength, psi (MPa)	W/C G/S*	Plain		PDA		Plain		PDA	
			Water* G/Y	CF* S/Y	Water G/Y	CF S/Y	Water G/Y	CF S/Y	Water G/Y	CF S/Y
70	3250 (22)	7.0	36	5.1	32	4.5	38	5.4	34	4.8
80	3450 (24)	6.75	36	5.3	32	4.7	38	5.6	34	5.0
90	3700 (26)	6.25	36	5.8	32	5.1	38	6.1	34	5.4
99	4600 (32)	5.5	36	6.5	32	5.8	38	6.9	34	6.2
Maximum size aggregate—¾″ (19 mm)										
70	3250 (22)	7.0	40	5.7	36	5.1	42	6.0	38	5.4
80	3450 (24)	6.75	40	5.9	36	5.3	42	6.25	38	5.6
90	3700 (26)	6.25	40	6.4	36	5.8	42	6.7	38	6.1
99	4600 (32)	5.5	40	7.3	36	6.5	42	7.6	38	6.9

*1 gal/sack = 3.785 L/sack; 1 gal/cu. yd. = 4.95 L/m³; 1 sack/cu. yd. = 1.31 sacks/m³. Based on 94 lb (43 kg) per sack.

Table 1.10(d)

Required compressive strength—3500 psi (24 MPa) Coefficient of variation—15% Maximum size aggregate—1½″ (38 mm)									
Percent of Tests above Req. Str.	Required Design Strength, psi (MPa)	W/C G/S*	3–4″ (76–102 mm) Slump				5–6″ (127–152 mm) Slump		
			Plain		PDA		Plain		PDA	
			Water* G/Y	CF* S/Y	Water G/Y	CF S/Y	Water G/Y	CF S/Y	Water G/Y	CF S/Y
70	3800 (26)	6.25	36	5.75	32	5.1	38	6.1	34	5.45
80	4000 (28)	6.0	36	6.0	32	5.35	38	6.35	34	5.65
90	4300 (30)	5.75	36	6.25	32	5.6	38	6.6	34	5.9
99	5400 (37)	4.75	36	7.6	32	6.75	38	8.0	34	7.0
Maximum size aggregate—¾″ (19 mm)										
70	3800 (26)	6.25	40	6.4	36	5.75	42	6.7	38	6.1
80	4000 (28)	6.0	40	6.7	36	6.0	42	7.0	38	6.36
90	4300 (30)	5.75	40	7.0	36	6.25	42	7.3	38	6.6
99	5400 (37)	4.75	40	8.4	36	7.6	42	8.85	38	7.9

* 1 gal/sack = 3.785 L/sack; 1 gal/cu. yd. = 4.95 L/m³; 1 sack/cu. yd. = 1.31 sacks/m³. Based on 94 lb (43 kg) per sack.

Table 1.11   Concrete Mixes for Small Jobs

Maximum size of aggregate, in. (mm)	Mix designation	Approximate sacks of cement* per cu. yd. of concrete	Aggregate, lb per 1-sack batch			
			Sand		Gravel or crushed stone	Iron blast furnace slag
			Air-entrained concrete	Concrete without air		
½ (13)	A	7.0	235	245	170	145
	B	6.9	225	235	190	165
	C	6.8	225	235	205	180
¾ (19)	A	6.6	225	235	225	195
	B	6.4	225	235	245	215
	C	6.3	215	225	265	235
1 (25)	A	6.4	225	235	245	210
	B	6.2	215	225	275	240
	C	6.1	205	215	290	255
1½ (38)	A	6.0	225	235	290	245
	B	5.8	215	225	320	275
	C	5.7	205	215	345	300
2 (51)	A	5.7	225	235	330	270
	B	5.6	215	225	360	300
	C	5.4	205	215	380	320

* Based on 94 lb (43 kg) per sack. One sack/cu. yd. = 1.31 sacks/m³.

APPROXIMATE METHODS FOR MIX DESIGN FOR CONCRETE

The scientific methods presented for design of concrete using the trial batch method cannot always be followed because of the lack of trained personnel and equipment. When such is the case, a conservative method is available. The proportions shown in Table 1.11, although not precise, will prove to be satisfactory for small work and for work at locations where the more scientific process cannot be used.

Fig. 1-1 shows the usual range in proportions of materials used in concrete. These ranges can be used as a guide for concrete mix design.

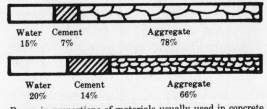

Range in proportions of materials usually used in concrete. Upper bar represents lean mix of stiff consistency with large aggregate. Lower bar represents rich mix of wet consistency with small aggregate.

Fig. 1-1

Mix B is the starting point. If the mix appears to be *under-sanded*, change to mix A. If the mix is *over-sanded*, change to mix C. In all cases the mixes refer to dry materials and there must be adjustments for water in the aggregates.

The water-cement ratio may be obtained from Table 1.3 for a given strength of concrete.

WATER IN THE AGGREGATES

The water-cement ratio must be carefully preserved if the results are to be reproducible from one batch to another. For this reason, *free water* in the aggregates must be considered as part of the mixing water. Since the free water varies from time to time in a given stock pile of aggregates, it is necessary to determine the free water content several times each day, with the added water adjusted accordingly. A definite procedure is available for determining the free water in the aggregates.

A representative sample of aggregates is weighed. The surfaces of the particles are then dried to a *saturated-surface dry state* in an oven or pan, or by pouring alcohol on the aggregates and setting afire. The dried aggregates are then weighed. The percentage moisture (by weight or density of the aggregate) is obtained using the equation

$$p = 100(w_w - w_D)/w_D \qquad (1.1)$$

in which p = percentage of moisture by dry weight or density, w_w = wet weight or density of the material, and w_D = dry weight or density of the material. The percentage of *surface water* is deducted from the total water required in order to obtain the desired water-cement ratio.

QUANTITY OF CONCRETE OBTAINED

The quantity of concrete or *yield* of a trial batch is used to predict the quantity of concrete to be obtained from the job mix. The yield is predicted using the *absolute volume method* to obtain the absolute volume of each of the component materials (i.e. gravel, sand, cement and water) using the equation

$$V_a = (\text{weight of loose material})/(\text{SG})(w_u) \qquad (1.2)$$

in which V_a = absolute volume of material, ft^3 (m^3); SG = specific gravity of the material; and w_u = density of water (62.4 lb/ft^3 or 1000 kg/m^3).

MAKING, CURING AND TESTING SPECIMENS

Making Cylinders

Test cylinders are made using the trial batch method in order to obtain the true strength of the manufactured material. The cylinders are also used during construction to insure that the strength of the concrete is maintained at the desired level.

The fresh concrete is placed in a cylindrical mold in 3 layers. Each layer is rodded 25 times. The concrete is troweled smooth at the top surface and the cylinder allowed to air cure for 24 hours, after which it is placed in a *damp room* under controlled humidity and temperature to age for 28 days. After the curing period has elapsed, the cylinders are tested in a testing machine to determine the compressive strength f'_c, the modulus of elasticity E_c, and often the complete stress-strain diagram.

Dimensions of test cylinders (and cubes) are discussed under *Compressive strength of concrete*, page 1. It should be mentioned here that the Canadian Standards recognize a method that predicts the 28 day strength after the concrete is only one day in place. The cylinders are *boiled in water* and then tested for compressive strength. Empirical equations predict the 28 day strength within about 10 percent, plus or minus.

Curing Cylinders

Proper curing of specimens is vitally important to the strength of the concrete. *Moist curing* is the most desirable method, as shown in Fig. 1-2.

Temperature of curing also affects concrete strength, as seen from Fig. 1-3, since strength and other desirable properties of concrete improve more rapidly at normal temperature than at low temperatures.

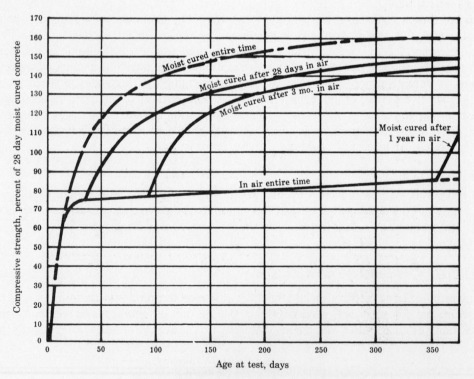

Fig. 1-2 Effects of Moist Curing on the Strength of Concrete

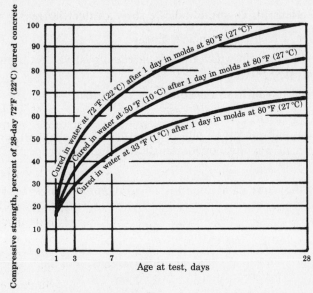

Fig. 1-3 Effects of Temperature on the Strength of Concrete

STRUCTURAL DESIGN CONSIDERATIONS

The *compression stress-strain diagram* provides the most important single factor for use in deriving equations for designing structural elements of reinforced concrete. The stress-strain diagrams are plotted using data obtained from the 28 day tests of concrete cylinders. During the loading process, loads in pounds (kilograms) and the corresponding strains in inches per inch (mm/mm) are recorded. The loads are transformed into direct stresses (P/A) and the stress-strain diagram is then plotted.

Fig. 1-4 shows a series of typical stress-strain diagrams obtained using different strengths of concrete.

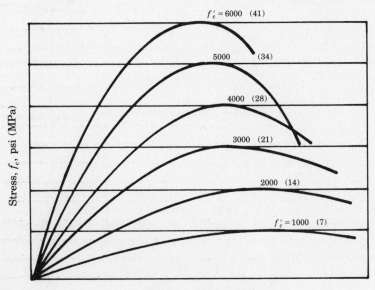

Fig. 1-4 Typical Stress-Strain Diagrams for Concrete

The stress-strain diagrams for concrete indicate three distinct ranges:

(1) The initial range, which is very nearly linear.

(2) The intermediate range, in which there is increasing curvature, ultimately reaching a point of maximum stress, f'_c.

(3) The final range, in which strain continues to increase while the load-carrying capacity decreases.

Fig. 1-5 presents the usually accepted plot of the stress-strain diagram for concrete subjected to axial load and flexure which was developed statistically. Certain properties of the curve must be described:

(1) The tangent to the curve at its origin is called the *initial tangent modulus of elasticity*, E_{ci}, psi (MPa).

(2) A line drawn from the origin to a point on the curve at which $f_c = 0.45f'_c$ is called the *secant modulus of elasticity*, E_{cs}, psi (MPa).

(3) For low strength concrete E_{ci} and E_{cs} differ widely. For high strength concrete there is practically no difference between the two values.

(4) For lightweight aggregate concrete the initial slope is somewhat less than that for normal weight concrete. The maximum stress occurs for larger strain values for lightweight concrete when f'_c is the same for both types of concrete.

(5) Definitions: ϵ_c = unit strain in the concrete for any stress, f_c

                            ϵ_o = strain corresponding to maximum stress, f'_c

                            ϵ_u = ultimate strain at rupture

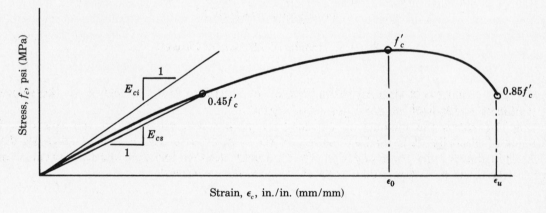

Fig. 1-5 Stress-Strain Diagram

MODULUS OF ELASTICITY OF CONCRETE

In general practice the *secant modulus of elasticity* is usually used and is simply referred to as E_c. Numerous experimental equations have been proposed for obtaining the value of E_c. One of the most recent empirical equations was presented by Pauw and was adopted by the ACI Code Committee. This equation represents a modification of Pauw's equation which was obtained from statistical correlation of test data.

The modified equation adopted by the ACI Code Committee is

$$E_c = 33w_c^{1.5}\sqrt{f'_c} \qquad (0.043w_c^{1.5}\sqrt{f'_c}) \qquad\qquad (1.3)$$

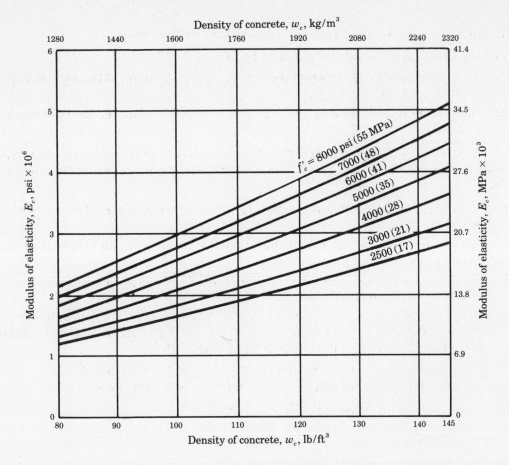

Fig. 1-6 Modulus of Elasticity of Concrete

in which E_c = modulus of elasticity, lb/in.^{2} (MPa); w_c = density of the concrete, lb/ft^3 (kg/m^3); and f'_c = ultimate strength of the concrete, lb/in.^{2} (MPa).

Fig. 1-6 shows the modulus of elasticity for various concrete strengths and densities. The modulus of elasticity E_c is used extensively in the *working stress design method* (*alternate design method*) in almost every phase of design. The modulus is used for deflection and stability calculations in both the alternate design method and the *strength design method*.

STRUCTURAL DESIGN AND THE STRESS-STRAIN DIAGRAM

In order to derive equations for the design of reinforced concrete structural elements it is necessary to describe the stress-strain function mathematically. After establishing an equation for the stresses in the concrete, it is only necessary to integrate the equation to obtain the magnitude and point of application of the total compression force.

Fig. 1-7 illustrates the various stages of development of the stress-strain function as the loads are increased. Selection of a design method depends on which stage is to be used as a basis of proportioning the structural elements.

Fig. 1-7(*a*) is used in the working stress design method, limiting f_c to a maximum value $0.45f'_c$. The diagram is considered to be linear, and the relationship between the elastic moduli of steel and concrete is used to transform the steel into equivalent concrete. The *modular ratio* is $n = E_s/E_c$.

Fig. 1-7(*d*) is used in the strength design method, using $0.85f'_c$ as the maximum stress. This is based on experimental data, using statistical methods for correlation.

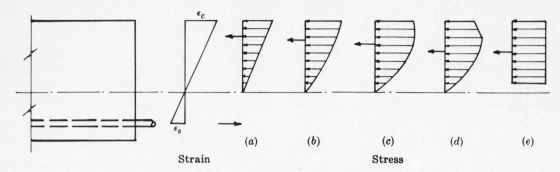

Fig. 1-7

In general practice an *equivalent rectangular stress-block* (*e*) is used to approximate the effects of the true ultimate stress-strain diagram. The limits of the block are defined so as to provide a total force in compression identical to that developed by the true diagram and to locate the force at its true point of application.

The 1983 ACI Code permits the use of any mathematical expression for describing the stress-strain diagram, providing the resulting equations will be in general agreement with comprehensive test results. The rectangular stress block fulfills these requirements.

TENSILE STRENGTH OF CONCRETE

The tensile strength of plain concrete is rather small compared to the compressive strength, and tension in the concrete is always neglected in design practice. It is important, however, to give some consideration to the tensile strength with regard to combined stresses which cause *diagonal tension failure*.

Beam tests have been utilized for obtaining the tensile strength of concrete in past years. A method has been developed for obtaining a measure of the tensile strength, called the *split-cylinder test*. Standard concrete cylinders are loaded along the sides until the cylinder splits. The stress at which splitting occurs is designated as the *split cylinder strength*, f_{sp}. This value is used to determine a design factor

$$F_{sp} = f_{sp}/\sqrt{f'_c} \qquad (1.4)$$

This factor is used in connection with *shear stresses* in concrete design to guard against diagonal tension failure.

REINFORCING STEEL

Steel reinforcing for concrete consists of bars, wires and welded wire fabric, all of which are manufactured in accord with ASTM specifications. The most important properties of reinforcing steel are:

(1) modulus of elasticity, E_s, psi (MPa)

(2) tensile strength, psi (MPa)

(3) yield point stress, f_y, psi (MPa)

(4) steel grade designation (yield strength)

(5) size or diameter of the bar or wire.

Fig. 1-8 illustrates properties (1), (2) and (3). Tables 1.12 and 1.13 provide information pertaining to properties (4) and (5).

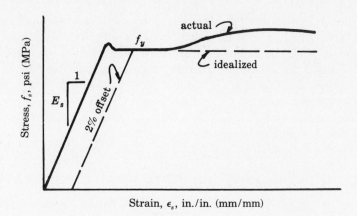

Fig. 1-8   Stress-Strain Diagram for Reinforcing Steel

The stress-strain diagram is *idealized* by assuming that stress is constant in the *plastic region* and equal to f_y.

The yield point is established by drawing a line parallel to the elastic region at some offset therefrom, usually about 0.002 in./in. (mm/mm).

The idealized stress-strain diagram consists of an elastic portion and a purely plastic portion, i.e. a sloping line and a horizontal line.

The modulus of elasticity of reinforcing bars is considered to be 29,000,000 psi (200 000 MPa).

Reinforcing bars are manufactured as *plain or deformed bars*. Deformed bars have ribbed projections which grip the concrete in order to provide better *bond* between the two materials. In the United States, main bars are always deformed. Plain bars are used for spirals and ties in columns. Practices differ in other countries, depending on the availability of deformed bars.

Formerly, bars were manufactured in both round and square shapes. In modern practice, however, square bars are not used. In place of the square bars, *equivalent round bars* are manufactured.

Bars are no longer designated by diameter or side dimension as in the past. Bar numbers are used in modern practice. For bars up to No. 8, relative to U.S. customary units, the number coincides with the number of eighths of an inch in the bar diameter; for larger bars the numbers are used merely for designation purposes. For metric (SI) reinforcing bars, there is no correspondence between size number and diameter.

Table 1.12

BAR SIZE DESIGNATION	WEIGHT POUNDS PER FOOT	NOMINAL DIMENSIONS – ROUND SECTIONS		
		DIAMETER INCHES	CROSS SECTIONAL AREA - SQ. INCHES	PERIMETER INCHES
#3	.376	.375	.11	1.178
#4	.668	.500	.20	1.571
#5	1.043	.625	.31	1.963
#6	1.502	.750	.44	2.356
#7	2.044	.875	.60	2.749
#8	2.670	1.000	.79	3.142
#9	3.400	1.128	1.00	3.544
#10	4.303	1.270	1.27	3.990
#11	5.313	1.410	1.56	4.430
#14	7.65	1.693	2.25	5.32
#18	13.60	2.257	4.00	7.09

ASTM STANDARD REINFORCING BARS

Table 1.13

METRIC REINFORCING BARS

BAR DESIG-NATION	MASS WEIGHT kg/m	NOMINAL DIMENSIONS		
		DIAMETER mm	CROSS SECTIONAL AREA mm²	PERIMETER mm
10M	0.785	11.3	100	35.5
15M	1.570	16.0	200	50.1
20M	2.355	19.5	300	61.3
25M	3.925	25.2	500	79.2
30M	5.495	29.9	700	93.9
35M	7.850	35.7	1000	112.2
45M	11.775	43.7	1500	137.3
55M	19.625	56.4	2500	177.2

The nominal dimensions of a deformed bar are equivalent to those of a plain round bar having the same mass per metre as the deformed bar.

Tables 1.12 and 1.13 indicate the sizes, numbers and various properties of currently used reinforcing bars; Table 1.13 has been provided by the Reinforcing Steel Institute of Canada.

The bars listed in Tables 1.12 and 1.13 are used primarily for main reinforcement in beams, slabs, columns, footings, walls and other structural elements. Bar sizes Nos. 3, 4 and 5 (10M, 15M) are often used for *spirals and lateral ties* in columns and as *stirrups* in beams. Spiral bars are usually plain, without deformations.

Reinforcing steel rods having diameters less than $\frac{1}{4}''$ (6 mm) are referred to as *wires*, and the sizes are designated by the AW & S gage number.

BENDING OF REINFORCING BARS FOR STRUCTURAL CONSIDERATIONS

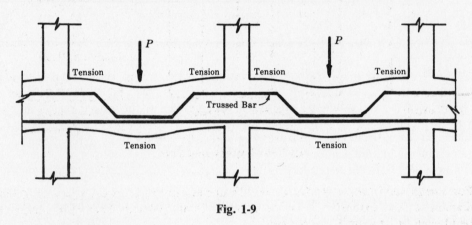

Fig. 1-9

Reinforcement is usually placed in the *tension zone* since the concrete is assumed to be incapable of resisting tension. Bars are, however, also often used in the compression zone. It is economical to use *bent-up bars* in such cases. Further, in order to anchor the bars properly to resist *bond stresses*, hooks are often provided at the ends of bars as shown in Fig. 1-10. Bars must be embedded properly in order to provide *anchorage*. Bending or *trussing* of bars is often used to accomplish this purpose.

Ties or spirals are always used in columns to hold the main vertical bars in place and to confine longitudinal bars. *Stirrups* are often used in beams to hold the longitudinal steel in place, but the most important function of stirrups is to resist *diagonal tension* stresses. The concrete acting alone cannot always resist all of the shear stress which is present, thus the stirrups are used.

Fig. 1-10 shows typical examples of bar bending for various purposes.

Numerous shapes of bent bars are used. Some are used so frequently that pattern numbers have been standardized. Bent bars are therefore often called for by pattern number, indicating the various center to center dimensions in a tabular form.

Radii of bend have been standardized and are specified in the ACI Code. The bend radii and extensions are usually specified in terms of the diameter of the size bar being bent, with minimum extensions beyond the bend being specified.

Bars should be *bent cold* in a fabricating plant. Bending bars on the job is usually not permitted except in special cases. Under no circumstances should a bar be heated since this introduces residual stresses into the bar. Torches should never be used to cut bars to the desired length. Bending of bars which have been partially embedded in concrete is prohibited except when the engineer states that such bars may be bent.

Standard hooks consist of one of the following:

(a) A 180 degree turn plus an extension of at least *4 bar diameters*. The extension must be at least $2\frac{1}{2}$ in. (64 mm) long.

(b) A 90 degree turn plus an extension of at least *12 bar diameters*.

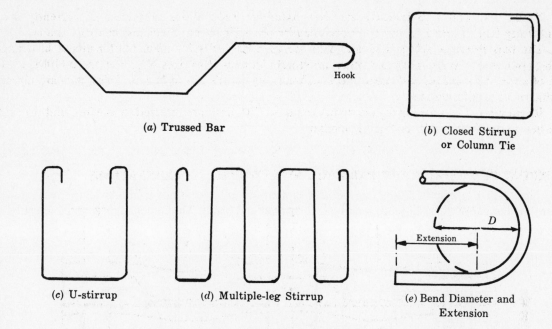

(a) Trussed Bar

(b) Closed Stirrup or Column Tie

(c) U-stirrup　　　　　(d) Multiple-leg Stirrup　　　　　(e) Bend Diameter and Extension

Fig. 1-10　Typical Bent Bars

(c)　For *stirrups and ties*, a 90 degree or 135 degree turn plus an extension of at least 6 bar diameters. The extension must be at least $2\frac{1}{2}$ in. (64 mm) long. The inside radius of bend shall be at least 1 bar diameter.

For *structural and intermediate grade steel*, bars in the size range No. 6 through No. 11 must have a *radius of bend* which is not less than $2\frac{1}{2}$ bar diameters. For steel other than structural and intermediate grades, and for bars other than stirrups or ties, the minimum bend diameter must conform to Table 1.14 or 1.15.

Table 1.14

Bar Size	Minimum Bend Diameter, bar diameters
#3–#8	6
#9, #10, #11	8
#14, #18	10

Table 1.15

Bar Size	Minimum Bend Diameter, bar diameters
10M–25M	6
30M, 35M	8
45M, 55M	10

The diameter of bend is closely controlled in order to insure that high *residual stresses* will not develop in the bar as a result of the bending operations. This is the primary reason for discouraging bending of bars at the job site and for requiring that bars should be bent *cold*.

Reinforcement should be bent and placed as accurately as possible. For flexural members, walls and columns, the variations permitted in placement of bars (as compared to specified dimensions) are given in Table 1.16.

Longitudinal locations of bends or ends of bars may be displaced ±2 in. (±50 mm) from the specified locations. Proper minimum clearances and cover must be maintained, however. Minimum cover is specified for fireproofing purposes and to protect the steel from corrosion due to atmospheric conditions. The appropriate minimum cover is stated in the paragraphs which follow.

Table 1.16

Effective depth d, in. (mm)	Tolerance on d, in. (mm)	Tolerance on minimum cover on reinforcements, in. (mm)
≤ 8 (200)	$\frac{3}{8}$ (10)	$-\frac{3}{8}$ (−10)
>8 (200)	$\frac{1}{2}$ (12)	$-\frac{1}{2}$ (−12)

BAR SPACING AND CONCRETE COVER FOR STEEL

Figs. 1-11 through 1-15 illustrate the key dimensions related to cover and spacing for reinforcement for the various cast-in-place members illustrated and listed. The dimensions X and Z must be at least equal to those stated.

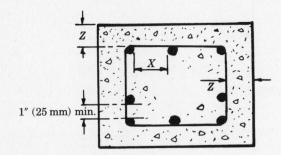

Fig. 1-11   Bar Spacing and Cover for Beams or Girders

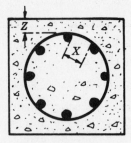

Fig. 1-12   Bar Spacing and Cover for Columns

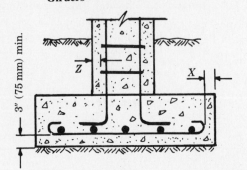

Fig. 1-13   Bar Spacing and Cover for Footings

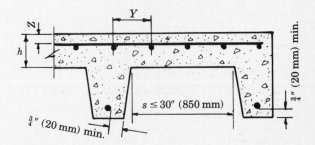

Fig. 1-14   Bar Spacing and Cover for Slabs and Joints

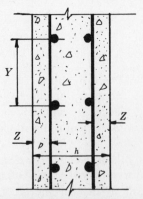

Fig. 1-15   Walls (Including Retaining Walls)

For Beams or Girders

$X \geq$ diameter of main bars

$X \geq 1\frac{1}{3} \times$ maximum aggregate size

$X \geq 1''$ (25 mm)

$Z \geq$ diameter of main bars, always

When not exposed to weather or earth,

$Z \geq 1\frac{1}{2}''$ (40 mm)

When exposed to weather or earth,

$Z \geq 1\frac{1}{2}''$ (40 mm), bars No. 5 (15M) or smaller

$Z \geq 2''$ (50 mm), bars larger than No. 5 (15M)

Columns with Spirals or Ties

$X \geq 1\frac{1}{2}$ diameters of vertical bars

$X \geq 1\frac{1}{3} \times$ maximum aggregate size

$X \geq 1\frac{1}{2}''$ (40 mm)

$Z \geq$ diameter of vertical bars, always

When not exposed to weather or earth,

$Z \geq 1\frac{1}{2}''$ (40 mm)

$Z \geq 1\frac{1}{3} \times$ maximum aggregate size

When exposed to weather or earth,

$Z \geq 2''$ (50 mm), unless the local building code requires greater clearance

Footings

$X \geq 2''$ (50 mm) when side forms are used

$X \geq 3''$ (75 mm) without side forms

When exposed to weather or earth,

$Z \geq 2''$ (50 mm) (formed surfaces)

Z or X is always greater than the bar diameter and exceeds $1\frac{1}{3}$ times the maximum aggregate size

Slabs, Walls and Joists

$Y \leq 3h$

$Y \leq 18''$ (460 mm)

$Y \geq$ bar diameter

$Y \geq 1\frac{1}{3} \times$ maximum aggregate size

$Y \geq 1''$ (25 mm)

When exposed to weather,

$Z \geq 1\frac{1}{2}''$ (40 mm), bars No. 5 (15M) or smaller

$Z \geq 2''$ (50 mm), for bars larger than No. 5 (15M)

When not exposed to weather,

$Z \geq \frac{3}{4}''$ (20 mm), bars No. 11 (35M) or smaller

$Z \geq 1\frac{1}{2}''$ (40 mm), bars No. 14 and No. 18 (45M and 55M)

When $S > 30''$ (750 mm), consider a joist as a beam.

Note. These provisions also apply to solid slabs without joists, but do not apply to *flat slab* construction.

Walls

$Y \leq 5h$ (Temperature Reinforcement)

$Y \leq 3h$ (Main Reinforcement)

$Y \leq 18''$ (460 mm)

$Y \geq$ bar diameter

$Y \geq 1\frac{1}{3} \times$ maximum aggregate size

$Y \geq 1''$ (25 mm)

When not exposed to weather or earth,

$Z \geq \frac{3}{4}''$ (20 mm)

When exposed to weather or earth,

$Z \geq 1\frac{1}{2}''$ (40 mm), bars No. 5 (15M) or smaller

$Z \geq 2''$ (50 mm), bars larger than No. 5 (15M)

IMPORTANT COMPONENTS OF STRUCTURES

In order to fully understand the Building Code provisions and to design structural elements, it is necessary to have a thorough understanding of the types of structures used and their component parts.

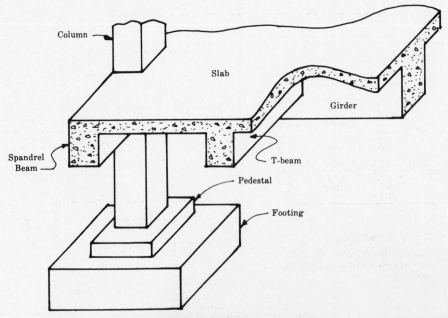

Fig. 1-16 Beam and Slab Construction

Figs. 1-16 through 1-18 illustrate the various types of elements usually encountered in building design and construction.

Notes for Fig. 1-16:

(1) The slab shown delivers load primarily to the T-beams and is called a *one-way slab*.

(2) The girder receives loads primarily from the T-beams as concentrated loads, then delivers the loads to the columns.

(3) Pedestals are used to spread the load over a large area of the footing. The pedestal receives the column loads.

(4) Spread footings bear directly on the earth.

(5) The girder is also a T-beam (in shape, even if not considered so in structural action).

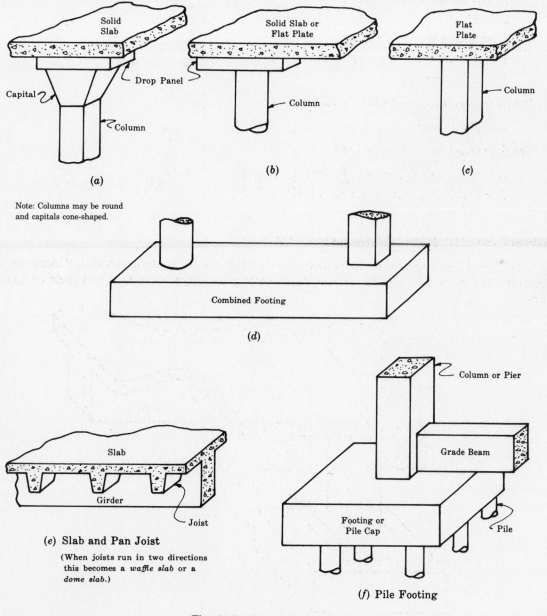

Fig. 1-17 Structural Systems

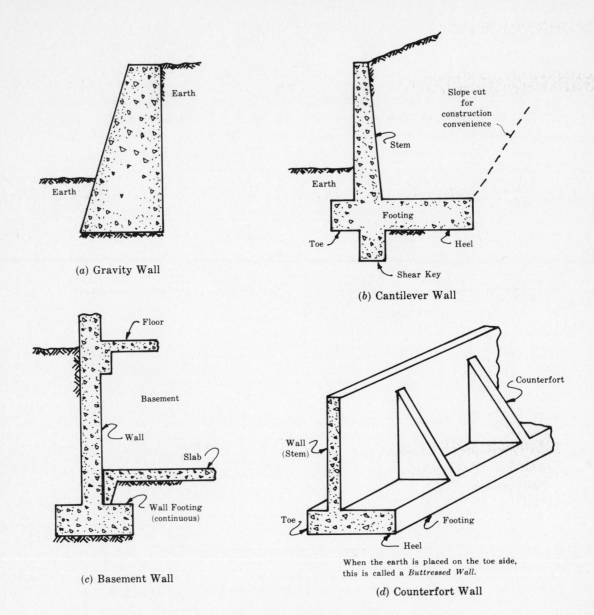

(a) Gravity Wall

(b) Cantilever Wall

(c) Basement Wall

When the earth is placed on the toe side, this is called a *Buttressed Wall.*

(d) Counterfort Wall

Fig. 1-18   Retaining Walls

Because of their shape, conditions of connection and purpose in the structure, the various elements act together in receiving and distributing the loads and eventually delivering those loads to the foundations.

Types of loads imposed on structures and the method of distribution of those loads are discussed in Chapters 2 and 3.

Chapter 2

Gravity Loads

FORCES, SHEAR, MOMENTS AND REACTIONS

NOTATION

b = width of a beam or column, in. (mm)

C = moment coefficient of $w(L')^2$

h = total depth of a member, in. (mm)

I = section moment of inertia, in.^{4} (mm^4)

K = stiffness factor, in.^{3} (mm^3)

L = any span length, usually center-to-center of supports, ft (m)

L = long span for two-way slabs, ft (m)

L' = clear span length (average of 2 adjacent spans for negative M), ft (m)

m = ratio of short span to long span, two-way slabs

M = bending moment, ft-kips or ft-pounds (kN·m)

P = any concentrated load, kips or pounds (kN)

P_{DL} = dead load concentrated load, kips or pounds (kN)

P_{LL} = live load concentrated load, kips or pounds (kN)

P_{TL} = total load concentrated load, kips or pounds (kN)

q = any uniformly distributed load, kips/ft (kN/m)

R = any reaction, kips or pounds (kN)

S = short span length, two-way slabs, ft (m)

V = shear force, kips or pounds (kN)

w = any uniformly distributed load, kips/ft or kips/ft^2 (kN/m or kN/m^2)

w_{DL} = uniformly distributed dead load

w_{LL} = uniformly distributed live load

w_{TL} = uniformly distributed total load

STRUCTURAL ANALYSIS AND DESIGN

The analysis of structures deals with the determination of loads, reactions, shear and bending moments. Structural design deals with the proportioning of members to resist the applied forces. The sequence involved in creating a structure, then, involves analysis first and then design.

The 1983 ACI Code requires that analysis be made using the *elastic theory*, whereas structural design may be accomplished using either the *alternate design method* or the *strength design method*.

EXACT AND APPROXIMATE METHODS OF ANALYSIS

There exist methods which provide for an exact mathematical analysis of structures. Such methods as slope-deflection and moment distribution may *always* be used to analyze concrete structures. In some cases it is absolutely necessary to use the exact methods. In the most common cases, however, it is sufficiently accurate to use approximate methods.

The 1983 ACI Code contains approximate coefficients for calculating shears and moments, which can be used when (and only when) specified conditions have been satisfied.

Since exact methods are studied in *statically indeterminate structures*, a prerequisite to the study of reinforced concrete, only the approximate methods will be discussed in this text. An exception exists in the case of the cantilever moment distribution method, which is an exact mathematical method when certain conditions are satisfied.

CONTINUOUS BEAMS AND SLABS

Approximate coefficients of shear and bending moment may be utilized when the following conditions are satisfied:

(1) Adjacent *clear* spans may not differ in length by more than 20% of the shorter span.

(2) The ratio of live load to dead load may not exceed 3.

(3) The loads *must be* uniformly distributed.

Beam and Slab Coefficients

When conditions (1), (2) and (3) are satisfied, the following listed approximate formulas which are stated in the ACI Code may be used to determine shear forces and bending moments in continuous beams and *one-way slabs*. (A one-way slab is one which distributes its load to two end supports only. A two-way slab distributes its load to four supports, one support existing along each of the ends and sides.)

For Positive Moment

End spans:

If discontinuous end is unrestrained	$w(L')^2/11$	(2.1)
If discontinuous end is integral with the support	$w(L')^2/14$	(2.2)
Interior spans	$w(L')^2/16$	(2.3)

For Negative Moment

Negative moment at exterior face of first interior support:

Two spans	$w(L')^2/9$	(2.4)
More than two spans	$w(L')^2/10$	(2.5)
Negative moment at other faces of interior supports	$w(L')^2/11$	(2.6)

Negative moment at face of all supports for (*a*) slabs with spans not exceeding 10 ft (3 m) and (*b*) beams and girders where the ratio of sum of column stiffnesses to beam stiffness exceeds 8 at each end of the span .... $w(L')^2/12$ (2.7)

Negative moment at interior faces of exterior supports, for members built integrally with their supports:

Where the support is a spandrel beam or girder	$w(L')^2/24$	(2.8)
Where the support is a column	$w(L')^2/16$	(2.9)

Shear Forces

Shear in end members at first interior support	$1.15w(L')/2$	(2.10)
Shear at all other supports	$w(L')/2$	(2.11)

End Reactions

Reactions to a supporting beam, column or wall are obtained as the sum of shear forces acting on both sides of the support.

Integral and unrestrained supports are illustrated in Fig. 2-1. Exterior and interior supports are shown in Fig. 2-2.

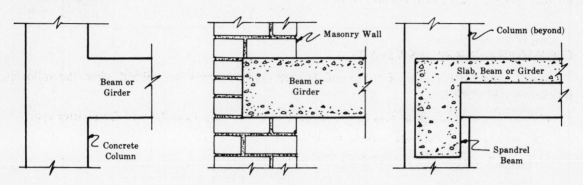

Fig. 2-1

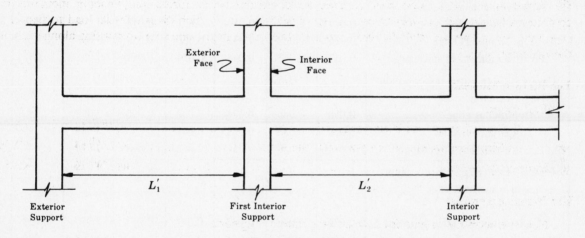

Fig. 2-2

TWO-WAY SLABS

The 1963 ACI Code provided three separate approximate methods for use in determining shear and moments in slabs which distribute their loads to four supports. While those methods are not given in the text of the 1983 ACI Code, the Code permits the use of methods that have proved satisfactory over a period of many years. All three methods of the 1963 ACI Code are discussed in Chapter 13. Two-way systems designed according to the *Direct Design Method* and the *Equivalent Frame Method* of the 1983 ACI Code are covered in Chapter 14.

Method 2 of the 1963 ACI Code has been used extensively in engineering practice. This method has been devised considering the *theory of elasticity* and the results of experiments. The method applies *only* when (a) the loads are uniformly distributed and (b) the ratio of live load to dead load does not exceed 3.

The following notes relative to the analysis of two-way slabs are reproduced from the ACI Code:

Table 2.1   Moment Coefficients for Two-Way Slabs (Method 2 of 1963 Code)

Moments	Short span						Long span, all values of m
	Values of m						
	1.0	0.9	0.8	0.7	0.6	0.5 and less	
Case 1—Interior panels							
Negative moment at—							
Continuous edge	0.033	0.040	0.048	0.055	0.063	0.083	0.033
Discontinuous edge	—	—	—	—	—	—	—
Positive moment at midspan	0.025	0.030	0.036	0.041	0.047	0.062	0.025
Case 2—One edge discontinuous							
Negative moment at—							
Continuous edge	0.041	0.048	0.055	0.062	0.069	0.085	0.041
Discontinuous edge	0.021	0.024	0.027	0.031	0.035	0.042	0.021
Positive moment at midspan	0.031	0.036	0.041	0.047	0.052	0.064	0.031
Case 3—Two edges discontinuous							
Negative moment at—							
Continuous edge	0.049	0.057	0.064	0.071	0.078	0.090	0.049
Discontinuous edge	0.025	0.028	0.032	0.036	0.039	0.045	0.025
Positive moment at midspan	0.037	0.043	0.048	0.054	0.059	0.068	0.037
Case 4—Three edges discontinuous							
Negative moment at—							
Continuous edge	0.058	0.066	0.074	0.082	0.090	0.098	0.058
Discontinuous edge	0.029	0.033	0.037	0.041	0.045	0.049	0.029
Positive moment at midspan	0.044	0.050	0.056	0.062	0.068	0.074	0.044
Case 5—Four edges discontinuous							
Negative moment at—							
Continuous edge	—	—	—	—	—	—	—
Discontinuous edge	0.033	0.038	0.043	0.047	0.053	0.055	0.033
Positive moment at midspan	0.050	0.057	0.064	0.072	0.080	0.083	0.050

C = moment coefficient for two-way slabs as given in Table 2.1

m = ratio of short span to long span for two-way slabs

S = length of short span for two-way slabs. The span shall be considered as the center-to-center distance between supports or the clear span plus twice the thickness of slab, whichever value is the smaller. (Since the beam width is not known for certain at the outset, it is sufficient and safe to use the center-to-center span.)

w = total uniform load, lb/ft^2 or kips/ft^2 (kN/m^2)

Limitations

These recommendations are intended to apply to slabs (solid or ribbed), isolated or continuous, supported on all four sides by walls or beams, in either case built monolithically with the slabs.

A two-way slab shall be considered as consisting of strips in each direction as follows: (a) A middle strip one-half panel in width, symmetrical about the panel centerline and extending through the panel in the direction in which moments are considered and (b) a column strip one-half panel in width, occupying the two quarter-panel areas outside the middle strip.

Where the ratio of short to long span is less than 0.5, the middle strip in the short direction shall be considered as having a width equal to the difference between the long and short span with the remaining area representing the two column strips.

The critical sections for moment calculations are referred to as the principal design sections and are located as follows: (a) For negative moment, along the edges of the panel at the faces of the supporting beams and (b) for positive moment, along the centerlines of the panels.

Bending Moments

The bending moments for the middle strips shall be computed using the formula

$$M = CwS^2 \qquad (2.12)$$

The average moments per foot (meter) of width in the column strip shall be two-thirds of the corresponding moments in the middle strip. In determining the spacing of the reinforcement in the column strip, the moment may be assumed to vary from a maximum at the edge of the middle strip to a minimum at the edge of the panel.

Where the negative moment on one side of a support is less than 80 percent of the moment on the other side, two-thirds of the difference shall be distributed in proportion to the relative stiffnesses of the slabs.

Shear

The shear stresses in the slab may be computed on the assumption that the load is distributed to the supports in accordance with equations (2.13) or (2.14) given below.

Supporting Beams

The loads on the supporting beams for a two-way rectangular panel may be assumed as these loads contained within the tributary areas of the panel bounded by the intersection of 45-degree lines from the corners with the median line of the panel parallel to the long side. (See Fig. 2-3.)

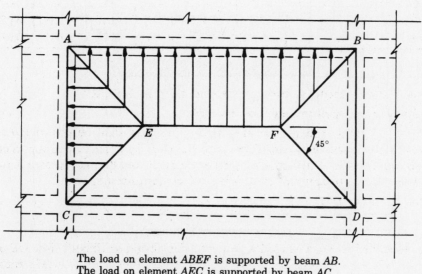

The load on element *ABEF* is supported by beam *AB*.
The load on element *AEC* is supported by beam *AC*.

Fig. 2-3

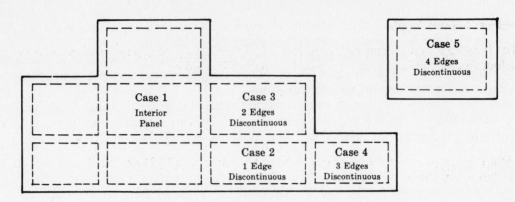

Fig. 2-4

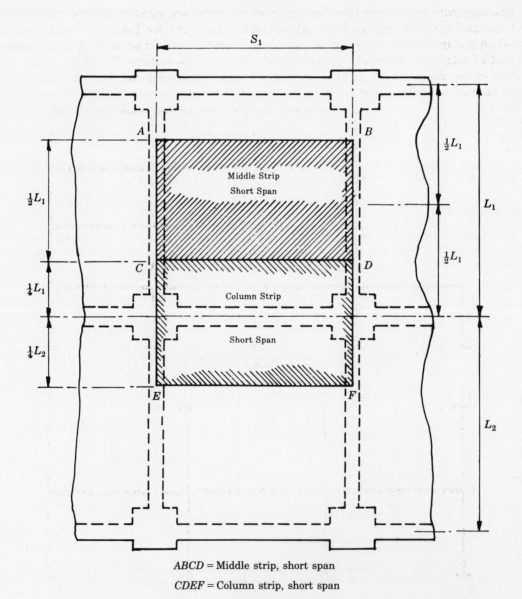

$ABCD$ = Middle strip, short span

$CDEF$ = Column strip, short span

Fig. 2-5

Equivalent Uniform Loads

The bending moments may be determined approximately by using an *equivalent uniform load* per lineal foot (meter) of beam for each panel supported as follows:

$$\text{For the short span:} \qquad \frac{wS}{3} \tag{2.13}$$

$$\text{For the long span:} \qquad \frac{wS}{3}\,\frac{(3-m^2)}{2} \tag{2.14}$$

Note that a *concentrated load* may not be transformed via (*2.13*) or (*2.14*); any such load must appear "as is" in the loading diagram.

FRAME ANALYSIS

The approximate methods previously discussed apply to *usual conditions* of construction, where loads are uniformly distributed, spans are nearly equal and the live loads are not excessively high.

When the approximate methods do not apply, mathematically exact methods must be used. The 1983 ACI Code does, however, permit the utilization of simplifications. For example, the complete analysis of the frame shown in Fig. 2-6 would be time-consuming, unless an electronic computer would be used for the analysis.

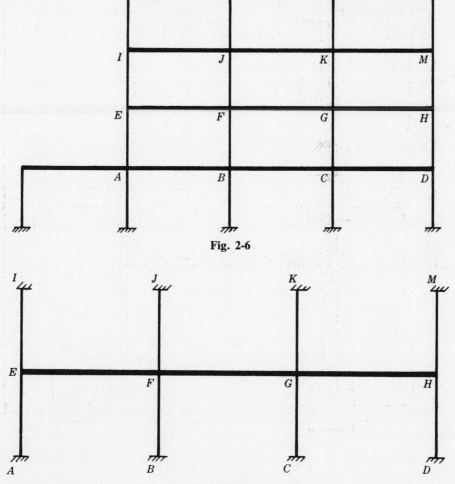

Fig. 2-6

Fig. 2-7

In analyzing the beams for level *EFGH*, the 1983 ACI Code permits the use of the substitute frame shown in Fig. 2-7. The columns may be assumed to be *fixed* at the floors above and below the level in question.

Maximum Design Values

The maximum shear force, reaction and bending moment *do not occur* when the entire frame is loaded. This becomes apparent by observing the deflections of a loaded frame.

Fig. 2-8 shows the deflection diagrams due to four different loading conditions, each of which must be considered in determining the maximum shear, moment and reactions. In each case, *all spans are loaded with the dead load* which consists of the weight of the materials.

Those loading conditions provide the maximum negative and positive moments at the supports and centers of spans, maximum column moments, maximum column reactions and maximum shear forces for the beams.

Methods of complete analysis of such frames are discussed in texts concerning statically indeterminate structures and will not be discussed in this book.

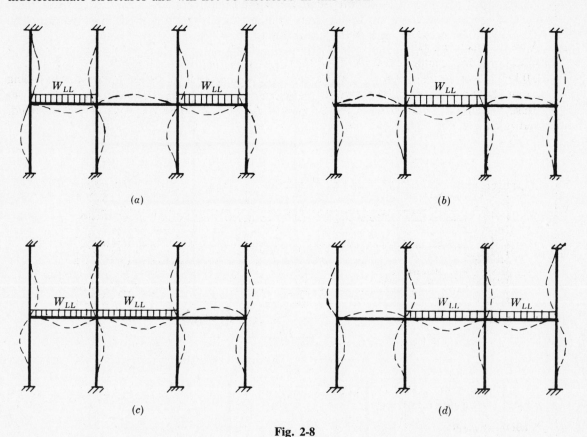

Fig. 2-8

Statics

In the study of reinforced concrete a thorough knowledge of statics is assumed. Nevertheless, a number of problems involving statics are provided for the purpose of review.

Column Forces and Moments

Axial loads and moments in columns due to gravity loads are properly determined using the principles of continuity and methods of solving statically indeterminate structures. Under normal circumstances, column moments on interior columns are relatively small when span lengths are nearly equal. However, the exterior columns may have reasonably large moments accompanied by relatively small axial loads.

When the approximate methods are used for obtaining beam moments, the column moments are also determined by approximate methods. The differences in the beam moments are distributed to the columns above and below the joint in the ratio of the *stiffnesses* of the columns.

It must be emphasized here, however, that approximate methods usually yield excessively large structural members. Approximate methods should not be used when an exact mathematical analysis is available.

Solved Problems

2.1. A two span beam (Fig. 2-9) is supported by spandrel beams at the outer edges and by a column in the center. Dead load (including beam weight) is 1.0 kip/ft and live load is 2.0 kips/ft on both beams. Calculate all critical *service-load* shear forces and bending moments for the beams. The torsional resistance of the spandrel beam is not sufficient to cause restraint of beam *ABC* at the masonry walls. (*Service loads* are loads actually applied, and not increased by use of load factors.)

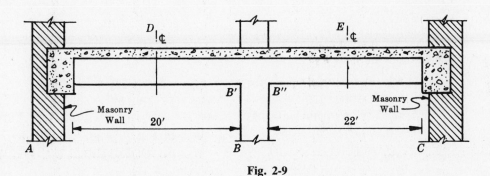

Fig. 2-9

(*a*) Loads are uniformly distributed, (*b*) $LL/DL = 2 < 3$, (*c*) $(L_2' - L_1')/L_1' = (22 - 20)/20 = 0.1 < 0.2$. ACI coefficients apply.

Bending Moments

$M_{AB} = -3(20)^2/24 = -50$ ft-kips $M_{BA} = -(3/9)[(20 + 22)/2]^2 = -147$ ft-kips $= M_{BC}$

$M_{CB} = -3(22)^2/24 = -60.5$ ft-kips $M_D = 3(20)^2/11 = 109$ ft-kips $M_E = 3(22)^2/11 = 132$ ft-kips

Shear Forces

$$V_A = 3(20)/2 = 30 \text{ kips} \qquad V_{B'} = 1.15(3)(20)/2 = 34.5 \text{ kips}$$
$$V_C = 3(22)/2 = 33 \text{ kips} \qquad V_{B''} = 1.15(3)(22)/2 = 37.95 \text{ kips}$$

Reactions

$$R_A = V_A = 30 \text{ kips} \qquad R_B = V_{B'} + V_{B''} = 72.45 \text{ kips} \qquad R_C = V_C = 33 \text{ kips}$$

2.2. Given, conditions identical to those of Problem 2.1, except that A and B are built integrally with the columns. Determine the critical service-load moments, shears and column reactions.

Refer to equations (2.2), (2.3), (2.4) and (2.9).

$$M_{AB'} = -3(20)^2/16 = -75 \text{ ft-kips} \qquad M_E = 3(22)^2/14 = 103.7 \text{ ft-kips}$$
$$M_{BA'} = -3(21)^2/9 \; = -147 \text{ ft-kips} \qquad M_D = 3(20)^2/14 = 85.7 \text{ ft-kips}$$

All shear forces and reactions are identical to those for Problem 2.1 for the corresponding sections.

2.3. Given, a five span beam as shown in Fig. 2-10. Total dead load is 1.5 kips/ft and total live load is 2.5 kips/ft. Calculate the critical moments, shear forces and column reactions. The beam is built into a girder at A and a column at E. The masonry wall at A does not offer restraint to beam AB.

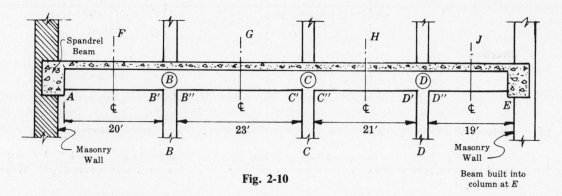

Fig. 2-10

Investigation indicates that the criteria are satisfied for the 1983 ACI Code coefficients. Note that the average spans are used for negative moments at supports. Total load $w_{TL} = 1.5 + 2.5 = 4.0$ kips/ft. Moments in foot kips are:

$$M_{BA} = -(4/10)(21.5)^2 = -184.9 \qquad M_{CB} = -(4/11)(22)^2 = -176.0$$
$$M_{DC} = -(4/11)(20)^2 = -145.4 \qquad M_{DE} = -(4/10)(20)^2 = -160.0$$
$$M_{ED} = -(4/16)(19)^2 = -90.25 \qquad M_{AB} = -(4/24)(20)^2 = -66.6$$
$$M_F = (4/11)(20)^2 = 145.4 \qquad M_G = (4/16)(23)^2 = 132.25$$
$$M_H = (4/16)(21)^2 = 110.0 \qquad M_J = (4/14)(19)^2 = 103.0$$
$$M_{CD} = -(4/11)(22)^2 = -176.0 \qquad M_{BC} = -(4/11)(19)^2 = -131.27$$

Short cuts may be introduced by making use of common quantities in the moment equations. For example,

$$M_H = (21/23)^2(M_G) = 110.0 \qquad \text{and} \qquad M_{DE} = (11/10)(M_{DC}) = -160.0$$

Shear forces in kips

$$V_A = 4(20/2) = 40.0 \qquad V_{B'} = 1.15 V_A = 46.0 \qquad V_{B''} = 4(23/2) = 46.0 = V_{C'}$$
$$V_E = 4(19/2) = 38.0 \qquad V_{D''} = 1.15 V_E = 43.7 \qquad V_{C''} = 4(21/2) = 42.0 = V_{D'}$$

Reactions in kips

$$R_A = V_A = 40.0 \qquad R_C = V_{C'} + V_{C''} = 88.0 \qquad R_E = V_E = 38.0$$
$$R_B = V_{B'} + V_{B''} = 92.0 \qquad R_D = V_{D'} + V_{D''} = 85.7$$

2.4. Fig. 2-11 shows the cross-section of a 4.5-inch-thick one-way slab. Live load is 100 lb/ft^2, and the floor covering weighs 1.5 lb/ft^2. Determine the shear forces and the reactions delivered to the supporting beams. The concrete weighs 150 lb/ft^3, or 12.5 lb/ft^2 per inch of thickness.

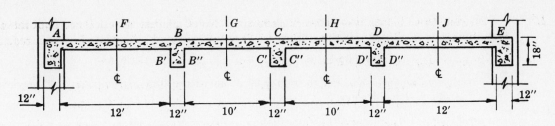

Fig. 2-11

A check will show that the 1983 ACI coefficients may be used.

$$w_{TL} = 100 + 1.5 + (4.5)(12.5) = 157.75 \text{ (use 158 lb/ft}^2)$$

Shear Forces

$$V_A = V_E = 158(12/2) = 948 \text{ lb/ft} \qquad V_{D''} = V_{B'} = 158(12/2)(1.15) = 1090 \text{ lb/ft}$$
$$V_{B''} = V_{D'} = 158(10/2) = 790 \text{ lb/ft} \qquad V_{C'} = V_{C''} = 158(10/2) = 790 \text{ lb/ft}$$

Reactions

$$R_D = R_B = V_{B'} + V_{B''} = 1880 \text{ lb/ft} \qquad R_C = V_{C'} + V_{C''} = 1580 \text{ lb/ft}$$
$$R_A = R_E = V_A = V_E = 948 \text{ lb/ft}$$

2.5. Determine the moments at A, B', B'', C', F and G for the slabs of Problem 2.4.

Beams A and E are spandrel beams, so $M = -w(L')^2/24$ in the slab at A and E. For negative moments at supports use average of adjacent clear spans. For positive moment use clear span. Moments in ft-lb are:

$$M_A = -158(12)^2/24 = -948 \qquad M_{B'} = -158(11)^2/10 = -1912$$
$$M_{B''} = (10/11)(M_{B'}) = -1738 \qquad M_{C'} = -158(10)^2/11 = -1437$$
$$M_F = 158(12)^2/14 = 1625 \qquad M_G = 158(10)^2/16 = 988$$

2.6. If beam B of Fig. 2-11 is 20 ft long and is simply supported on brick walls, determine the end reactions and the moment at the center of the beam. (Dead load + live load of beam $B = 225$ lb/ft.)

End reactions are $w_{TL}(L/2) = 2205(20/2) = 22{,}050$ lb to the wall.

M at center $= wL^2/8 = 2205(20)^2/8 = 110{,}250$ ft-lb.

2.7. Fig. 2-12 shows a continuous slab supported on intermediate beams. Live load is 100 lb/ft^2 and dead load is 50 lb/ft^2. Determine all slab shear forces and moments and the reactions to the beams.

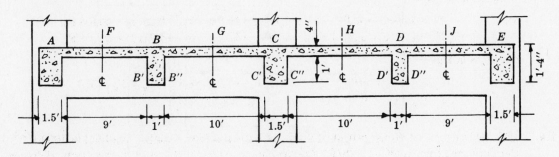

Fig. 2-12

A check shows that the 1983 ACI coefficients apply. Note symmetry and calculate half of the values. Use moments in ft-lb/ft of supporting beam.

$$\text{Total load } w_{TL} = w_{LL} + w_{DL} = 100 + 50 = 150 \text{ lb/ft}^2$$

For negative moments, note that the slab spans do not exceed 10 ft, and equation (2.7) applies. Use average of adjacent span lengths.

$$M_{AB} = -150(9)^2/12 = -1013 \qquad M_{ED} = M_{AB} \qquad M_{DE} = M_{BA}$$
$$M_{BA} = -150(9.5)^2/12 = -1128 \qquad M_{CD} = M_{CB} \qquad M_{DC} = M_{BC}$$
$$M_{CB} = -150(10)^2/12 = -1250$$

For positive moments, ends supported on girders, or integral with supports when considering slabs:

$$M_F = M_J = 150(9)^2/14 = 868 \qquad M_G = M_H = 150(10)^2/16 = 938$$

Shear forces in lb/ft of supporting beams:

$$V_A = V_E = 150(9/2) = 675 \qquad V_{B''} = V_{C'} = 150(10/2) = 750$$
$$V_{B'} = V_{D''} = 1.15V_A = 776 \qquad V_{C''} = V_{D'} = 150(10/2) = 750$$

Reactions in lb/ft of supporting beam:

$$R_A = R_E = V_A = 675 \qquad R_B = R_D = V_{B'} + V_{B''} = 1526 \qquad R_C = V_{C'} + V_{C''} = 1500$$

2.8. Determine the total design loads for beams A, C and E of Problem 2.7.

Loads are delivered from the slab in lb/ft of beam. Clear spans were used for slab analysis, so added loads over the beams must be determined.

Added $DL = (4+12)(1)(150)/12 = 200$ lb/ft^2. Added $LL = (100)(1)(1) = 100$ lb/ft^2.
Added total load $= 200 + 100 = 300$ lb/ft^2.

Beams A, C and E: added $w_{TL} = 1.5(300) = 450$ lb/ft

$$w_A = w_E = 450 + 675 = 1125 \text{ lb/ft} \qquad w_C = 450 + 1500 = 1950 \text{ lb/ft}$$

2.9. The two-way slab shown in Fig. 2-13 is an interior slab and is 4″ thick. The live load is 50 lb/ft^2. Determine the slab moments for designing the middle strip in the short direction at D, C and E. Assume all adjacent slabs are identical to the one shown. Use Method 2 of the 1963 ACI Code.

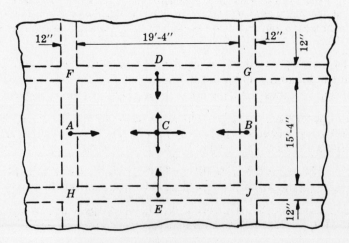

Fig. 2-13

L is lesser of: $(19' - 4'') + (2)(4'') = (20' - 0'')$ and $(19' - 4'') + (2)(12'')/2 = (20' - 4'')$. S is lesser of: $(15' - 4'') + (2)(4'') = (16' - 0'')$ and $(15' - 4'') + (2)(12'')/2 = 16' - 4''$. Then $L = 20'$ and $S = 16'$, so $m = S/L = 0.8$.

From Table 2.1, Case 1, $m = 0.8$ obtain: for negative M, continuous edge, $C = 0.048$; for positive M, $C = 0.036$.

Slab weight $= w_{DL} = 4(12.5) = 50$ lb/ft^2. $w_{TL} = w_{DL} + w_{LL} = 100$ lb/ft^2.

At D and E: $M = -Cw_{TL}S^2 = -0.048(100)(16)^2 = -1229$ ft-lb/ft.

At C: $M = +Cw_{TL}S^2 = 0.036(100)(16)^2 = 922$ ft-lb/ft.

2.10. Determine the moments at A, B and C for the middle strip, long direction, for the slab of Fig. 2-13. Use Method 2 of the 1963 ACI Code.

For the long direction, for all values of m, $C = 0.033$ for negative moment and $C = 0.025$ for positive moment (Table 2.1).

At A and B: $M = -Cw_{TL}S^2 = -0.033(100)(16)^2 = -845$ ft-lb/ft.

At C: $M = +Cw_{TL}S^2 = 0.025(100)(16)^2 = 640$ ft-lb/ft.

Note. The *design moments* may be reduced to those values which exist at the *face of the support*, in accord with the 1963 ACI Code. The reductions in moments are usually small and the calculations are lengthy. Thus the reduction is usually unwarranted considering the fact that the design coefficients are somewhat approximate.

2.11. Use the solutions obtained in Problem 2.9 to determine the moments in the column strips, short direction.

The column strip width is $(2)(20/4) = 10$ ft. The 1963 ACI Code required that the *average moments* in the column strip shall be two-thirds of the moments in the adjacent middle strip. That Code permitted varying the moment from a maximum at the edge of the middle strip to a minimum at the support. Two solutions are possible:

(*a*) Using average moment of 2/3 of middle strip moment:

At support (both sides of G or J), $M = -0.667(1229) = -819$ ft-lb/ft.

At center (both sides of B), $M = -0.667(922) = -615$ ft-lb/ft.

(*b*) If the average column strip moments are 2/3 of the middle strip moments, the column strip moments may be varied from $(3/3)M$ to $(1/3)M$.

At G, $M = -0.333(1229) = -409$ ft-lb/ft. At B, $M = 0.333(922) = 307$ ft-lb/ft.

If (*a*) is used, the reinforcement is spaced uniformly over the column strip. If (*b*) is used, the reinforcement spacing is *gradually increased* from the middle strip to the support. The latter provides better load distribution, but the former is more practical from the viewpoint of simplicity in construction.

2.12. Determine the design moment for point D, middle strip, long span for the slab shown in Fig. 2-14. Total dead load including the slab weight is 75 lb/ft^2 and live load is 100 lb/ft^2. The slab has 6″ constant thickness. Beams are 12″ wide. Use Method 2 of the 1963 ACI Code.

Center to center spans are identical to clear spans plus twice the slab thickness. The values of C (for $M = Cw_{TL}S^2$) are independent of $m = S/L$ for the long span. Moments are in ft-lb/ft width of slab. From Table 2.1: Slab A, $C_D = 0.033$; slab B, $C_D = 0.058$.

$$M'_D = -0.033(175)(20)^2 = -2310 \qquad M''_D = -0.058(175)(20)^2 = -4060$$
$$\text{Ratio } M'_D/M''_D = 2310/4060 = 0.0569 < 0.80; \text{ distribution necessary}$$

Stiffness I/L is the same for the two slabs. Thus distribution factors are 0.5 for each side. Difference in moments $= |M''_D| - |M'_D| = 1750$. Distribute 2/3 of the difference to each side.

$$M''_D = -4060 + (2/3)(1750)(0.5) = -3477 \qquad M'_D = -2310 - (2/3)(1750)(0.5) = -2893$$

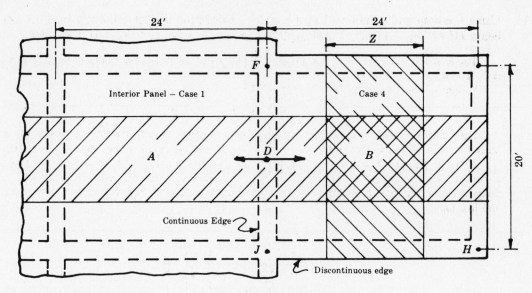

Fig. 2-14

2.13. Determine the width of the middle strips (X_A and X_B) for slabs A and B in the short direction and the width of the column strip (Y) between the two middle strips for the slabs shown in Fig. 2-15. Use Method 2 of the 1963 ACI Code.

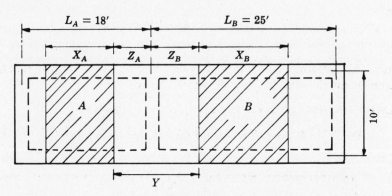

Fig. 2-15

For slab A, $m = 10/18 = 0.555 > 0.5$ For slab B, $m = 10/25 = 0.4 < 0.5$

For slab A, $X_A = 18/2 = 9'$ For slab B, $X_B = 25 - 10 = 15'$

For slab A, $Z_A = 18/4 = 4.5'$ For slab B, $Z_B = (25 - 15)/2 = 5'$

$$Y = 4.5 + 5.0 = 9.5'$$

2.14. Determine the end shear for design of the middle strip in the long and short directions for slab B in Fig. 2-14. Use data and solution for Problem 2.12 (along with Method 2 of the 1963 ACI Code).

$m = S/L = 20/24 = 0.833$

Short beam: $w'_s = w_{TL}S/3 = 175(20/3) = 1167$ lb/ft.

Long beam: $w'_L = (w_{TL}S/3)(3 - m^2)/2 = 1167[3 - (0.833)^2]/2 = 1346$ lb/ft.

The end shear force for the *short slab span* = w'_L since that force is delivered to the long beam. In a like manner, the end shear force for the *long slab span* = w'_s.

2.15. Using the data and solution of Problem 2.14, determine the total uniform load for designing beam JH of Fig. 2-14. The total depth of the beam is 24″ and the slab is 6″ thick.

The load to beam $JH = w'_L$ + weight of the shaded portion of the beam shown in Fig. 2-16. The area is $2(1) - (0.5)(0.5) = 1.75$ ft^2.

Weight of shaded portion $= 1.75(150) = 263$ lb/ft.

$$w_{TL} = 1346 + 263 = 1609 \text{ lb/ft}$$

Note that additional live load was not added since the wall rests on the slab and live load cannot be placed on this 6″ width. The weight of the wall would also be added to the load on the beam.

Note. In this problem and other problems in this chapter the dead load and live load have been combined. This procedure is normal for use with the *alternate design method*. When using the *strength design method*, the dead load and live load are treated separately since *load factors* differ for dead load and live load.

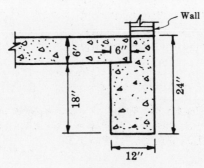

Fig. 2-16

2.16. Determine the uniform load for designing beam FJ of Fig. 2-14. Use data from Problem 2.14. The beam below the slab is 12″ wide and 18″ deep.

The beam receives load from two sides, as shown in Fig. 2-17.

$$w_{TL} = 2w'_S + \text{weight of beam below slab} = 2(1167) + (1.5)(1)(150) = 2559 \text{ lb/ft}$$

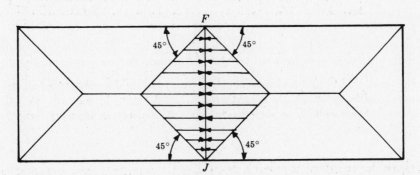

Fig. 2-17

2.17. Derive equation (2.13) for the equivalent uniform load w'_s delivered to the short beam (AC) due to a uniform load w on the slab. Do not include the weight of the beam. See Fig. 2-18.

The load is delivered to beam AC from area AEC. The maximum load is delivered along line EG and equals $wS/2$. Fig. 2-19(a) shows the actual loading condition and Fig. 2-19(b) shows the *equivalent* loading condition. (Load from one side only considered.)

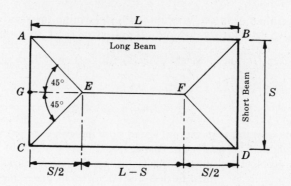

Fig. 2-18

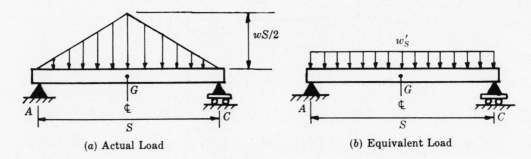

(a) Actual Load   (b) Equivalent Load

Fig. 2-19

Using the actual load condition, $R_A = (wS/2)(S/2)(1/2) = wS^2/8$.

The moment $M_G = (wS^2/80)(S/2) - (wS^2/8)(S/6) = wS^3/24$.

Using the equivalent load condition, $M_G = w'_s(S^2/8)$.

Considering M_G identical for both conditions, $wS^3/24 = w'_s(S^2/8)$ or $w'_s = wS/3$ (valid in any consistent system of units).

2.18. Derive equation (2.14) for the equivalent uniform load w'_L for the *long beam* (CD) shown in Fig. 2-18, due to a uniform load w. Do not include the weight of the beam.

The actual load condition is shown in Fig. 2-20(a) and the equivalent load condition is shown in Fig. 2-20(b). (Load from one side only considered.)

Using statics to obtain reactions, the moment M_G can be obtained.

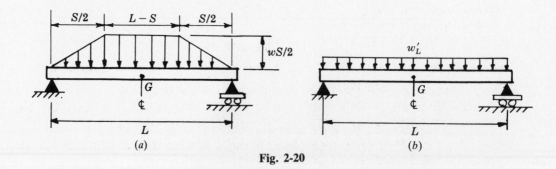

Fig. 2-20

Due to actual load conditions, $M_G = (wS^2/8)(S/3) + (wS/2)[(L-S)/2][(L+S)/4]$. Substituting $m = S/L$, obtain $M_G = (wSL^2)(3-m^2)/48$.

Due to the equivalent load conditions, $M_G = w_L'(L^2/8)$.

Equating the actual moment and the equivalent moment, obtain $w_L' = (wS/3)(3-m^2)/2$ (valid in any consistent system of units).

2.19. Fig. 2-21 shows the joints of a frame and the beam end moments.

$$M_{AB} = -110 \text{ ft-kips} \qquad M_{BA} = -150 \text{ ft-kips} \qquad M_{BC} = -100 \text{ ft-kips}$$

Stiffness factors are shown on the figure. Determine the column moments.

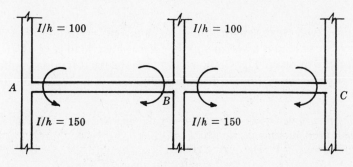

Fig. 2-21

At joint A: For column above, $M = 110[100/(100+150)] = 44$ ft-kips
                 For column below, $M = 110[150/(100+150)] = 66$ ft-kips

At joint B: For column above, $M = (150-100)[100/(100+150)] = 20$ ft-kips
                 For column below, $M = (150-100)[150/(100+150)] = 30$ ft-kips

The axial loads in the columns are obtained by summing the shear forces on both sides of the column at all floor levels above the joint in question. The weight of the column above the joint must also be added thereto.

2.20. Refer to Fig. 2-9 of Problem 2.1. Assume span AB is 7 meters long and span BC is 8 meters long. Dead load (DL), including the weight of the beam, is 5 kN/m and live load (LL) is 10 kN/m. Calculate the *factored-load* moments and shears using the approximate coefficients given in the 1983 ACI Code. All other conditions are identical to those in Fig. 2-9. The loads given have been *factored* using 1.4DL and 1.7LL.

The Code criteria for using the approximate coefficients are: (*a*) loads are uniformly distributed; (*b*) the ratio of live load to dead load does not exceed 3; (*c*) the ratio of adjacent clear spans, (Long span − Short span)/Short span, does not exceed 0.2. (The long span is designated as L_2' and the short span as L_1'. For negative moments at exterior ends of exterior spans and for positive moments, the particular span length is L'.)

Thus: (*a*) Loads are uniform   (O.K.)
       (*b*) LL/DL = 10/5 = 2 < 3   (O.K.)
       (*c*) $(L_2' - L_1')/L_1' = (8-7)/8 = 0.125 < 0.2$ (O.K.)

The approximate coefficients may be used. With $W = $ LL + DL, the bending moments are:

$$M_{AB} = -w(L')^2/24 = -(15)(7)^2/24 = -30.63 \text{ kN} \cdot \text{m}$$
$$M_{BA} = -w[(L_1' + L_2')/2]^2/9 = -(15/9)[(7+8)/2]^2 = 93.75 \text{ kN} \cdot \text{m}$$
$$M_{CB} = -w(L')^2/24 = -(15)(8)^2/24 = -40.0 \text{ kN} \cdot \text{m}$$
$$M_D = +w(L')^2/11 = +(15)(7)^2/11 = +66.82 \text{ kN} \cdot \text{m}$$
$$M_E = +w(L')^2/11 = +(15)(8)^2/11 = +87.27 \text{ kN} \cdot \text{m}$$

The shear forces are:

$$V_A = wL'/2 = (15)(7)/2 = 52.5 \text{ kN}$$
$$V'_B = 1.15 \, wL'/2 = (1.15)(15)(7)/2 = 60.38 \text{ kN}$$
$$V''_B = 1.15 \, wL'/2 = (1.15)(15)(8)/2 = 69.0 \text{ kN}$$

The reactions are:

$$R_A = V_A = 52.5 \text{ kN}$$
$$R_B = V'_B + V''_B = 60.38 + 69.0 = 129.38 \text{ kN}$$
$$R_C = V_C = 60.0 \text{ kN}$$

2.21. Refer to Fig. 2-10. Span lengths are $AB = 8$ m, $BC = 9$ m, $CD = 8$ m and $DE = 7$ m. The total *factored* dead load is 8 kN/m and the *factored* live load is 12 kN/m. The other conditions are identical to those in Fig. 2-10. Determine the factored-load (ultimate) design bending moments. (Load factors for ultimate moments are 1.4 times dead load and 1.7 times live load.)

Checking first to determine whether or not the approximate coefficients apply:
(*a*) Loads are uniformly distributed   (O.K.)
(*b*) LL/DL = 8/7 = 1.143 < 3.0   (O.K.)
(*c*) $(L'_2 - L'_1)/L'_1 = (9 - 8)/9 = 0.11 < 0.2$ for spans CE and DE (O.K.)
Therefore the coefficients apply, and the design bending moments, in kN · m, are:

$$M_{BA} = -(1/10)(15)(8)^2 = -96.0 \qquad M_{CD} = -(1/11)(15)(8.5)^2 = -98.52$$
$$M_{CB} = -(1/11)(15)(8.5)^2 = -98.52 \qquad M_{BC} = -(1/11)(15)(8.5)^2 = -98.52$$
$$M_{DC} = -(1/11)(15)(7.5)^2 = -7.67 \qquad M_F = +(1/11)(15)(8)^2 = +87.27$$
$$M_{DE} = -(1/10)(15)(7.5)^2 = -84.38 \qquad M_H = +(1/16)(15)(8)^2 = +60.0$$
$$M_{ED} = -(1/16)(15)(7)^2 = -45.94 \qquad M_G = +(1/16)(15)(9)^2 = +75.94$$
$$M_{AB} = -(1/24)(15)(8.5)^2 = 45.16 \qquad M_J = +(1/14)(15)(7)^2 = +52.5$$

2.22. Refer to Fig. 2-14. The span lengths in the long direction are 8 m, and the spans in the short direction are 7 m. The dead load (not including the weight of the slab) is 2.47 kN/m^2 and the live load is 5.0 kN/m^2. The beams are 300 mm wide and the slab thickness is 150 mm. Use Method 2 of the 1963 ACI Code to determine the *service-load* moment for point D, middle strip, long span. Assume that reinforced concrete (with normal-density aggregates) has a weight density of 23.5 kN/m^3.

The criteria for use of Method 2 of the 1963 ACI Code are: (*a*) loads are uniformly distributed; (*b*) ratio of live load to dead load (LL/DL) does not exceed 3; (*c*) adjacent clear spans may not differ in length by more than 20 percent of the short span; (*d*) design span lengths are the smaller of the center-to-center spans and the clear spans plus twice the slab thickness.
Checking the criteria:
(*a*) Loads are uniformly distributed   (O.K.)
(*b*) The slab weight per unit area is (150/1000)(23.5) = 3.53 kN/m^2, so total dead load = 3.53 + 2.47 = 6.0 kN/m^2. Thus, LL/DL = 5.0/6.0 = 0.833 < 3.0   (O.K.)
(*c*) Adjacent spans are the same length   (O.K.)
(*d*) Design span lengths center-to-center and clear span plus twice the slab thickness are identical   (O.K.)
It should also be noted that when the *aspect ratio* (ratio of long span to perpendicular short span) exceeds 2.0, two-way slab action will not occur. In such cases, the slab should be designed as a one-way slab in the short direction. (Temperature and shrinkage reinforcement must then be provided in the long direction.)

The bending moments using equation (2.12) are equal to CwS^2. For the long span, Table 2.1 shows that the coefficients C are independent of the inverse of the aspect ratio, $m = S/L$. So, for slab A, an interior panel, C_D (at point D) = 0.333, and for slab B, with three edges discontinuous, $C_D = 0.058$. The total load (D + L) is $6.0 + 5.0 = 11.0$ kN/m^2. The moments at D are:

$$M'_D = -CwS^2 = -(0.033)(11.0)(7)^2 = -17.79 \text{ kN} \cdot \text{m}$$
$$M''_D = -(0.058)(11.0)(7)^2 = -31.26 \text{ kN} \cdot \text{m}$$

The ratio $M'_D/M''_D = 17.79/31.26 = 0.569 < 0.8$. Thus, the moments on the two sides of point D must be distributed in accord with the *stiffness factors* on each side. The stiffness $E_C I/L$ is the same for both sides, so the distribution factors are 0.5. Thus,

$$|M''_D| - |M'_D| = 31.26 - 17.79 = 13.47 \text{ kN} \cdot \text{m}$$

Distribute 2/3 of the difference to each side in accord with their stiffness, or $(2/3)(1/2) = 2/6 = 1/3$. Hence,

$$M'_D = -17.79 - (1/3)(13.47) = -22.28 \text{ kN} \cdot \text{m}$$
$$M''_D = -31.26 + (1/3)(13.47) = -26.77 \text{ kN} \cdot \text{m}$$

2.23. Determine the *equivalent uniform load* superimposed on beam JF of Fig. 2-14 using the data given in Problem 2.22.

For the short span JF, length = 7 m, and the load condition shown in Fig. 2-19 applies.

Weight density of slab = (23.5)(150/1000) =	3.53 kN/m^2
Superimposed dead load	= 2.47
Total uniform dead load	= 6.0
Total uniform live load	= 5.0
Total uniform load = w	= 11.0　kN/m^2

Then, by equation (2.13), $w'_s = wS/3 = (11.0)(7)/3 = 25.67$ kN/m.

The dead load of the beam must be added to this. For example, if the beam width is 300 mm and the beam web extends 600 mm below the slab, the lineal weight density of the web (or stem) is

$$(300/1000)(600/1000)(23.5) = 4.23 \text{ kN/m}$$

The total service live load plus dead load for the beam design would then be

$$(2)(25.67) + 4.23 = 55.57 \text{ kN/m}$$

2.24. Determine the equivalent uniform load superimposed on long beam JH in Fig. 2-14 using the data given in Problem 2.22. Assume that the web of the beam extends 500 mm below the slab.

For the long span JH, length = 8 m, and the load condition shown in Fig. 2-20 applies. From equation (2.14),

$$w'_L = (wS/3)(3 - m^2)/2$$

where $m = S/L$. Note that beam JH receives slab load from one side only.

$$m = S/L = 7/8 = 0.875$$

From Problem 2.23, the total uniform load (L + D) on the slab is 11.0 kN/m^2. Thus,

$$w'_L = [(11)(7)/3][3.0 - (0.875)^2]/2 = 28.67 \text{ kN/m}$$

Since beam weight = $(300/1000)(500/1000)(23.5) = 3.53$ kN/m, the equivalent uniform load for designing the beam is

$$w = 28.67 + 3.53 = 32.2 \text{ kN/m}$$

Supplementary Problems

2.25. Determine slab shear force at A, B', B'' and C' in Fig. 2-22. Express answers in kips/ft of supporting beam.   *Ans.* $V_A = 2.25$, $V_{B'} = = 2.59$, $V_{B''} = 2.38$, $V_{C'} = 2.38$

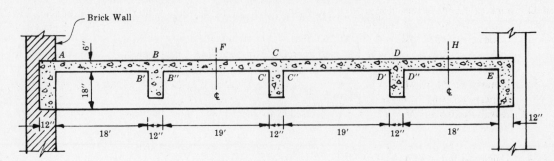

Fig. 2-22

2.26. Determine the negative moments in the slab of Fig. 2-22 at A and B'. Express answers in ft-kips/ft of supporting beam. Use coefficients of the 1983 ACI Code.   *Ans.* $M_A = -3.38$, $M_{B'} = -8.56$

2.27. Determine the positive moments in the slab of Fig. 2-22 at F and H. Express answers in ft-kips/ft of supporting beam. Use coefficients of the 1983 ACI Code.   *Ans.* $M_F = +5.64$, $M_H = +5.79$

2.28. Determine the shear forces and bending moments in the slab of Fig. 2-22 at D'' and E. Use coefficients of the 1983 ACI Code.   *Ans.* $V_{D''} = 2.59$, $M_{D''} = -8.56$, $V_E = 2.25$, $M_E = -3.38$

2.29. Determine the width (X) of the middle strip, short direction for each slab shown in Fig. 2-23. Express answers in feet. Use Method 2 of the 1963 ACI Code.
Ans. $X_I = 12.0$, $X_{II} = 10.0$, $X_{III} = 10.0$, $X_{IV} = 10.0$, $X_V = 10.0$

   Data for Fig. 2-23:
   (1) Slab thickness $= 6''$
   (2) All beams are $12''$ wide and project $16''$ below the slab
   (3) $w_{LL} = 150$ lb/ft^2
   (4) All beams are built into columns

2.30. Determine the width (Y) of the column strips along DE and HJM for the slabs in Fig. 2-23. Express answers in feet. Use Method 2 of the 1963 ACI Code.   *Ans.* $Y_{DE} = 5.75$, $Y_{HJM} = 8.25$

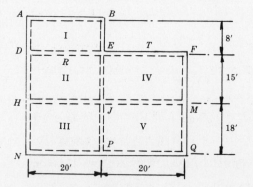

Fig. 2-23

2.31. Determine the negative moments for the middle strip, long span at the intersection of slabs II and IV in Fig. 2-23. Express answers in ft-kips/ft. Use Method 2 of the 1963 ACI Code.
Ans. $M_{II} = -2.07$, $M_{IV} = -2.48$

2.32. Determine the negative moments for the middle strip, short span at the intersection of slabs I and II in Fig. 2-23. (*Hint.* Distribution is necessary.) Express answers in ft-kips/ft. Use Method 2 of the 1963 ACI Code.   *Ans.* $M_I = -2.09$, $M_{II} = -2.60$

2.33. Determine the positive moments for the middle strips (long and short spans) for slab V in Fig. 2-23. Express answers in ft-kips/ft. Use Method 2 of the 1963 ACI Code.
Ans. Long span $M = +2.70$, Short span $M = +3.13$

2.34. Calculate the design dead and live equivalent uniform loads delivered from the slab to beams AD and DH in Fig. 2-23. Express answers in kips/ft. Use Method 2 of the 1963 ACI Code.
Ans. Beam AD: $w'_{DL} = 0.20$, $w'_{LL} = 0.40$, Beam DH: $w'_{DL} = 0.375$, $w'_{LL} = 0.75$

2.35. Calculate the total load shear forces and moments for beams DE and EF in Fig. 2-23. Points R and T lie at the centers of the respective spans. Include necessary portions of the beam weight. Express shear in kips, moments in ft-kips. Use Method 2 of the 1963 ACI Code.
Ans. $V_D = 24.26$, $V_{ED} = 27.90$, $V_{EF} = 18.5$, $V_F = 16.08$, $M_{DE} = -54.5$,
$M_{ED} = -97.5$, $M_{EF} = -64.33$, $M_{FE} = -36.19$, $M_R = +62.5$, $M_T = +41.35$

2.36. Calculate the moments at the column ends for columns GD and DA in Fig. 2-24.
Ans. $M_{GD} = 90$ ft-kips, $M_{DG} = 36$ ft-kips, $M_{DA} = 60$ ft-kips

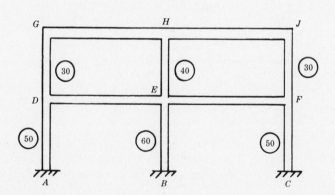

Beam	DE	ED	EF	FE	GH	HG	HJ	JH
Moment, ft-kips	−96	−100	−130	−88	−90	−100	−120	−70

Stiffness factors (I/L) are shown in circles on the figure.

Fig. 2-24

2.37. Calculate the moments at the column ends for columns HE and EB in Fig. 2-24.
Ans. $M_{HE} = 20$ ft-kips, $M_{EH} = 12$ ft-kips, $M_{EB} = 18$ ft-kips

2.38. Calculate the moments at the column ends for columns JF and FC in Fig. 2-24.
   Ans. $M_{JF} = 70$ ft-kips, $M_{FJ} = 33$ ft-kips, $M_{FC} = 55$ ft-kips

2.39. Solve Problem 2.20 if span AB is 8 m and span BC is 9 m.
   Ans. $M_{AB} = -40.0$ kN·m $\quad M_{BA} = -120.4$ kN·m $\quad M_{CB} = -50.6$ kN·m
   $\quad M_D = +87.3$ kN·m $\quad\ M_E = +110.4$ kN·m
   $\quad V_A = 60.0$ kN $\qquad V_B' = 69.0$ kN $\qquad V_B'' = 77.6$ kN
   $\quad R_A = 60.0$ kN $\qquad R_B = 146.6$ kN $\qquad R_C = 60.0$ kN

2.40. Solve Problem 2.21 if all spans are 10 m long and both dead load and live load are 10 kN/m.
   Ans. $M_{DE} = -100.0$ kN·m $\quad M_{BA} = -100$ kN·m $\quad M_{CB} = -90.0$ kN·m $\quad M_{DC} = -90.0$ kN·m
   $\quad M_{ED} = -62.5$ kN·m $\quad M_{AB} = -41.7$ kN·m $\quad M_{BC} = -90.0$ kN·m $\quad M_{CD} = -90.0$ kN·m
   $\qquad M_F = +90.0$ kN·m $\qquad M_G = +62.5$ kN·m $\qquad M_H = +62.5$ kN·m $\qquad M_J = +71.4$ kN·m

2.41. Determine the middle-strip short-span moments in the slab of Fig. 2-14 if the long and short spans are 10 m and 8 m, respectively. Use other data from Problem 2.22.
   Ans. $M_D' = -39.9$ kN·m, $M_D'' = -46.0$ kN·m

2.42. Solve Problem 2.24 for a long span of 10 m, a short span of 8 m and a web extension below the beam of 700 mm.   *Ans.* $w_L' = 39.55$ kN/m

Chapter 3

Lateral Loads

FORCES, SHEAR, MOMENTS AND REACTIONS

NOTATION

CK = modified stiffness factor, in.-kips (kN $\cdot$ m)

h = column height, ft (m)

H_i = lateral load at any joint i, kips (kN)

I = section moment of inertia, in.^{4} (mm^4)

K = stiffness factor, $E_c I/L$, in.-kips (kN $\cdot$ m)

L = length of beam, ft (m)

M = bending moment, ft-kips (kN $\cdot$ m)

M^F = fixed-end moment, ft-kips (kN $\cdot$ m)

V = story shear due to lateral loads, kips (kN)

GENERAL NOTES

Forces due to wind, earthquakes or soil pressure must be considered in the analysis of structures. Methods used for determining the shears, moments and reactions due to loads caused by such conditions are studied in courses dealing with *statically indeterminate structures*. Such procedures include the slope-deflection method and the moment distribution method, a knowledge of which is prerequisite to the study of reinforced concrete structural analysis.

Courses pertaining to statically indeterminate structures provide only the basic principles one must master in order to analyze reinforced concrete structures for lateral loads. These principles must be expanded in the study of reinforced concrete analysis and design, particularly with reference to lateral loads. Therefore several methods for such analysis will be discussed and illustrated in this chapter.

Use of the modified stiffness factor $CK = CE_c I/L$ (not simply I/L) is necessary because: (1) current practice is to use stronger concrete in columns than in beams, so that E_c becomes variable; (2) end conditions may vary from member to member, which is reflected in the variability of the numerical parameter C. In this chapter constant cross-sectional moments of inertia are assumed; see Chapter 14 for problems involving variable moments of inertia.

PORTAL METHOD

In using the portal method for lateral load analysis one assumes that a *point of contraflexure* exists at the midpoint of every member. The structure becomes *statically determinate*, and only the principles of statics are involved in obtaining the shears, moments and axial forces.

This method provides excellent results for the intermediate stories of high-rise structures. The solutions obtained for the upper two stories and the lower two stories are often in error by as much as 50 percent. However, for the purpose of illustration of the method, simple structural frameworks will be utilized in this text.

Assumptions in the Portal Method

(1) A point of contraflexure (zero moment) occurs at the midpoint of all members.

(2)   The total horizontal force resisted at any level is the sum of all horizontal forces applied above that level.

(3)   The total horizontal force at any level is distributed so that the interior columns resist twice as much horizontal force as the exterior columns.

(4)   Vertical forces are obtained by statics using assumptions 1, 2 and 3.

The *free-body diagram* for the analysis of the structure shown in Fig. 3-1(*a*) is illustrated in Fig. 3-1(*b*).

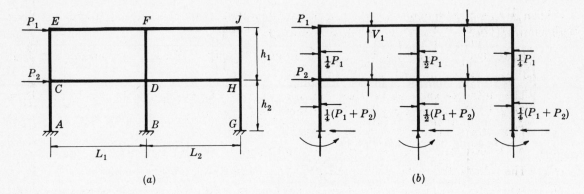

Fig. 3-1

The vertical forces are obtained using statics, as shown in Fig. 3-2. Summing moments about E, we obtain

$$V_1 = P_1 h_1 / 4L_1 \tag{3.1}$$

If a section is cut horizontally at E, we obtain from statics

$$M_E = P_1 h_1 / 8L_1 \tag{3.2}$$

Similar equations are used for the interior columns, with $H = P/2$ rather than $P/4$ as for the exterior columns.

Since the applied forces P_1 and P_2 are known and the lengths of all members are known, the horizontal and vertical forces and the joint moments may be obtained using statics.

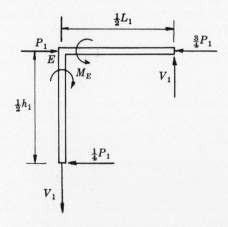

Fig. 3-2

Inaccuracies in the Portal Method

The portal method provides somewhat inaccurate results due to a number of factors, most of which are related to the disregard of the stiffness of the members. Regardless of the inaccuracies involved, the method is important for use in the preliminary stages of a design in order to obtain approximate sizes of members for use with more accurate methods.

SWAY (CANTILEVER) MOMENT DISTRIBUTION

When certain requirements are fulfilled, the *sway moment distribution method* provides an *exact* mathematical solution. The requirements for an exact solution are:

(1) In any two-column bay and at any floor level, the column stiffness factors must be identical.

(2) Loads must be applied laterally at the joints.

(3) The stiffness factors K for the girders are modified by multiplying by $C = 6$, i.e. $CK = 6E_cI/L$. The girder carryover factors then become *zero*.

(4) The stiffness factors for the columns are $K = E_cI/h$.

(5) All column carryover factors are -1.0 and beam carryover factors are zero.

(6) The structure must be symmetrical about a vertical axis (the centerline).

(7) When the structure consists of more than one bay, the structure must be separated into several single bay structures for which (1) is satisfied. The shear is divided in proportion to the relative stiffnesses of the columns of the single bay structures. The final solution consists of superimposing the separate solutions for the single bay structures, one upon another, in order to obtain the results for the complete original structure.

(8) The fixed-end moment, M^F, on any column is obtained from the equation

$$M^F = -Vh/4 \qquad (3.3)$$

where $V =$ sum of the horizontal forces on the single bay structure, above the midheight of the story in question
$h =$ height of the story in question, center-to-center of girders

PRELIMINARY AND FINAL ANALYSES

It is always necessary to make a preliminary analysis using assumed dimensions and stiffness factors when employing the cantilever moment distribution; the same is true for computer or any other type of solution.

The portal method provides a simple means of determining approximate moments and axial loads for use in making a preliminary design. Thereafter, the more precise methods may be employed to obtain an accurate solution for the lateral load problem.

Solved Problems

3.1. Use the *portal method* to determine the moments and reactions at all joints in the upper story of the frame shown in Fig. 3-3.

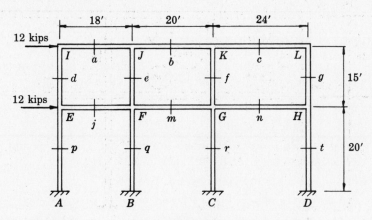

Fig. 3-3

The *free-body diagrams* for the elements of the upper story are shown in Fig. 3-4. Points of contraflexure are assumed to exist at the midpoints of all members. The upper story shear force (12 kips) is divided so that the exterior columns resist 2 kips each and the interior columns resist 4 kips each.

From (*a*), ΣM about $J = 0$. $V_a = 2(7.5)/9 = 1.67$ kips.

From (*b*), ΣM about $J = 0$. $4(7.5) - 1.67(9) - 10V_b = 0$ and $V_b = 1.5$ kips.

From (*c*), ΣM about $K = 0$. $4(7.5) - 1.5(10) - 12V_c = 0$ and $V_c = 1.25$ kips.

From (*d*), ΣM about $c = 0$. $2(7.5) - 12V_g = 0$ and $V_g = 1.25$ kips.

The moments may now be computed using statics. Moments are in ft-kips.

$M_{Ia} = 9(1.67) = 15.0$ $M_{Lc} = 1.25(12) = 15.0$ $M_{Jb} = 1.5(10) = 15.0$ $M_{Je} = -4(7.5) = -30.0$

$M_{Ja} = 9(1.67) = 15.0$ $M_{Id} = -2(7.5) = -15.0$ $M_{Lg} = -2(7.5) = -15.0$

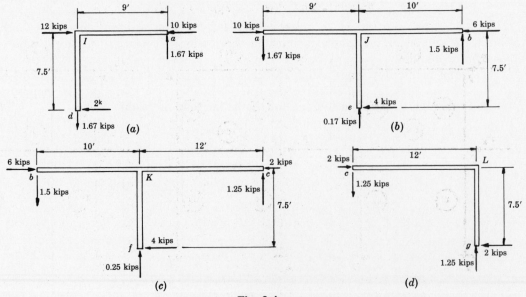

Fig. 3-4

3.2. Given a 3-bay, 2-story structural frame as shown in Fig. 3-5, with lateral loads of 12 kips at the second floor and roof joints; encircled numbers are the stiffness factors K for the members. Determine all moments using the *sway moment distribution* method.

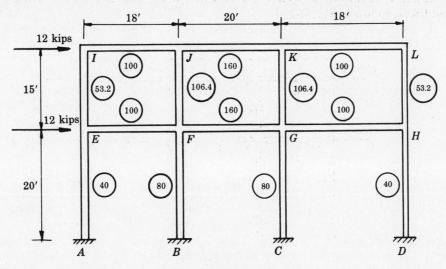

Fig. 3-5

(1) Separate the frame into three single bay frames having *symmetrical column stiffness factors* at any floor level. This is shown in Fig. 3-6.

(2) Calculate the load distribution constants at the roof level:

$$P_I = P_J = P_K = \frac{2(53.2)}{6(53.2)} (12) = 4.0 \text{ kips}$$

(3) Calculate load distribution constants at second floor level:

$$P_E = P_F = P_G = \frac{2(40)}{6(40)} (12) = 4.0 \text{ kips}$$

(4) Calculate fixed-end moments on columns, $M^F = -Vh/4$:

Frame 1 Col's. *EI, FJ*; $M^F = -(4.0)(15.4) = -15.0$ ft-kips

Col's *AE, BF*; $M^F = -(4.0 + 4.0)(20/4) = -40.0$ ft-kips

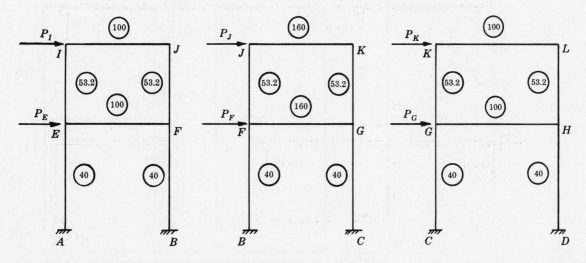

Fig. 3-6

Frame 2 Col's. FJ, GK; $M^F = -15.0$ ft-kips
          Col's. BF, CG; $M^F = -40.0$ ft-kips
Frame 3 Col's. GK, HL; $M^F = -15.0$ ft-kips
          Col's. CG, DH; $M^F = -40.0$ ft-kips

(5) Calculate modified (sway) stiffness factors (CK) and distribution factors (D):

$$\text{Columns:} \quad CK = K = E_c I/h$$
$$\text{Girders:} \quad CK = 6K = 6E_c I/L$$

Frame 1				
$CK_{EA} =$	40.0	$D_{EA} =$	$40/693.2 =$	0.0577
$CK_{EI} =$	53.2	$D_{EI} =$	$53.2/693.2 =$	0.0765
$CK_{EF} = (6)(100) =$	600.0	$D_{EF} =$	$600/693.2 =$	0.8658
$\Sigma\, CK_E =$	693.2		$\Sigma =$	1.0000
$CK_{IE} =$	53.2	$D_{IE} =$	$53.2/653.2 =$	0.0814
$CK_{IJ} = (6)(100) =$	600.0	$D_{IJ} =$	$600/653.2 =$	0.9186
$\Sigma\, CK_I =$	653.2		$\Sigma =$	1.0000

For corresponding locations on Frames 2 and 3, the distribution factors are identical to those on Frame 1. Carryover factors are -1.0 for all columns and 0 for all beams. Therefore, distribute moments only on the columns; the sum of column moments at a joint gives the beam moment. The moment distribution is indicated in Fig. 3-7.

A		E		I	
0.0	0.0577	0.0765	0.0814		D.F.
−40.00	−40.00	−15.00	−15.00		
	−3.17 ← +3.17	+4.21 → −4.21			F.E.M. (M^F)
		−1.56 ← +1.56			
	−0.09 ← +0.09	+0.12 → −0.12			
		+0.01			
−43.26	−36.74	−12.23	−17.76		SUM

Fig. 3-7

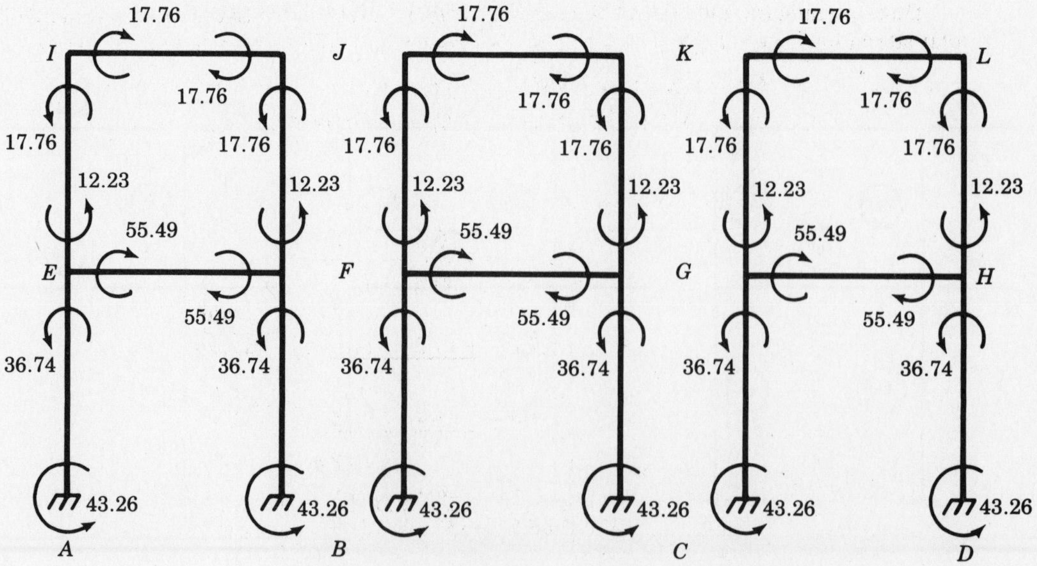

Fig. 3-8

The beam moments are obtained as the algebraic sum of the column moments at the joint in question, as shown in Fig. 3-8.

The final moments are obtained by putting the separate bents together (i.e. superposition) and adding the corresponding moments algebraically.

The final solution is shown in Fig. 3-9.

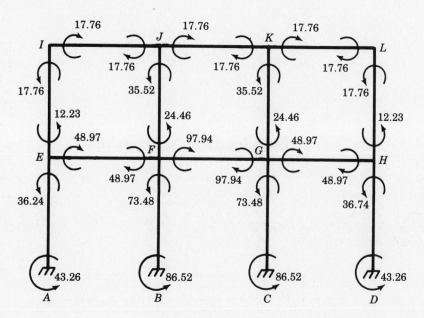

Fig. 3-9

3.3. Refer to Fig. 3-11. Determine the moments at the ends of all columns if the beam span lengths are 10 m, the column heights are 4 m and the applied lateral loads at D and G are each 50 kN. The encircled stiffness factors are in kN·m. Use the cantilever (or sway) moment distribution method.

Due to symmetry, only one of the bays needs to be solved. Each bay is subjected to one-half of the total lateral load, i.e. 25 kN, at each joint. For frame $AB-GH$ the distribution factors are

$$DA = DG = (40)(1)/[(40)(1) + (20)(6) + (40)(1)] = 40/200 = 0.2$$
$$GD = (40)(1)/[(40)(1) + (10)(6)] = 40/100 = 0.4$$

	A		D		G
D.F.		0.2	0.2		0.4
	−50.0	−50.0	−25.0		−25.0
			−10.0 ←		+10.0
	−17.0 ←	+17.0	+17.0	→	−17.0
			−6.8 ←		+6.8
	−1.4 ←	+1.4	+1.4	→	−1.4
			−0.6 ←		+0.6
	−0.1 ←	+0.1	+0.1		
SUM	−68.5	−31.5	−23.9		−26.0

Fig. 3-10

The fixed-end moments are

$$M_{GD}^F = Vh/4 = -(25)(4)/4 = -25.0 \text{ kN} \cdot \text{m}$$
$$M_{DG}^F = M_{DA}^F = -(50)(4/4) = -50.0 \text{ kN} \cdot \text{m}$$

The carryover factors are all -1.0. The moment distribution is shown in Fig. 3-10.
The final column moments, in kN·m, are

$$M_{AD} = M_{CF} = -68.5 \qquad M_{BE} = (2)(-68.5) = -137.0$$
$$M_{DA} = M_{FC} = -31.5 \qquad M_{EB} = (2)(-31.5) = -63.0$$
$$M_{DG} = M_{FI} = -23.9 \qquad M_{EH} = (2)(-23.9) = -47.8$$
$$M_{GD} = M_{IF} = -26.0 \qquad M_{HE} = (2)(-26.0) = -52.0$$

Supplementary Problems

3.4.   Use the *portal method* to find the moments in all of the columns and girders of the frame shown in Fig. 3-11. The answers, in ft-kips, are listed in Table 3.1.

3.5.   Use the *cantilever moment distribution method* to find the moments in all of the columns and girders of the frame shown in Fig. 3-11. The answers, in ft-kips, are given in Table 3.1.

Table 3.1

Moment	Portal Method	Cantilever Moment Distribution Method
AD	−18.75	−24.72
DA	−18.75	−12.78
DG	−6.25	−5.14
GD	−6.25	−7.33
BE	−37.50	−49.44
EB	−37.50	−25.56
EH	−12.50	−10.28
HE	−12.50	−14.64
CF	−18.75	−24.72
FC	−18.75	−12.78
FI	−6.25	−5.14
IF	−6.25	−7.33
DE	+25.00	+17.92
ED	+25.00	+17.92
EF	+25.00	+17.92
FE	+25.00	+17.92
GH	+6.25	+7.33
HG	+6.25	+7.33
HI	+6.25	+7.33
IH	+6.25	+7.33

3.6.   Find all end moments in the frame shown in Fig. 3-12 using the *portal method*. Answers, in ft-kips, are listed in Table 3.2.

3.7.   Solve Problem 3.6 using the *cantilever moment distribution method*. Answers, in ft-kips, are listed in Table 3.2.

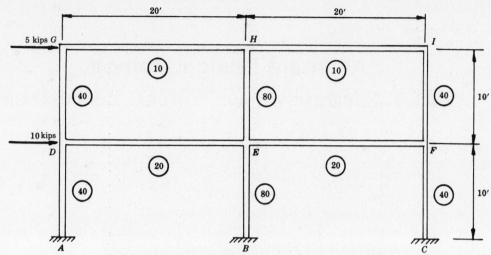

Stiffness factors are shown in circles on each member.
Fig. 3-11

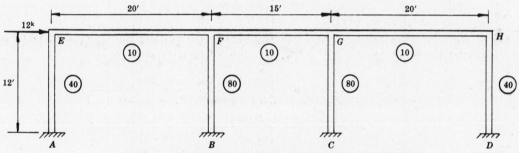

Stiffness factors are shown in circles on each member.
Fig. 3-12

Table 3.2

Moment	Portal Method	Cantilever Moment Distribution Method
AE	−12.0	−16.8
EA	−12.0	−7.4
BF	−24.0	−33.6
FB	−24.0	−14.8
CG	−24.0	−33.6
GC	−24.0	−14.8
DH	−12.0	−16.8
HD	−12.0	−7.4
EF	+12.0	+7.4
FE	+12.0	+7.4
FG	+12.0	+7.4
GF	+12.0	+7.4
GH	+12.0	+7.4
HG	+12.0	+7.4

3.8. In Fig. 3-11, if beam span lengths are 8 m, column heights are 3 m and lateral loads at D and G are each 80 kN, find the moments at the ends of all columns. K-values in kN · m are shown on the figure. *Partial Ans.* $M_{AD} = -82.1$ kN · m, $M_{DA} = -37.9$ kN · m, $M_{DG} = -28.7$ kN · m, $M_{GD} = -31.2$ kN · m

Chapter 4

Alternate Design Method

GENERAL REQUIREMENTS AND FLEXURAL COMPUTATIONS

NOTATION

A_g = gross area of concrete cross section, in.^{2} (mm^2)

A_s = area of tension reinforcement, in.^{2} (mm^2)

A_s' = area of compression reinforcement, in.^{2} (mm^2)

A_v = total area of web reinforcement for shear, V_s, measured in a direction parallel to the longitudinal reinforcement, in.^{2} (mm^2)

α = angle between inclined stirrups or bent-up bars and longitudinal axis of a member, degrees (radians)

b = width of compression face of a flexural member, and smaller dimension of a column cross section, in. (mm)

b_0 = perimeter of critical section for punching shear in slabs and footings, in. (mm)

b_w = width of web of a T-beam, I-beam or spandrel beam (inverted L-beam), in. (mm)

β_c = aspect ratio of a slab or column cross section, long side/short side

C = a factor modifying the stiffness K of a member

C_c = compression force in concrete, kips (kN)

C_s = compression force in reinforcement, kips (kN)

d = distance from extreme compression fiber to centroid of tension steel, in. (mm)

d' = distance from extreme compression fiber to centroid of compression steel, in. (mm)

E_c = modulus of elasticity of concrete, psi (MPa)

E_s = modulus of elasticity of steel (29×10^6 psi or 200 000 MPa)

f_b = bearing stress, psi (MPa)

f_s = stress in tension reinforcement or web reinforcement, psi (MPa)

f_s' = stress in compression reinforcement, psi (MPa)

f_y = yield strength of reinforcement, psi (MPa)

h = total thickness or total depth of a slab or beam, and larger dimension of a column cross section, in. (mm)

h_f = flange thickness for T-beams, I-beams and spandrel beams (inverted L-beam), in. (mm)

I = moment of inertia of a cross section, in.^{4} (mm^4)

k = a factor related to the location of the neutral axis of a cross section

K = stiffness factor, $E_c I/L$, in.-kips (kN $\cdot$ m)

L = span length, ft (m)

M = bending moment, in.-kips (kN $\cdot$ m)

n = modular ratio, E_s/E_c

N = load normal to a cross section (axial load), to be taken as positive for compression and negative for tension, psi (MPa)

p = steel ratio for tension reinforcement, A_s/bd

p' = steel ratio for compression reinforcement, A'_s/bd

p_f = steel ratio for tension reinforcement required to balance the compression force in overhanging flanges of T-beams, I-beams or spandrel beams (inverted L-beams), $A_{sf}/b_w d$

p_w = steel ratio for tension reinforcement required to balance the compression force in the webs of T-beams, I-beams or spandrel beams (inverted L-beams), $A_{sw}/b_w d$

INTRODUCTION

The vast majority of reinforced concrete structures built in America before 1963 were proportioned based on a straight-line theory called working stress design (referenced as the *alternate design method* in the 1983 ACI Code). Although strength design techniques are rapidly supplanting working stress design, the student or the designer should be proficient in both. Hence the fundamentals of the straight-line theory are presented here. Strength design will be discussed thoroughly in other chapters.

When using working stress design techniques, members are proportioned so that they may sustain the anticipated real loads induced (working or service loads) without the stresses in the concrete or reinforcing exceeding the proportional limits of the individual materials. Although the stress-strain diagram of concrete does not exhibit an initial straight-line portion, it is still assumed that Hooke's law does apply to concrete at stresses below $0.45f'_c$.

This leads to the basic assumptions in working stress design required for the following derivations and discussion:

(1) Plane sections before bending remain plane after bending.

(2) Both the concrete and reinforcing steel obey Hooke's law in the regions considered.

(3) Strain is proportional to the distance from the neutral axis. See (1) above.

(4) The tensile strength of the concrete is neglected.

(5) Perfect bond or adhesion is developed between the concrete and reinforcing steel so that there is no slippage between the two materials.

(6) The other basic assumptions concerning deformation and flexure of homogeneous members are valid.

(7) The modulus of elasticity of concrete is given by equation (*1.3*) of Chapter 1.

(8) The modulus of elasticity of steel is as given above.

Unfortunately the ordinary formulas for computing stresses and deformations in members of homogeneous, isotropic materials do not apply to the design of reinforced concrete. Many rather complex expressions can be developed for the members which are a composite of concrete and reinforcing steel.

It will become apparent that the developed expressions are rather lengthy and would prove to be quite cumbersome in the routine design of reinforced concrete. Tables, charts and other design aids will be discussed and derived which will greatly facilitate the solution of the derived formulas. However, so that there will be a thorough understanding of the principles involved, the fundamentals will be studied first and then we will consider the proportioning of members by using various design aids.

ALLOWABLE STRESSES

In working stress design all of the margin of safety is provided for by the fact that the calculated stresses in the members are such that they are considerably below the yield stress or ultimate stess of the various materials when the member is subjected to the anticipated service loads of the structure.

Paraphrased below are the portions of the 1983 ACI Code pertaining to allowable stresses under the alternate design method. As usual, expressions in U.S. customary units are followed by their SI equivalents in parentheses (except when one form holds for both systems). It should be emphasized that the SI forms were obtained by "hard conversion"—exact, or "soft," conversion followed by smoothing—of the U.S. customary forms. For example, the exact SI equivalent of $1.1\sqrt{f'_c}$ is, to five decimals, $0.91340\sqrt{f'_c}$, which is smoothed into

$$(0.9090\cdots)\sqrt{f'_c} = \frac{1}{11}\sqrt{f'_c}$$

Permissible service load stresses

Stresses in concrete shall not exceed the following:

Flexure

   Extreme fiber in compression . . . . $0.45f'_c$

Shear

   Beams and one-way slabs and footings

   Shear carried by concrete, v_c . . . . $1.1\sqrt{f'_c}$ $(\sqrt{f'_c}/11)$

   Maximum shear carried by concrete
   plus shear reinforcement . . . . $v_c + 4.4\sqrt{f'_c}$ $(v_c + 3\sqrt{f'_c}/8)$

   Joists
   Shear carried by concrete, v_c . . . . $1.2\sqrt{f'_c}$ $(\sqrt{f'_c}/10)$

   Two-way slabs and footings

   Shear carried by concrete, v_c . . . . $\left(1 + \dfrac{2}{\beta_c}\right)\sqrt{f'_c}$ $\left(\left(1 + \dfrac{2}{\beta_c}\right)\sqrt{f'_c}/12\right)$

   but not greater than $2\sqrt{f'_c}$ $(\sqrt{f'_c}/6)$

Bearing

   On loaded area . . . . $0.3f'_c$

Tensile stress in reinforcement f_s shall not exceed the following:

Grade 40 or Grade 50 (Grade 300M)
reinforcement. . . . . 20,000 psi   (140 MPa)

Grade 60 (Grade 400M) reinforcement or greater
and welded wire fabric (smooth or deformed). . . . . 24,000 psi   (170 MPa)

For flexural reinforcement, 3/8 in. (10 mm) or
less in diameter, in one-way slabs of not
more than 12 ft (4 m) span. . . . . $0.50f_v$
                           but not greater than 30,000 psi   (200 MPa)

Development and splices of reinforcement

   M_n shall be taken as computed moment capacity assuming all positive moment tension reinforcement at the section to be stressed to the permissible tensile stress f_s, and V_u shall be taken as unfactored shear force at the section.

Flexure

For investigation of stresses at service loads, straight-line theory (for flexure) shall be used with the following assumptions:

(1) Strains vary linearly as the distance from the neutral axis, except, for deep flexural members with overall depth-span ratios greater than 2/5 for continuous spans and 4/5 for simple spans, a nonlinear distribution of strain shall be considered.

(2) The stress-strain relationship of concrete is a straight line under service loads within permissible service load stresses.

(3) In reinforced concrete members, concrete resists no tension.

(4) The modular ratio, $n = E_s/E_c$, may be taken as the nearest whole number (but not less than 6). Except in calculations for deflections, the value of n for low-density concrete shall be assumed to be the same as for normal density concrete of the same strength.

(5) In doubly reinforced flexural members, an effective modular ratio of $2E_s/E_c$ shall be used to transform compression reinforcement for stress computations. Compressive stress in such reinforcement shall not exceed permissible tensile stress.

Because effects due to wind loads and earthquake forces are intermittent and of short duration, the 1983 ACI Code permits members to be proportioned for stresses one-third greater than those stated above when the applied stresses are due to wind or earthquake forces. However, the section so proportioned must be capable of resisting the combinations of dead and live loads if the allowable stresses are based on 100% of the values tabulated.

EFFECT OF REINFORCEMENT IN CONCRETE

As previously stated, the tensile strength of concrete is small compared to the compressive strength (normal density concrete has a tensile strength of $7.5\sqrt{f'_c}$ $(0.7\sqrt{f'_c})$). Hence it is assumed in structural design that this tensile strength is nil. Therefore it is necessary to strengthen or reinforce concrete members where they are subjected to tensile stresses. This reinforcement is usually accomplished by the embedment of steel bars or rods which must then be assumed to resist 100% of the tensile forces.

Fig. 4-1 is a schematic elevation of a reinforced concrete beam. As the magnitude of the vertical loads is increased, the elongation of the bottommost fibers of the beam exceeds the ultimate tensile strain of the concrete and it cracks. Since the strain in the section is proportional to the distance from the neutral axis, as the magnitude of the loads continues to increase, the tensile stresses increase and

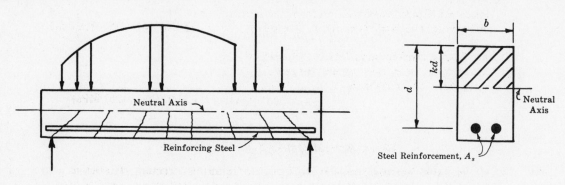

Fig. 4-1   Fig. 4-2

the cracks continue to increase in size and number and spread upward toward the neutral axis. These cracks should be perpendicular to the direction of the maximum principal tensile stress in the concrete. Hence the inclination of these cracks is a function of the flexural, shear and axial stresses to which the section is subjected.

Obviously, when the concrete is cracked it is no longer capable of transmitting or resisting tensile forces. Then the tensile forces in the bottom of the beam must be resisted by the reinforcement, and the compressive forces at the top are resisted by the concrete. Thus the effective cross section in resisting flexure is shown in Fig. 4-2. The shaded area above the neutral axis is the compression zone, and the steel reinforcement, A_s, is all that resists the tensile stresses below the neutral axis.

It is important to remember that because the tensile strength of concrete is so low, these cracks form at a load level such that the compressive stresses in the concrete and the tensile stresses in the reinforcement are still well below the ultimate or yield strengths. Also, the strains or deflections associated with the formation of these cracks are small enough that the appearance and serviceability of the structure are not impaired.

RECTANGULAR BEAMS

Fig. 4-3 shows the assumed distribution of strain and stress for a rectangular beam.

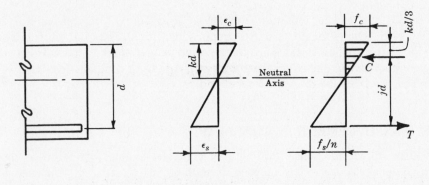

Fig. 4-3

As a matter of convenience, the principal expressions derived in Problem 4.1 are summarized below:

$$f_c = \frac{f_s k}{n(1-k)} \tag{4.1}$$

$$f_s = \frac{nf_c(1-k)}{k} \tag{4.2}$$

$$k = \sqrt{2pn + (pn)^2} - pn \tag{4.3}$$

$$j = 1 - k/3 \tag{4.4}$$

$$R = f_c jk/2 \tag{4.5}$$

$$M = Rbd^2 \tag{4.6}$$

$$d = \sqrt{M/Rb} \tag{4.7}$$

$$p = kf_c/2f_s \tag{4.8}$$

$$A_s = M/f_s jd \tag{4.9}$$

With the above equations it is possible to design a reinforced rectangular beam with tensile reinforcement. It will be shown later that aids can be developed for these equations which will greatly facilitate their solution.

TRANSFORMED SECTION

The foregoing expressions were developed assuming that the reinforced concrete beam was not homogeneous but was a composite of concrete and reinforcing steel. Although this method complies with the laws of mechanics, it can prove to be somewhat tedious when analyzing sections more complex in makeup than the previous simple example. A method will now be developed which will "transform" the composite section into an equivalent homogeneous beam. The transformed section is effected by substituting a certain cross-sectional area of concrete for the area of reinforcing steel. See Problem 4.2.

To transform the composite section of concrete and steel into an equivalent homogeneous section, the area of reinforcing steel is replaced by an area of concrete n times as great. The transformed section of Fig. 4-2 would then be as shown in Fig. 4-4, with $n = E_s/E_c$.

Thus, with the section in Fig. 4-4, the standard formulas for flexure of a homogeneous beam may be used. And equations (4.1) through (4.9) may be derived beginning with the flexure formula $f = Mc/I$.

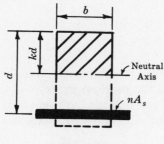

Fig. 4-4

T-BEAMS

In much reinforced concrete construction and particularly in floor systems, a concrete slab is cast monolithically with and connected to rectangular beams forming a T-beam. Fig. 4-5 below is a cross section of a concrete floor system which is composed of beams and slabs.

It is assumed, within certain limitations which will be discussed later, that the rectangular beam which has a width b_w acts structurally with the slab, b wide and h_f thick. This then forms a T-beam shown shaded in the figure.

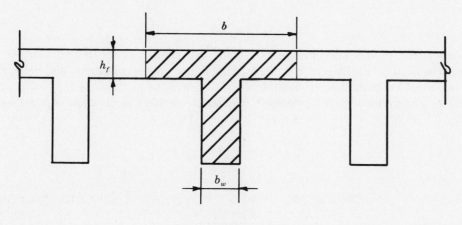

Fig. 4-5

If the neutral axis of the section falls within the slab or flange of the T-beam as shown in Fig. 4-6, then the analysis for the section is the same as for a rectangular beam. Because the effect of the concrete below the neutral axis has been neglected, the value for b_w could be anything and equations (4.1) through (4.9) would still be valid. The beam would be analyzed as a simple rectangular one with width b.

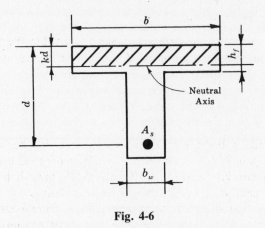

Fig. 4-6

If, however, the area of the flange is not capable of resisting the compressive force, then the neutral axis falls below the slab and the previous expressions are not applicable.

Fig. 4-7 is similar to Fig. 4-3 except that the location of the centroid of the compressive force is not as readily determined. Hence the derivations are somewhat more complex.

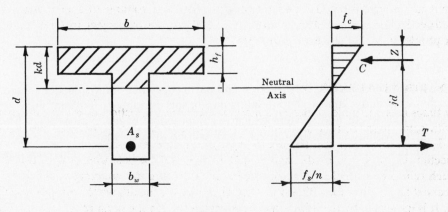

Fig. 4-7

The derivations of the formulas for a T-beam are similar to the rectangular beam. In the derivations contained in this chapter, the effect of the flange only is considered and then added to the rectangular portion or stem. See Problem 4.3.

Again, as a convenience, the principal expressions derived in Problem 4.3 are summarized below:

$$f_c = \frac{f_s k}{n(1-k)} \tag{4.10}$$

$$f_s = \frac{n f_c (1-k)}{k} \tag{4.11}$$

$$k = \frac{pn + \frac{1}{2}(h_f/d)^2}{pn + (h_f/d)} \tag{4.12}$$

$$j = \frac{6 - 6(h_f/d) + 2(h_f/d)^2 + (h_f/d)^3(\frac{1}{2}pn)}{6 - 3(h_f/d)} \qquad (4.13)$$

$$R = f_c j(1 - h_f/2kd)(h_f/d) \qquad (4.14)$$

$$M = Rbd^2 \qquad (4.15)$$

$$d = \sqrt{M/Rb} \qquad (4.16)$$

$$A_s = M/f_s jd \qquad (4.17)$$

Later, design charts or tables will be developed which will aid in the solution of equations (4.10) through (4.17). These same equations could be developed using the homogeneous or transformed section.

BALANCED DESIGN

A term frequently used in the alternate design method for flexural members is "balanced design." Sometimes this is referred to as balanced reinforcement. If a member is so proportioned that the stresses in both the concrete and the reinforcing steel reach their maximum allowable values at the same time, then the section is said to have balanced design.

Balanced design means that there is exactly enough reinforcement to develop the maximum allowable compressive stress in the concrete. If there is a lesser amount of steel, then the concrete compressive strength cannot be developed and the section is said to be "underreinforced." If there is more reinforcement than required to develop the concrete strength, the section is "overreinforced."

At first it would appear that balanced design would prove to be the most economical solution to a design. This is not always true. In fact the greatest economy is almost always attained using underreinforced sections. The reinforcing steel is usually the most expensive component in the section, and in underreinforced members the ratio of steel volume to total volume is made less. Underreinforced members are deeper and stiffer and are not as subject to immediate or long-term deflection problems as the shallower members are.

COMPRESSION REINFORCEMENT

Many times it is desirable or perhaps mandatory to have a section of minimum depth in order to comply with some architectural or structural requirement. In fact, sometimes the section based on balanced design is too deep and the depth must be made even less.

If a section is arbitrarily made shallower than required for balanced design, there will be more than enough reinforcing steel to develop the concrete. Hence the concrete is overstressed when the steel is stressed to its allowable. When this condition exists, the allowable compressive force must be increased. It is increased by the addition of reinforcing steel in the compression zone. The reinforcing steel is capable of resisting compressive stresses many times greater than concrete can. Compression steel is always provided at the supports in continuous members because Section 12.11 of the 1983 ACI Code requires that one-fourth of the midspan tension steel must be extended into the support.

When compressive steel is present, the section is doubly reinforced and can withstand a greater moment and/or be made shallower. When this is done, the expressions previously derived are no longer valid because it was assumed that there was tension reinforcement only. Therefore with compressive reinforcement Fig. 4-2 would now look like Fig. 4-8.

Fig. 4-3 would now appear like Fig. 4-9. ϵ_c, ϵ_s', C_c, and C_s' are the unit strains and compressive forces in the concrete and steel respectively.

Again as a summary, the expressions used in the working stress design of doubly reinforced beams as derived in Problem 4.4 are:

$$f_c = \frac{f_s k}{n(1 - k)} \qquad (4.18)$$

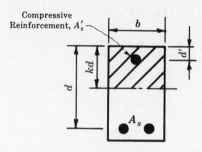

Fig. 4-8

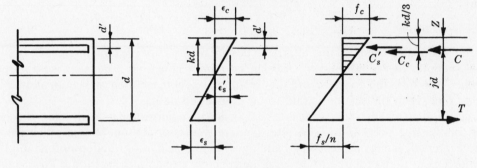

Fig. 4-9

$$f_s = nf_c\left(\frac{1-k}{k}\right) \tag{4.19}$$

$$f'_s = 2nf_c\,\frac{k-d'/d}{k} \tag{4.20}$$

$$k = \begin{cases} \sqrt{2n(p+2p'd'/d)+n^2(p+2p')^2}-n(p+2p')\,, \\ \text{when compression steel is transformed by } 2n \\[4pt] \sqrt{[np+(2n-1)p']^2+2[np+(2n-1)p'd'/d]}-[np+(2n-1)p']\,, \\ \text{when compression steel is transformed by } 2n-1 \end{cases} \tag{4.21}$$

$$j = 1 - z/d \tag{4.22}$$

$$M_c = Rbd^2 \tag{4.23}$$

$$M_s = A'_s f'_s (d - d') \tag{4.24}$$

$$d = \sqrt{M/Rb} \tag{4.25}$$

$$A_s = M/f_s jd \tag{4.26}$$

$$A'_s = \frac{M - Rbd^2}{f'_s(d-d')} \tag{4.27}$$

$$z = \begin{cases} \dfrac{k^3 d/3 + 2np'd'(k-d'/d)}{k^2 + 2np'(k-d'/d)}\,, \\ \text{when compression steel is transformed by } n \\[6pt] \dfrac{k^3 d/3 + 4np'd'(k-d'/d)}{k^2 + 4np'(k-d'/d)}\,, \\ \text{when compression steel is transformed by } 2n-1 \end{cases} \tag{4.28}$$

There also will be design aids developed for the solution of equations (4.18) through (4.28).

MODULUS OF ELASTICITY OF CONCRETE

Equation (1.3) of Chapter 1 gives the empirical formula for E_c adopted by the 1983 ACI Code. The formula is assumed valid for concrete densities ranging from 90 lb/ft^3 (1500 kg/m^3) to 155 lb/ft^3 (2500 kg/m^3).

It is obvious that the modulus of elasticity of concrete varies with the density. The modular ratio then varies inversely as the modulus of elasticity of the concrete. Hence it would seem that the value for the modular ratio for lightweight (low density) structural concrete would be much greater than that for normal density concrete of equal cylinder strength.

Except in deflection calculations, the 1983 ACI Code specifies that n for low density concrete be taken as the same as normal density concrete of equal strength. This should require an explanation.

Assuming that $f_c' = 3000$ psi, then for 145 lb/ft^3 density,

$$E_c = 33(145)^{1.5}\sqrt{3000} = 3{,}160{,}000 \text{ psi} \qquad \text{and} \qquad n = \frac{E_s}{E_c} = \frac{29{,}000{,}000}{3{,}160{,}000} = 9.2$$

The Code permits use of the nearest whole number.

For $w_c = 100$ lb/ft^3 and $f_c' = 3000$ psi, $n = 16$.

If $f_s = 20{,}000$ psi, then for $n = 9.2$

$$k = \frac{1}{1 + f_s/nf_c} = \frac{1}{1 + 20{,}000[9.2(3000)(0.45)]} = 0.383 \qquad \text{and} \qquad j = 1 - k/3 = 0.872$$

If $n = 16$,

$$k = \frac{1}{1 + 20{,}000/[16(3000)(0.45)]} = 0.52 \qquad \text{and} \qquad j = 1 - 0.52/3 = 0.827$$

The balanced steel ratios would be

$$p = f_c k/2f_s = 0.0129 \qquad \text{for} \qquad n = 9.2 \qquad \text{and} \qquad p = 0.0175 \qquad \text{for} \qquad n = 16$$

(similar values would be obtained from an SI calculation.)

This means that balanced reinforcement for the lightweight concrete would be approximately 35% greater than balanced reinforcement for the normal density concrete. Strength design theory (Chapter 5) will show that the apparent ultimate capacity of the lightweight section would be approximately 30% greater than that of the normal density section.

If the same values of n are compared for an underreinforced section, as an example for $p = 0.008$, then $k = \sqrt{2np + (np)^2} - np = \sqrt{2(9.2)(0.008) + (9.2)^2(0.008)^2} - 9.2(0.008) = 0.318$ for $n = 9.2$. And $k = 0.393$ for $n = 16$. Then $j = 0.894$ for $n = 9.2$, and $j = 0.869$ for $n = 16$. If $M = f_s j d A_s$, the moment capacity for the normal density concrete would be approximately 3% greater than that for the lightweight section. The ultimate capacities would be identical.

Although the above comparison is merely an arbitrarily selected example, it does show what the results are when n is varied. Other examples would bear out this trend.

With balanced reinforcement in the alternate design method, the lightweight concrete section has ostensibly a much greater ultimate capacity if the two n values are used. For underreinforced sections the working stress resisting moments are almost equal and the ultimate resisting moments are identical regardless of what n is used. Therefore in flexural computations it is logical that the value of n should be taken as that determined for normal density concrete regardless of the actual density of the concrete.

DESIGN AIDS AND TABLES

Two methods have been presented with which the flexural design of reinforced concrete can be accomplished when the straight-line theory is used: the analysis (1) as a composite non-homogeneous section and (2) as a homogeneous transformed section. With either of these two methods the laws of mechanics may be applied to a free body of the section, or values may be substituted directly into the derived equations (4.1) through (4.28). It is obvious that these methods are not suitable for rapid investigation or design and that they do not lend themselves to use as routine techniques for the design office.

As previously discussed, many design aids have been developed based on the derivations contained in this chapter. These have been in the form of charts, curves, tables, nomographs, etc. However, it seems that the solution of these equations by the use of tables has proved to be the most satisfactory for the design office.

The American Concrete Institute and others have developed series of tables which are universally accepted, and these will be the ones used herein.

Table 4.1, which is reproduced with permission of the ACI, is a series of coefficients developed for use in solving equations (4.1) through (4.9) when U.S. customary units are employed. The tables are divided into sections depending on the allowable steel stress f_s which varies from 16,000 psi to 33,000 psi. At the left-hand column of each of the tables is the concrete cylinder strength f'_c and the modular ratio n. Also at the left is the allowable compressive stress in the concrete f_c. The body of the tables contains the appropriate values of the coefficient $K = R$; the ratio of the distance from the compression face to neutral axis to the distance from the compression face to centroid of the tension reinforcement, k; the ratio of the distance from the centroid of the compressive force to centroid of the tension reinforcement to the distance from the compression face to the centroid of the tension reinforcement, j; and the percentage of tension reinforcement, p.

Knowing f_s and f_c [equations (4.1) and (4.2)], Table 4.1 solves directly for k, j, R and p [equations (4.3), (4.4), (4.5) and (4.8)]. Knowing K or R, the value for the resisting moment M may then be determined by equation (4.6). Or knowing K or R and M, the effective depth d of the section is determined by equation (4.7). With the value of j and d, the area of tension reinforcement A_s is computed using equation (4.9).

The procedure outlined in the above paragraph is the one that would be followed when designing. In an investigation of an established section, this procedure may be reversed.

Table 4.1 indicates that for a given steel stress and over wide ranges of concrete stress, the value for j varies a small amount. If a new coefficient $a = f_s j / 12,000$ is determined using the average of j for a given steel stress, then equation (4.9) is rewritten

$$A_s = M/ad \qquad (4.29)$$

where M is in foot-kips, d in inches and A_s in square inches. Values of a are included at the top of each table and based on the average value of j.

As a matter of comparison, for $f_s = 20,000$ psi, the maximum value of j is 0.898 and the minimum is 0.828. Then

$$a(\text{max.}) = \frac{20,000}{12,000}\,(0.898) = 1.50 \qquad a(\text{min.}) = \frac{20,000}{12,000}\,(0.828) = 1.38$$

The average value of a is given as 1.44, which is approximately 4% different from both the maximum and minimum values. For a given concrete strength and modular ratio, the variation between the maximum and minimum values is even less. For $f_s = 20,000$ psi and $f'_c = 5000$ psi, the value of j varies from 0.872 to 0.828 and a would then vary from 1.45 to 1.38. Hence the assumption of an average value of a and j is well within the precision required for ordinary design practices.

In the flexural design of a slab, the moment capacity may be expressed in terms of a unit width. If in equation (4.5), $b = 12''$, then a table has been developed for the moment resisted by the concrete for various depths of slabs.

Knowing f'_c, f_c, f_s and d, the body of Table 4.2 contains the resisting moment. If the required moment capacity is known, then the effective depth is found at the top of the table. Table 4.2 is reproduced with permission of the ACI.

Equations (4.13) and (4.14) show that j and K or R are functions of the ratio h_f/d. Hence in the design of T-sections, the tables previously developed are not valid and a new set is needed which

Table 4.1 Coefficients (K, k, j, p) for Rectangular Sections

f'_c, psi and n	f_c, psi	K	k	j	p	K	k	j	p
		$f_s = 16{,}000$ psi		$a = 1.13$		$f_s = 18{,}000$ psi		$a = 1.29$	
2500	875.	137.	.356	.881	.0097	128.	.329	.890	.0080
	1000.	169.	.387	.871	.0121	158.	.359	.880	.0100
	1125.	201.	.415	.862	.0146	190.	.387	.871	.0121
10.1	1250.	235.	.441	.853	.0172	222.	.412	.863	.0143
	1500.	306.	.486	.838	.0228	291.	.457	.848	.0190
3000	1050.	173.	.376	.875	.0124	162.	.349	.884	.0102
	1200.	212.	.408	.864	.0153	199.	.380	.873	.0127
	1350.	252.	.437	.854	.0184	238.	.408	.864	.0153
9.2	1500.	294.	.463	.846	.0217	278.	.434	.855	.0181
	1800.	380.	.509	.830	.0286	362.	.479	.840	.0240
4000	1400.	249.	.412	.863	.0180	234.	.384	.872	.0149
	1600.	303.	.444	.852	.0222	286.	.416	.861	.0185
	1800.	359.	.474	.842	.0266	341.	.444	.852	.0222
8.0	2000.	417.	.500	.833	.0313	397.	.471	.843	.0261
	2400.	536.	.545	.818	.0409	513.	.516	.828	.0344
5000	1750.	327.	.437	.854	.0239	309.	.408	.864	.0199
	2000.	397.	.470	.843	.0294	376.	.441	.853	.0245
	2250.	468.	.500	.833	.0351	446.	.470	.843	.0294
7.1	2500.	542.	.526	.825	.0411	518.	.497	.835	.0345
	3000.	694.	.571	.810	.0535	666.	.542	.819	.0452

$$p = \frac{A_s}{bd}$$

$$k = \frac{1}{1 + f_s/nf_c} \qquad j = 1 - \tfrac{1}{3}k$$

$$p^* = \frac{f_c}{2f_s} \times k \qquad K = \frac{f_c}{2}kj$$

$$a = \frac{f_s}{12{,}000} \times (\text{av. } j\text{-value})$$

for use in

$$A_s = \frac{M}{ad} \quad \text{or} \quad A_s = \frac{NE}{adi}$$

f_c, psi	K	k	j	p	K	k	j	p	K	k	j	p
	$f_s = 20{,}000$ psi		$a = 1.44$		$f_s = 22{,}000$ psi		$a = 1.60$		$f_s = 24{,}000$ psi		$a = 1.76$	
2500, 10.1												
875.	120.	.306	.898	.0067	113.	.287	.904	.0057	107.	.269	.910	.0049
1000.	149.	.336	.888	.0084	141.	.315	.895	.0072	133.	.296	.901	.0062
1125.	179.	.362	.879	.0102	170.	.341	.886	.0087	161.	.321	.893	.0075
1250.	211.	.387	.871	.0121	200.	.365	.878	.0104	191.	.345	.885	.0090
1500.	277.	.431	.856	.0162	264.	.408	.864	.0139	253.	.387	.871	.0121
3000, 9.2												
1050.	152.	.326	.891	.0085	144.	.305	.898	.0073	136.	.287	.904	.0063
1200.	188.	.356	.881	.0107	178.	.334	.889	.0091	169.	.315	.895	.0079
1350.	226.	.383	.872	.0129	214.	.361	.880	.0111	204.	.341	.886	.0096
1500.	265.	.408	.864	.0153	252.	.385	.872	.0131	240.	.365	.878	.0114
1800.	346.	.453	.849	.0204	331.	.429	.857	.0176	317.	.408	.864	.0153
4000, 8.0												
1400.	221.	.359	.880	.0126	210.	.337	.888	.0107	199.	.318	.894	.0093
1600.	272.	.390	.870	.0156	258.	.368	.877	.0134	246.	.348	.884	.0116
1800.	324.	.419	.860	.0188	309.	.396	.868	.0162	295.	.375	.875	.0141
2000.	379.	.444	.852	.0222	362.	.421	.860	.0191	347.	.400	.867	.0167
2400.	492.	.490	.837	.0294	472.	.466	.845	.0254	454.	.444	.852	.0222
5000, 7.1												
1750.	292.	.383	.872	.0168	278.	.361	.880	.0144	265.	.341	.886	.0124
2000.	358.	.415	.862	.0208	341.	.392	.869	.0178	326.	.372	.876	.0155
2250.	426.	.444	.852	.0250	407.	.421	.860	.0215	390.	.400	.867	.0187
2500.	496.	.470	.843	.0294	475.	.447	.851	.0254	456.	.425	.858	.0221
3000.	641.	.516	.828	.0387	617.	.492	.836	.0335	595.	.470	.843	.0294

f_c, psi	K	k	j	p	K	k	j	p	K	k	j	p
	$f_s = 27{,}000$ psi		$a = 2.00$		$f_s = 30{,}000$ psi		$a = 2.24$		$f_s = 33{,}000$ psi		$a = 2.48$	
2500, 10.1												
875.	99.	.247	.918	.0040	92.	.228	.924	.0033	86.	.211	.930	.0028
1000.	124.	.272	.909	.0050	115.	.252	.916	.0042	108.	.234	.922	.0036
1125.	150.	.296	.901	.0062	140.	.275	.908	.0052	132.	.256	.915	.0044
1250.	178.	.319	.894	.0074	167.	.296	.901	.0062	157.	.277	.908	.0052
1500.	237.	.359	.880	.0100	224.	.336	.888	.0084	211.	.315	.895	.0072
3000, 9.2												
1050.	126.	.264	.912	.0051	117.	.244	.919	.0043	110.	.226	.925	.0036
1200.	157.	.290	.903	.0064	147.	.269	.910	.0054	138.	.251	.916	.0046
1350.	190.	.315	.895	.0079	178.	.293	.902	.0066	168.	.273	.909	.0056
1500.	225.	.338	.887	.0094	211.	.315	.895	.0079	199.	.295	.902	.0067
1800.	299.	.380	.873	.0127	282.	.356	.881	.0107	267.	.334	.889	.0091
4000, 8.0												
1400.	185.	.293	.902	.0076	173.	.272	.909	.0063	162.	.253	.916	.0054
1600.	230.	.322	.893	.0095	215.	.299	.900	.0080	203.	.279	.907	.0068
1800.	277.	.348	.884	.0116	260.	.324	.892	.0097	246.	.304	.899	.0083
2000.	326.	.372	.876	.0138	308.	.348	.884	.0116	291.	.327	.891	.0099
2400.	430.	.416	.861	.0185	407.	.390	.870	.0156	387.	.368	.877	.0134
5000, 7.1												
1750.	247.	.315	.895	.0102	231.	.293	.902	.0085	218.	.274	.909	.0073
2000.	305.	.345	.885	.0128	287.	.321	.893	.0107	271.	.301	.900	.0091
2250.	366.	.372	.876	.0155	346.	.347	.884	.0130	327.	.326	.891	.0111
2500.	430.	.397	.868	.0184	407.	.372	.876	.0155	386.	.350	.883	.0132
3000.	564.	.441	.853	.0245	537.	.415	.862	.0208	511.	.392	.869	.0178

*"Balanced steel ratio" applies to problems involving bending only.

Table 4.2   Resisting Moments of Rectangular Sections 1 ft Wide (Slabs)

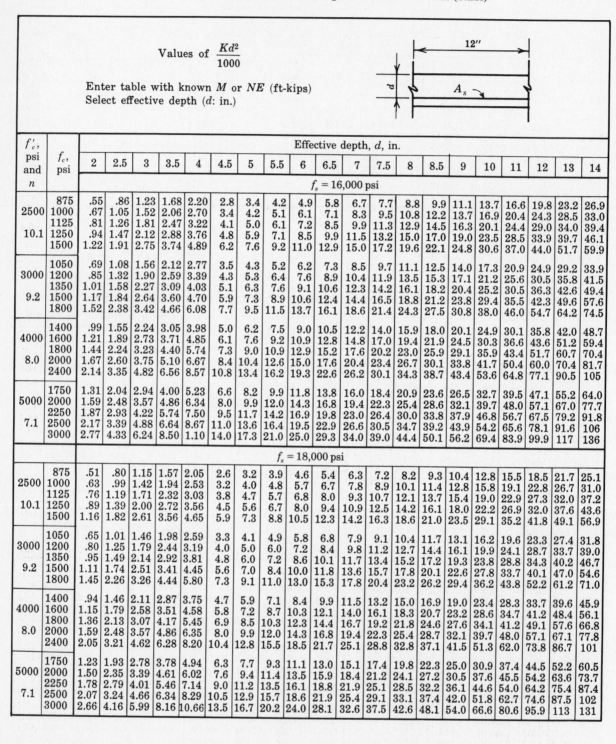

Values of $\dfrac{Kd^2}{1000}$

Enter table with known M or NE (ft-kips)
Select effective depth (d: in.)

f'_c, psi and n	f_c, psi	Effective depth, d, in.																			
		2	2.5	3	3.5	4	4.5	5	5.5	6	6.5	7	7.5	8	8.5	9	10	11	12	13	14
								$f_s = 16{,}000$ psi													
2500	875	.55	.86	1.23	1.68	2.20	2.8	3.4	4.2	4.9	5.8	6.7	7.7	8.8	9.9	11.1	13.7	16.6	19.8	23.2	26.9
	1000	.67	1.05	1.52	2.06	2.70	3.4	4.2	5.1	6.1	7.1	8.3	9.5	10.8	12.2	13.7	16.9	20.4	24.3	28.5	33.0
	1125	.81	1.26	1.81	2.47	3.22	4.1	5.0	6.1	7.2	8.5	9.9	11.3	12.9	14.5	16.3	20.1	24.4	29.0	34.0	39.4
10.1	1250	.94	1.47	2.12	2.88	3.76	4.8	5.9	7.1	8.5	9.9	11.5	13.2	15.0	17.0	19.0	23.5	28.5	33.9	39.7	46.1
	1500	1.22	1.91	2.75	3.74	4.89	6.2	7.6	9.2	11.0	12.9	15.0	17.2	19.6	22.1	24.8	30.6	37.0	44.0	51.7	59.9
3000	1050	.69	1.08	1.56	2.12	2.77	3.5	4.3	5.2	6.2	7.3	8.5	9.7	11.1	12.5	14.0	17.3	20.9	24.9	29.2	33.9
	1200	.85	1.32	1.90	2.59	3.39	4.3	5.3	6.4	7.6	8.9	10.4	11.9	13.5	15.3	17.1	21.2	25.6	30.5	35.8	41.5
	1350	1.01	1.58	2.27	3.09	4.03	5.1	6.3	7.6	9.1	10.6	12.3	14.2	16.1	18.2	20.4	25.2	30.5	36.3	42.6	49.4
9.2	1500	1.17	1.84	2.64	3.60	4.70	5.9	7.3	8.9	10.6	12.4	14.4	16.5	18.8	21.2	23.8	29.4	35.5	42.3	49.6	57.6
	1800	1.52	2.38	3.42	4.66	6.08	7.7	9.5	11.5	13.7	16.1	18.6	21.4	24.3	27.5	30.8	38.0	46.0	54.7	64.2	74.5
4000	1400	.99	1.55	2.24	3.05	3.98	5.0	6.2	7.5	9.0	10.5	12.2	14.0	15.9	18.0	20.1	24.9	30.1	35.8	42.0	48.7
	1600	1.21	1.89	2.73	3.71	4.85	6.1	7.6	9.2	10.9	12.8	14.8	17.0	19.4	21.9	24.5	30.3	36.6	43.6	51.2	59.4
	1800	1.44	2.24	3.23	4.40	5.74	7.3	9.0	10.9	12.9	15.2	17.6	20.2	23.0	25.9	29.1	35.9	43.4	51.7	60.7	70.4
8.0	2000	1.67	2.60	3.75	5.10	6.67	8.4	10.4	12.6	15.0	17.6	20.4	23.4	26.7	30.1	33.8	41.7	50.4	60.0	70.4	81.7
	2400	2.14	3.35	4.82	6.56	8.57	10.8	13.4	16.2	19.3	22.6	26.2	30.1	34.3	38.7	43.4	53.6	64.8	77.1	90.5	105
5000	1750	1.31	2.04	2.94	4.00	5.23	6.6	8.2	9.9	11.8	13.8	16.0	18.4	20.9	23.6	26.5	32.7	39.5	47.1	55.2	64.0
	2000	1.59	2.48	3.57	4.86	6.34	8.0	9.9	12.0	14.3	16.8	19.4	22.3	25.4	28.6	32.1	39.7	48.0	57.1	67.0	77.7
	2250	1.87	2.93	4.22	5.74	7.50	9.5	11.7	14.2	16.9	19.8	23.0	26.4	30.0	33.8	37.9	46.8	56.7	67.5	79.2	91.8
7.1	2500	2.17	3.39	4.88	6.64	8.67	11.0	13.6	16.4	19.5	22.9	26.6	30.5	34.7	39.2	43.9	54.2	65.6	78.1	91.6	106
	3000	2.77	4.33	6.24	8.50	1.10	14.0	17.3	21.0	25.0	29.3	34.0	39.0	44.4	50.1	56.2	69.4	83.9	99.9	117	136
								$f_s = 18{,}000$ psi													
2500	875	.51	.80	1.15	1.57	2.05	2.6	3.2	3.9	4.6	5.4	6.3	7.2	8.2	9.3	10.4	12.8	15.5	18.5	21.7	25.1
	1000	.63	.99	1.42	1.94	2.53	3.2	4.0	4.8	5.7	6.7	7.8	8.9	10.1	11.4	12.8	15.8	19.1	22.8	26.7	31.0
	1125	.76	1.19	1.71	2.32	3.03	3.8	4.7	5.7	6.8	8.0	9.3	10.7	12.1	13.7	15.4	19.0	22.9	27.3	32.0	37.2
10.1	1250	.89	1.39	2.00	2.72	3.56	4.5	5.6	6.7	8.0	9.4	10.9	12.5	14.2	16.1	18.0	22.2	26.9	32.0	37.6	43.6
	1500	1.16	1.82	2.61	3.56	4.65	5.9	7.3	8.8	10.5	12.3	14.2	16.3	18.6	21.0	23.5	29.1	35.2	41.8	49.1	56.9
3000	1050	.65	1.01	1.46	1.98	2.59	3.3	4.1	4.9	5.8	6.8	7.9	9.1	10.4	11.7	13.1	16.2	19.6	23.3	27.4	31.8
	1200	.80	1.25	1.79	2.44	3.19	4.0	5.0	6.0	7.2	8.4	9.8	11.2	12.7	14.4	16.1	19.9	24.1	28.7	33.7	39.0
	1350	.95	1.49	2.14	2.92	3.81	4.8	6.0	7.2	8.6	10.1	11.7	13.4	15.2	17.2	19.3	23.8	28.8	34.3	40.2	46.7
9.2	1500	1.11	1.74	2.51	3.41	4.45	5.6	7.0	8.4	10.0	11.8	13.6	15.7	17.8	20.1	22.6	27.8	33.7	40.1	47.0	54.6
	1800	1.45	2.26	3.26	4.44	5.80	7.3	9.1	11.0	13.0	15.3	17.8	20.4	23.2	26.2	29.4	36.2	43.8	52.2	61.2	71.0
4000	1400	.94	1.46	2.11	2.87	3.75	4.7	5.9	7.1	8.4	9.9	11.5	13.2	15.0	16.9	19.0	23.4	28.3	33.7	39.6	45.9
	1600	1.15	1.79	2.58	3.51	4.58	5.8	7.2	8.7	10.3	12.1	14.0	16.1	18.3	20.7	23.2	28.6	34.7	41.2	48.4	56.1
	1800	1.36	2.13	3.07	4.17	5.45	6.9	8.5	10.3	12.3	14.4	16.7	19.2	21.8	24.6	27.6	34.1	41.2	49.1	57.6	66.8
8.0	2000	1.59	2.48	3.57	4.86	6.35	8.0	9.9	12.0	14.3	16.8	19.4	22.3	25.4	28.7	32.1	39.7	48.0	57.1	67.1	77.8
	2400	2.05	3.21	4.62	6.28	8.20	10.4	12.8	15.5	18.5	21.7	25.1	28.8	32.8	37.1	41.5	51.3	62.0	73.8	86.7	101
5000	1750	1.23	1.93	2.78	3.78	4.94	6.3	7.7	9.3	11.1	13.0	15.1	17.4	19.8	22.3	25.0	30.9	37.4	44.5	52.2	60.5
	2000	1.50	2.35	3.39	4.61	6.02	7.6	9.4	11.4	13.5	15.9	18.4	21.2	24.1	27.2	30.5	37.6	45.5	54.2	63.6	73.7
	2250	1.78	2.79	4.01	5.46	7.14	9.0	11.2	13.5	16.1	18.8	21.9	25.1	28.5	32.2	36.1	44.6	54.0	64.2	75.4	87.4
7.1	2500	2.07	3.24	4.66	6.34	8.29	10.5	12.9	15.7	18.6	21.9	25.4	29.1	33.1	37.4	42.0	51.8	62.7	74.6	87.5	102
	3000	2.66	4.16	5.99	8.16	10.66	13.5	16.7	20.2	24.0	28.1	32.6	37.5	42.6	48.1	54.0	66.6	80.6	95.9	113	131

Table 4.2   (*cont.*)

f'_c, psi and n	f_c, psi	2	2.5	3	3.5	4	4.5	5	5.5	6	6.5	7	7.5	8	8.5	9	10	11	12	13	14
										Effective depth, d, in.											
								$f_s = 20{,}000$ psi													
2500	875	.48	.75	1.08	1.47	1.93	2.4	3.0	3.6	4.3	5.1	5.9	6.8	7.7	8.7	9.8	12.0	14.6	17.3	20.3	23.6
	1000	.60	.93	1.34	1.83	2.38	3.0	3.7	4.5	5.4	6.3	7.3	8.4	9.5	10.8	12.1	14.9	18.0	21.5	25.2	29.2
	1125	.72	1.12	1.61	2.20	2.87	3.6	4.5	5.4	6.5	7.6	8.8	10.1	11.5	12.9	14.5	17.9	21.7	25.8	30.3	35.1
10.1	1250	.84	1.32	1.90	2.58	3.37	4.3	5.3	6.4	7.6	8.9	10.3	11.9	13.5	15.2	17.1	21.1	25.5	30.3	35.6	41.3
	1500	1.11	1.73	2.49	3.39	4.43	5.6	6.9	8.4	10.0	11.7	13.6	15.6	17.7	20.0	22.4	27.7	33.5	39.9	46.8	54.3
3000	1050	.61	.95	1.37	1.87	2.44	3.1	3.8	4.6	5.5	6.4	7.5	8.6	9.8	11.0	12.3	15.2	18.4	21.9	25.8	29.9
	1200	.75	1.18	1.69	2.30	3.01	3.8	4.7	5.7	6.8	7.9	9.2	10.6	12.0	13.6	15.2	18.8	22.8	27.1	31.8	36.9
	1350	.90	1.41	2.03	2.76	3.61	4.6	5.6	6.8	8.1	9.5	11.1	12.7	14.4	16.3	18.3	22.6	27.3	32.5	38.1	44.2
9.2	1500	1.06	1.65	2.38	3.24	4.23	5.4	6.6	8.0	9.5	11.2	13.0	14.9	16.9	19.1	21.4	26.5	32.0	38.1	44.7	51.9
	1800	1.38	2.16	3.12	4.24	5.54	7.0	8.7	10.5	12.5	14.6	17.0	19.5	22.2	25.0	28.0	34.6	41.9	49.8	58.5	67.8
4000	1400	.88	1.38	1.99	2.71	3.54	4.5	5.5	6.7	8.0	9.3	10.8	12.4	14.2	16.0	17.9	22.1	26.8	31.9	37.4	43.4
	1600	1.09	1.70	2.44	3.33	4.35	5.5	6.8	8.2	9.8	11.5	13.3	15.3	17.4	19.6	22.0	27.2	32.9	39.1	45.9	53.2
	1800	1.30	2.03	2.92	3.97	5.19	6.6	8.1	9.8	11.7	13.7	15.9	18.2	20.7	23.4	26.3	32.4	39.2	46.7	54.8	63.5
8.0	2000	1.51	2.37	3.41	4.64	6.06	7.7	9.5	11.5	13.6	16.0	18.6	21.3	24.2	27.4	30.7	37.9	45.8	54.5	64.0	74.2
	2400	1.97	3.07	4.43	6.02	7.87	10.0	12.3	14.9	17.7	20.8	24.1	27.7	31.5	35.5	39.8	49.2	59.5	70.8	83.1	96.4
5000	1750	1.17	1.83	2.63	3.58	4.68	5.9	7.3	8.8	10.5	12.4	14.3	16.5	18.7	21.1	23.7	29.2	35.4	42.1	49.4	57.3
	2000	1.43	2.24	3.22	4.38	5.72	7.2	8.9	10.8	12.9	15.1	17.5	20.1	22.9	25.8	29.0	35.8	43.3	51.5	60.5	70.1
	2250	1.70	2.66	3.83	5.21	6.81	8.6	10.6	12.9	15.3	18.0	20.9	23.9	27.2	30.8	34.5	42.6	51.5	61.3	71.9	83.4
7.1	2500	1.98	3.10	4.46	6.07	7.93	10.0	12.4	15.0	17.8	20.9	24.3	27.9	31.7	35.8	40.1	49.6	60.0	71.4	83.8	97.1
	3000	2.56	4.00	5.77	7.85	10.25	13.0	16.0	19.4	23.1	27.1	31.4	36.0	41.0	46.3	51.9	64.1	77.5	92.2	108	126
								$f_s = 22{,}000$ psi													
2500	875	.45	.71	1.02	1.39	1.81	2.3	2.8	3.4	4.1	4.8	5.6	6.4	7.3	8.2	9.2	11.3	13.7	16.3	19.2	22.2
	1000	.56	.88	1.27	1.73	2.25	2.9	3.5	4.3	5.1	5.9	6.9	7.9	9.0	10.2	11.4	14.1	17.0	20.3	23.8	27.6
	1125	.68	1.06	1.53	2.08	2.72	3.4	4.2	5.1	6.1	7.2	8.3	9.6	10.9	12.3	13.8	17.0	20.5	24.5	28.7	33.3
10.1	1250	.80	1.25	1.80	2.45	3.20	4.1	5.0	6.1	7.2	8.5	9.8	11.3	12.8	14.5	16.2	20.0	24.2	28.8	33.8	39.2
	1500	1.06	1.65	2.38	3.24	4.23	5.4	6.6	8.0	9.5	11.2	13.0	14.9	16.9	19.1	21.4	26.4	32.0	38.1	44.7	51.8
3000	1050	.58	.90	1.30	1.76	2.30	2.9	3.6	4.4	5.2	6.1	7.1	8.1	9.2	10.4	11.7	14.4	17.4	20.7	24.3	28.2
	1200	.71	1.11	1.60	2.18	2.85	3.6	4.5	5.4	6.4	7.5	8.7	10.0	11.4	12.9	14.4	17.8	21.6	25.7	30.1	34.9
	1350	.86	1.34	1.93	2.62	3.43	4.3	5.4	6.5	7.7	9.1	10.5	12.1	13.7	15.5	17.4	21.4	25.9	30.9	36.2	42.0
9.2	1500	1.01	1.57	2.27	3.09	4.03	5.1	6.3	7.6	9.1	10.6	12.3	14.2	16.1	18.2	20.4	25.2	30.5	36.3	42.6	49.4
	1800	1.32	2.07	2.98	4.06	5.30	6.7	8.3	10.0	11.9	14.0	16.2	18.6	21.2	23.9	26.8	33.1	40.1	47.7	56.0	64.9
4000	1400	.84	1.31	1.89	2.57	3.35	4.2	5.2	6.3	7.5	8.9	10.3	11.8	13.4	15.1	17.0	21.0	25.4	30.2	35.4	41.1
	1600	1.03	1.61	2.32	3.16	4.13	5.2	6.5	7.8	9.3	10.9	12.7	14.5	16.5	18.7	20.9	25.8	31.2	37.2	43.6	50.6
	1800	1.24	1.93	2.78	3.79	4.95	6.3	7.7	9.4	11.1	13.1	15.1	17.4	19.8	22.3	25.0	30.9	37.4	44.5	52.2	60.6
8.0	2000	1.45	2.26	3.26	4.43	5.79	7.3	9.0	10.9	13.0	15.3	17.7	20.4	23.2	26.2	29.3	36.2	43.8	52.1	61.2	70.9
	2400	1.89	2.95	4.25	5.79	7.56	9.6	11.8	14.3	17.0	20.0	23.1	26.6	30.2	34.1	38.3	47.2	57.2	68.0	79.8	92.6
5000	1750	1.11	1.74	2.50	3.40	4.45	5.6	6.9	8.4	10.0	11.7	13.6	15.6	17.8	20.1	22.5	27.8	33.6	40.0	47.0	54.5
	2000	1.36	2.13	3.07	4.18	5.46	6.9	8.5	10.3	12.3	14.4	16.7	19.2	21.8	24.6	27.6	34.1	41.3	49.1	57.6	66.8
	2250	1.63	2.54	3.66	4.98	6.51	8.2	10.2	12.3	14.6	17.2	19.9	22.9	26.0	29.4	33.0	40.7	49.2	58.6	68.8	79.8
7.1	2500	1.90	2.97	4.28	5.82	7.60	9.6	11.9	14.4	17.1	20.1	23.3	26.7	30.4	34.3	38.5	47.5	57.5	68.4	80.3	93.1
	3000	2.47	3.86	5.55	7.56	9.87	12.5	15.4	18.7	22.2	26.1	30.2	34.7	39.5	44.6	50.0	61.7	74.6	88.8	104	121
								$f_s = 24{,}000$ psi													
2500	875	.43	.67	.96	1.31	1.71	2.2	2.7	3.2	3.9	4.5	5.3	6.0	6.9	7.7	8.7	10.7	13.0	15.4	18.1	21.0
	1000	.53	.83	1.20	1.64	2.14	2.7	3.3	4.0	4.8	5.6	6.5	7.5	8.5	9.6	10.8	13.3	16.2	19.2	22.6	26.2
	1125	.65	1.01	1.45	1.98	2.58	3.3	4.0	4.9	5.8	6.8	7.9	9.1	10.3	11.7	13.1	16.1	19.5	23.2	27.3	31.6
10.1	1250	.76	1.19	1.72	2.34	3.05	3.9	4.8	5.8	6.9	8.1	9.3	10.7	12.2	13.8	15.4	19.1	23.1	27.5	32.2	37.4
	1500	1.01	1.58	2.28	3.10	4.04	5.1	6.3	7.6	9.1	10.7	12.4	14.2	16.2	18.3	20.5	25.3	30.6	36.4	42.7	49.5
3000	1050	.55	.85	1.23	1.67	2.18	2.8	3.4	4.1	4.9	5.8	6.7	7.7	8.7	9.8	11.0	13.6	16.5	19.6	23.0	26.7
	1200	.68	1.06	1.52	2.07	2.71	3.4	4.2	5.1	6.1	7.1	8.3	9.5	10.8	12.2	13.7	16.9	20.5	24.4	28.6	33.2
	1350	.82	1.28	1.84	2.50	3.26	4.1	5.1	6.2	7.3	8.6	10.0	11.5	13.1	14.7	16.5	20.4	24.7	29.4	34.5	40.0
9.2	1500	.96	1.50	2.16	2.95	3.85	4.9	6.0	7.3	8.7	10.2	11.8	13.5	15.4	17.4	19.5	24.0	29.1	34.6	40.6	47.1
	1800	1.27	1.98	2.86	3.89	5.08	6.4	7.9	9.6	11.4	13.4	15.6	17.9	20.3	22.9	25.7	31.7	38.4	45.7	53.6	62.2
4000	1400	.80	1.24	1.79	2.44	3.19	4.0	5.0	6.0	7.2	8.4	9.8	11.2	12.7	14.4	16.1	19.9	24.1	28.7	33.6	39.0
	1600	.98	1.54	2.21	3.01	3.94	5.0	6.2	7.4	8.9	10.4	12.1	13.8	15.7	17.8	19.9	24.6	29.8	35.4	41.6	48.2
	1800	1.18	1.85	2.66	3.62	4.73	6.0	7.4	8.9	10.6	12.5	14.5	16.6	18.9	21.3	23.9	29.5	35.7	42.5	49.9	57.9
8.0	2000	1.39	2.17	3.12	4.25	5.55	7.0	8.7	10.5	12.5	14.6	17.0	19.5	22.2	25.0	28.1	34.7	41.9	49.9	58.6	67.9
	2400	1.82	2.84	4.09	5.57	7.27	9.2	11.4	13.7	16.4	19.2	22.3	25.6	29.1	32.8	36.8	45.4	55.0	65.4	76.8	89.0
5000	1750	1.06	1.65	2.38	3.24	4.23	5.4	6.6	8.0	9.5	11.2	13.0	14.9	16.9	19.1	21.4	26.5	32.0	38.1	44.7	51.8
	2000	1.30	2.04	2.93	3.99	5.21	6.6	8.1	9.9	11.7	13.8	16.0	18.3	20.8	23.5	26.4	32.6	39.4	46.9	55.0	63.8
	2250	1.56	2.44	3.51	4.77	6.24	7.9	9.7	11.8	14.0	16.5	19.1	21.9	24.9	28.2	31.6	39.0	47.2	56.1	65.9	76.4
7.1	2500	1.82	2.85	4.11	5.59	7.30	9.2	11.4	13.8	16.4	19.3	22.4	25.7	29.2	33.0	36.9	45.6	55.2	65.7	77.1	89.4
	3000	2.38	3.72	5.35	7.29	9.52	12.0	14.9	18.0	21.4	25.1	29.1	33.5	38.1	43.0	48.2	59.5	72.0	85.6	101	117

Table 4.2 *(cont.)*

f'_c, psi and n	f_c, psi	2	2.5	3	3.5	4	4.5	5	5.5	6	6.5	7	7.5	8	8.5	9	10	11	12	13	14
		Effective depth, d, in.																			
		$f_s = 27{,}000$ psi																			
2500	875	.40	.62	.89	1.21	1.58	2.0	2.5	3.0	3.6	4.2	4.9	5.6	6.3	7.2	8.0	9.9	12.0	14.3	16.7	19.4
	1000	.50	.77	1.11	1.52	1.98	2.5	3.1	3.7	4.5	5.2	6.1	7.0	7.9	8.9	10.0	12.4	15.0	17.8	20.9	24.3
10.1	1125	.60	.94	1.35	1.84	2.40	3.0	3.8	4.5	5.4	6.3	7.4	8.4	9.6	10.8	12.2	15.0	18.2	21.6	25.4	29.4
	1250	.71	1.11	1.60	2.18	2.85	3.6	4.4	5.4	6.4	7.5	8.7	10.0	11.4	12.9	14.4	17.8	21.5	25.6	30.1	34.9
	1500	.95	1.48	2.14	2.91	3.80	4.8	5.9	7.2	8.5	10.0	11.6	13.3	15.2	17.1	19.2	23.7	28.7	34.2	40.1	46.5
3000	1050	.50	.79	1.14	1.55	2.02	2.6	3.2	3.8	4.5	5.3	6.2	7.1	8.1	9.1	10.2	12.6	15.3	18.2	21.3	24.7
	1200	.63	.98	1.42	1.93	2.52	3.2	3.9	4.8	5.7	6.6	7.7	8.8	10.1	11.4	12.7	15.7	19.0	22.6	26.6	30.8
9.2	1350	.76	1.19	1.71	2.33	3.05	3.9	4.8	5.8	6.9	8.0	9.3	10.7	12.2	13.8	15.4	19.0	23.0	27.4	32.2	37.3
	1500	.90	1.41	2.03	2.76	3.60	4.6	5.6	6.8	8.1	9.5	11.0	12.7	14.4	16.3	18.2	22.5	27.2	32.4	38.0	44.1
	1800	1.20	1.87	2.69	3.66	4.78	6.1	7.5	9.0	10.8	12.6	14.6	16.8	19.1	21.6	24.2	29.9	36.2	43.0	50.5	58.6
4000	1400	.74	1.16	1.67	2.27	2.96	3.7	4.6	5.6	6.7	7.8	9.1	10.4	11.9	13.4	15.0	18.5	22.4	26.7	31.3	36.3
	1600	.92	1.44	2.07	2.81	3.68	4.7	5.7	6.9	8.3	9.7	11.3	12.9	14.7	16.6	18.6	23.0	27.8	33.1	38.8	45.0
	1800	1.11	1.73	2.49	3.39	4.43	5.6	6.9	8.4	10.0	11.7	13.6	15.6	17.7	20.0	22.4	27.7	33.5	39.9	46.8	54.2
8.0	2000	1.30	2.04	2.93	3.99	5.22	6.6	8.1	9.9	11.7	13.8	16.0	18.3	20.9	23.5	26.4	32.6	39.4	46.9	55.1	63.9
	2400	1.72	2.69	3.87	5.26	6.87	8.7	10.7	13.0	15.5	18.2	21.1	24.2	27.5	31.0	34.8	43.0	52.0	61.9	72.6	84.2
5000	1750	.99	1.54	2.22	3.02	3.95	5.0	6.2	7.5	8.9	10.4	12.1	13.9	15.8	17.8	20.0	24.7	29.9	35.5	41.7	48.4
	2000	1.22	1.91	2.75	3.74	4.88	6.2	7.6	9.2	11.0	12.9	14.9	17.2	19.5	22.0	24.7	30.5	36.9	43.9	51.6	59.8
	2250	1.47	2.29	3.30	4.49	5.86	7.4	9.2	11.1	13.2	15.5	18.0	20.6	23.4	26.5	29.7	36.6	44.3	52.8	61.9	71.8
7.1	2500	1.72	2.69	3.87	5.27	6.88	8.7	10.8	13.0	15.5	18.2	21.1	24.2	27.5	31.1	34.9	43.0	52.1	62.0	72.7	84.3
	3000	2.26	3.53	5.08	6.91	9.03	11.4	14.1	17.1	20.3	23.8	27.6	31.7	36.1	40.8	45.7	56.4	68.3	81.3	95.4	111
		$f_s = 30{,}000$ psi																			
2500	875	.37	.58	.83	1.13	1.47	1.9	2.3	2.8	3.3	3.9	4.5	5.2	5.9	6.6	7.5	9.2	11.1	13.2	15.5	18.0
	1000	.46	.72	1.04	1.41	1.85	2.3	2.9	3.5	4.2	4.9	5.7	6.5	7.4	8.3	9.3	11.5	14.0	16.6	19.5	22.6
10.1	1125	.56	.88	1.26	1.72	2.25	2.8	3.5	4.2	5.1	5.9	6.9	7.9	9.0	10.1	11.4	14.0	17.0	20.2	23.7	27.5
	1250	.67	1.04	1.50	2.04	2.67	3.4	4.2	5.0	6.0	7.0	8.2	9.4	10.7	12.1	13.5	16.7	20.2	24.0	28.2	32.7
	1500	.89	1.40	2.01	2.74	3.58	4.5	5.6	6.8	8.0	9.4	11.0	12.6	14.3	16.1	18.1	22.4	27.0	32.2	37.8	43.8
3000	1050	.47	.73	1.06	1.44	1.88	2.4	2.9	3.6	4.2	5.0	5.8	6.6	7.5	8.5	9.5	11.7	14.2	16.9	19.9	23.0
	1200	.59	.92	1.32	1.80	2.35	3.0	3.7	4.4	5.3	6.2	7.2	8.3	9.4	10.6	11.9	14.7	17.8	21.2	24.8	28.8
9.2	1350	.71	1.11	1.61	2.18	2.85	3.6	4.5	5.4	6.4	7.5	8.7	10.0	11.4	12.9	14.4	17.8	21.6	25.7	30.1	35.0
	1500	.85	1.32	1.90	2.59	3.38	4.3	5.3	6.4	7.6	8.9	10.4	11.9	13.5	15.3	17.1	21.1	25.6	30.5	35.7	41.5
	1800	1.13	1.76	2.54	3.46	4.51	5.7	7.1	8.5	10.2	11.9	13.8	15.9	18.1	20.4	22.9	28.2	34.1	40.6	47.7	55.3
4000	1400	.69	1.08	1.56	2.12	2.77	3.5	4.3	5.2	6.2	7.3	8.5	9.7	11.1	12.5	14.0	17.3	20.9	24.9	29.2	33.9
	1600	.86	1.35	1.94	2.64	3.45	4.4	5.4	6.5	7.8	9.1	10.6	12.1	13.8	15.6	17.4	21.5	26.1	31.0	36.4	42.2
	1800	1.04	1.63	2.34	3.19	4.17	5.3	6.5	7.9	9.4	11.0	12.8	14.6	16.7	18.8	21.1	26.0	31.5	37.5	44.0	51.0
8.0	2000	1.23	1.92	2.77	3.77	4.92	6.2	7.7	9.3	11.1	13.0	15.1	17.3	19.7	22.2	24.9	30.8	37.2	44.3	52.0	60.3
	2400	1.63	2.55	3.67	4.99	6.52	8.2	10.2	12.3	14.7	17.2	20.0	22.9	26.1	29.4	33.0	40.7	49.3	58.7	68.8	79.8
5000	1750	.92	1.45	2.08	2.83	3.70	4.7	5.8	7.0	8.3	9.8	11.3	13.0	14.8	16.7	18.7	23.1	28.0	33.3	39.1	45.3
	2000	1.15	1.79	2.58	3.51	4.59	5.8	7.2	8.7	10.3	12.1	14.1	16.1	18.4	20.7	23.2	28.7	34.7	41.3	48.5	56.2
	2250	1.38	2.16	3.11	4.23	5.53	7.0	8.6	10.5	12.4	14.6	16.9	19.4	22.1	25.0	28.0	34.6	41.8	49.8	58.4	67.7
7.1	2500	1.63	2.54	3.66	4.99	6.51	8.2	10.2	12.3	14.7	17.2	19.9	22.9	26.1	29.4	33.0	40.7	49.3	58.6	68.8	79.8
	3000	2.15	3.35	4.83	6.57	8.59	10.9	13.4	16.2	19.3	22.7	26.3	30.2	34.3	38.8	43.5	53.7	64.9	77.3	90.7	105
		$f_s = 33{,}000$ psi																			
2500	875	.34	.54	.77	1.05	1.37	1.7	2.1	2.6	3.1	3.6	4.2	4.8	5.5	6.2	7.0	8.6	10.4	12.4	14.5	16.8
	1000	.43	.68	.97	1.32	1.73	2.2	2.7	3.3	3.9	4.6	5.3	6.1	6.9	7.8	8.7	10.8	13.1	15.6	18.3	21.2
10.1	1125	.53	.82	1.19	1.61	2.11	2.7	3.3	4.0	4.7	5.6	6.5	7.4	8.4	9.5	10.7	13.2	15.9	19.0	22.3	25.8
	1250	.63	.98	1.41	1.92	2.51	3.2	3.9	4.7	5.7	6.6	7.7	8.8	10.0	11.3	12.7	15.7	19.0	22.6	26.5	30.8
	1500	.84	1.32	1.90	2.59	3.38	4.3	5.3	6.4	7.6	8.9	10.4	11.9	13.5	15.3	17.1	21.1	25.6	30.4	35.7	41.4
3000	1050	.44	.69	.99	1.35	1.76	2.2	2.7	3.3	4.0	4.6	5.4	6.2	7.0	7.9	8.9	11.0	13.3	15.8	18.6	21.5
	1200	.55	.86	1.24	1.69	2.21	2.8	3.4	4.2	5.0	5.8	6.8	7.8	8.8	10.0	11.2	13.8	16.7	19.8	23.3	27.0
9.2	1350	.67	1.05	1.51	2.06	2.68	3.4	4.2	5.1	6.0	7.1	8.2	9.4	10.7	12.1	13.6	16.8	20.3	24.2	28.4	32.9
	1500	.80	1.25	1.79	2.44	3.19	4.0	5.0	6.0	7.2	8.4	9.8	11.2	12.8	14.4	16.2	19.9	24.1	28.7	33.7	39.1
	1800	1.07	1.67	2.41	3.27	4.28	5.4	6.7	8.1	9.6	11.3	13.1	15.0	17.1	19.3	21.6	26.7	32.3	38.5	45.2	52.4
4000	1400	.65	1.02	1.46	1.99	2.60	3.3	4.1	4.9	5.8	6.9	8.0	9.1	10.4	11.7	13.2	16.2	19.7	23.4	27.4	31.8
	1600	.81	1.27	1.82	2.48	3.24	4.1	5.1	6.1	7.3	8.6	9.9	11.4	13.0	14.6	16.4	20.3	24.5	29.2	34.3	39.7
	1800	.98	1.54	2.21	3.01	3.93	5.0	6.1	7.4	8.8	10.4	12.0	13.8	15.7	17.8	19.9	24.6	29.7	35.4	41.5	48.2
8.0	2000	1.16	1.82	2.62	3.56	4.66	5.9	7.3	8.8	10.5	12.3	14.3	16.4	18.6	21.0	23.6	29.1	35.2	41.9	49.2	57.0
	2400	1.55	2.42	3.49	4.74	6.20	7.8	9.7	11.7	13.9	16.4	19.0	21.8	24.8	28.0	31.4	38.7	46.9	55.8	65.4	75.9
5000	1750	.87	1.36	1.96	2.66	3.48	4.4	5.4	6.6	7.8	9.2	10.7	12.2	13.9	15.7	17.6	21.8	26.3	31.3	36.8	42.6
	2000	1.08	1.69	2.44	3.32	4.33	5.5	6.8	8.2	9.7	11.4	13.3	15.2	17.3	19.6	21.9	27.1	32.8	39.0	45.7	53.1
	2250	1.31	2.04	2.94	4.01	5.23	6.6	8.2	9.9	11.8	13.8	16.0	18.4	20.9	23.6	26.5	32.7	39.6	47.1	55.3	64.1
7.1	2500	1.54	2.41	3.48	4.73	6.18	7.8	9.7	11.7	13.9	16.3	18.9	21.7	24.7	27.9	31.3	38.6	46.7	55.6	65.3	75.7
	3000	2.05	3.20	4.60	6.27	8.18	10.4	12.8	15.5	18.4	21.6	25.1	28.8	32.7	37.0	41.4	51.1	61.9	73.7	86.4	100

Table 4.3 Coefficients (a and K) for T-Sections

$$a = \frac{f_s}{12,000} \times \text{average value of } j$$

Average values of j are taken from Table 4.1; a and K are used in:

$$A_s = \frac{M}{ad} \quad \text{or} \quad A_s = \frac{NE}{adi};$$

$$A'_s = \frac{M - KF}{cd} \quad \text{or} \quad A'_s = \frac{NE - KF}{cd}$$

$$K = \frac{f_c}{2} \times \frac{h_f}{d} \times \left(2 - \frac{h_f}{d} - \frac{h_f}{kd} + \frac{2h_f^2}{3kd^2}\right)$$

f'_c, psi and n	f_c, psi	\multicolumn{11}{c}{$f_s = 16,000$ psi — h_f/d}											\multicolumn{11}{c}{$f_s = 18,000$ — h_f/d}										
		.10	.12	.14	.16	.18	.20	.24	.28	.32	.36	.40	.10	.12	.14	.16	.18	.20	.24	.28	.32	.36	.40
		\multicolumn{11}{c}{a}											\multicolumn{11}{c}{a}										
		1.26	1.25	1.24	1.23	1.22	1.21	1.19	1.17	1.16	1.14	1.12	1.42	1.41	1.40	1.39	1.38	1.37	1.34	1.32	1.30	1.28	1.26
		\multicolumn{11}{c}{K}											\multicolumn{11}{c}{K}										
2500	875	72	82	92	101	108	115	125	132	136	137	136	71	81	90	98	105	111	121	126	128	127	124
	1000	83	96	107	118	127	135	149	158	165	168	168	82	94	105	115	124	132	144	152	157	158	157
	1125	94	109	122	135	146	156	172	185	193	199	201	93	108	121	132	143	152	167	178	185	189	189
10.1	1250	106	122	138	152	164	176	195	211	222	229	234	105	121	136	149	162	172	191	204	214	220	222
	1500	128	149	168	186	202	217	242	263	279	291	299	127	147	166	183	199	213	237	257	271	281	287
3000	1050	87	100	112	123	132	141	154	164	170	173	172	86	99	110	120	129	137	149	157	161	162	160
	1200	100	116	130	143	155	165	182	195	204	209	212	99	114	128	141	152	161	177	188	195	199	199
	1350	114	132	148	163	177	189	210	227	238	246	251	113	130	146	161	174	186	205	220	230	236	238
9.2	1500	127	148	167	184	200	214	238	258	273	283	290	126	146	165	181	196	210	233	251	264	273	277
	1800	154	180	203	224	244	263	295	321	341	357	368	153	178	201	222	241	259	289	314	333	346	356
4000	1400	117	135	152	167	181	193	213	229	239	246	249	116	134	150	164	177	189	207	221	229	234	234
	1600	135	157	176	194	211	226	251	271	285	295	301	134	155	174	192	207	221	245	263	275	283	286
	1800	153	178	201	222	241	258	288	312	331	344	353	152	176	198	219	237	254	282	304	321	332	338
8.0	2000	171	199	225	249	271	291	326	354	376	393	405	170	197	223	246	267	286	320	346	366	381	391
	2400	207	242	273	303	330	356	400	438	468	492	510	206	240	271	300	327	351	394	430	458	479	495
5000	1750	148	171	192	212	230	246	273	294	309	319	325	146	169	190	209	226	241	266	285	298	306	309
	2000	170	197	223	246	267	286	320	346	366	381	390	169	196	220	243	263	281	313	337	355	367	374
	2250	193	224	253	280	304	327	366	398	423	442	456	191	222	250	276	300	322	359	389	412	429	439
7.1	2500	215	251	283	314	342	368	413	450	481	504	521	214	249	281	310	338	363	406	441	469	490	505
	3000	260	304	344	382	417	449	507	555	595	627	652	259	302	341	378	412	444	500	546	583	613	635

f'_c, psi and n	f_c, psi	\multicolumn{11}{c}{$f_s = 20,000$ psi — h_f/d}											\multicolumn{11}{c}{$f_s = 22,000$ psi — h_f/d}										
		.10	.12	.14	.16	.18	.20	.24	.28	.32	.36	.40	.10	.12	.14	.16	.18	.20	.24	.28	.32	.36	.40
		\multicolumn{11}{c}{a}											\multicolumn{11}{c}{a}										
		1.58	1.57	1.55	1.54	1.53	1.52	1.49	1.47	1.45	1.42	1.41	1.74	1.72	1.71	1.70	1.69	1.67	1.64	1.62	1.59	1.56	1.55
		\multicolumn{11}{c}{K}											\multicolumn{11}{c}{K}										
2500	875	70	80	89	96	103	108	116	120	120	118	112	69	78	87	94	100	105	111	113	112	108	101
	1000	81	93	104	113	121	128	139	146	149	148	145	80	92	102	111	118	125	134	139	141	139	134
	1125	92	106	119	130	140	149	162	172	177	179	178	91	105	117	128	137	145	158	166	169	169	166
10.1	1250	104	120	134	147	159	169	186	198	206	210	211	103	118	132	145	156	166	181	192	198	200	199
	1500	126	146	164	181	196	210	233	250	263	271	276	125	145	163	179	193	206	228	244	255	262	264
3000	1050	85	97	108	118	126	133	144	150	152	151	147	84	96	106	115	123	129	139	143	144	140	134
	1200	98	113	126	138	148	158	172	181	187	188	186	97	112	124	136	145	154	167	174	178	177	173
	1350	112	129	144	158	171	182	200	213	221	225	225	111	128	143	156	168	178	195	206	212	214	213
9.2	1500	125	145	163	179	193	206	228	244	255	262	264	124	143	161	176	190	203	223	237	246	251	252
	1800	152	177	199	220	238	255	284	307	324	336	343	151	175	197	217	235	251	279	300	315	325	330
4000	1400	115	132	148	161	174	184	201	213	219	221	219	114	130	145	159	170	180	195	205	209	209	205
	1600	133	153	172	189	204	217	239	255	265	270	271	132	152	170	186	200	213	233	247	255	258	257
	1800	151	175	196	216	234	249	276	296	311	320	324	150	173	194	213	230	245	270	288	301	307	309
8.0	2000	169	196	220	243	263	282	314	338	356	369	376	168	194	218	240	260	278	307	330	346	356	361
	2400	205	238	269	297	323	347	388	422	448	467	481	204	237	267	294	320	343	382	414	438	455	466
5000	1750	145	167	187	205	222	236	259	276	286	292	292	144	165	185	202	218	231	252	267	275	278	276
	2000	168	194	218	239	259	277	306	328	344	353	357	166	192	215	236	255	272	299	319	332	339	341
	2250	190	220	248	273	296	317	353	380	401	415	423	189	218	245	270	292	312	346	371	389	401	406
7.1	2500	213	247	278	307	334	358	399	432	458	476	488	211	245	276	304	330	353	393	424	447	462	472
	3000	258	300	339	375	408	439	493	537	572	599	619	257	298	336	372	404	434	486	528	561	585	602

Table 4.3 *(cont.)*

f'_c, psi and n	f_c, psi	$f_s = 24,000$ psi											$f_s = 27,000$ psi										
		h_f/d											h_f/d										
		.10	.12	.14	.16	.18	.20	.24	.28	.32	.36	.40	.10	.12	.14	.16	.18	.20	.24	.28	.32	.36	.40
		a											a										
		1.90	1.88	1.86	1.85	1.84	1.82	1.79	1.76	1.74	1.71	1.69	2.13	2.11	2.10	2.09	2.07	2.05	2.02	1.98	1.95	1.92	1.90
		K											K										
2500	875	68	77	85	92	97	101	106	107	104	98	89	67	75	82	88	93	96	99	98	92	84	72
	1000	79	90	100	109	116	121	130	133	133	129	122	78	88	98	105	111	116	122	124	121	114	105
10.1	1125	91	104	115	126	134	142	153	159	161	160	155	89	102	113	122	130	137	146	150	149	145	137
	1250	102	117	131	143	153	162	176	185	190	190	187	100	115	128	139	149	157	169	176	178	176	170
	1500	124	144	161	176	190	203	223	238	247	252	253	123	142	158	173	186	198	216	228	235	237	235
3000	1050	83	94	104	113	120	126	133	136	135	130	121	81	92	101	109	115	120	125	126	122	114	102
	1200	96	110	122	133	142	150	161	168	169	167	161	95	108	120	129	138	144	153	157	156	151	141
	1350	110	126	141	153	165	174	189	199	203	204	200	108	124	138	150	160	169	181	188	190	188	181
9.2	1500	123	142	159	174	187	199	217	230	238	240	239	122	140	156	170	182	193	210	220	225	224	220
	1800	150	174	195	215	232	248	274	293	306	314	317	149	172	192	211	227	242	266	282	293	298	298
4000	1400	112	129	143	156	167	176	189	197	199	197	190	111	126	140	151	161	169	180	185	184	178	168
	1600	131	150	167	183	197	208	227	239	245	246	242	129	148	164	179	191	202	218	227	230	227	220
	1800	149	171	192	210	226	241	264	280	291	295	294	147	169	188	206	221	234	255	268	275	277	272
8.0	2000	167	192	216	237	256	273	301	322	336	344	347	165	190	213	233	251	267	292	310	321	326	325
	2400	203	235	265	292	316	338	376	406	428	443	451	201	232	261	287	311	332	367	394	413	424	429
5000	1750	142	163	182	199	214	226	245	258	264	264	259	140	161	179	194	207	219	235	244	247	243	234
	2000	165	190	213	233	251	267	292	310	321	325	324	163	187	209	228	245	259	282	297	304	305	300
	2250	187	217	243	267	288	307	339	362	378	387	390	186	214	239	262	282	300	329	349	361	366	365
7.1	2500	210	243	273	301	326	348	386	415	435	448	455	208	240	270	296	320	341	376	401	418	428	430
	3000	255	296	334	369	400	429	479	519	549	571	586	253	293	330	364	394	422	469	506	532	551	561

f'_c, psi and n	f_c, psi	$f_s = 30,000$ psi											$f_s = 33,000$ psi										
		h_f/d											h_f/d										
		.10	.12	.14	.16	.18	.20	.24	.28	.32	.36	.40	.10	.12	.14	.16	.18	.20	.24	.28	.32	.36	.40
		a											a										
		2.37	2.35	2.33	2.32	2.30	2.28	2.24	2.21	2.17	2.14	2.11	2.61	2.59	2.56	2.55	2.53	2.51	2.46	2.43	2.39	2.35	2.32
		K											K										
2500	875	65	73	80	85	89	91	92	88	80	69	54	64	71	77	81	84	86	85	79	68	54	37
	1000	76	87	95	102	107	111	115	114	109	100	87	75	85	92	98	103	106	108	105	97	85	70
10.1	1125	88	100	110	119	126	132	139	140	137	130	120	86	98	107	115	122	126	131	131	125	116	102
	1250	99	113	125	136	145	152	162	166	166	161	152	98	111	123	132	140	147	155	157	154	147	135
	1500	122	140	156	170	182	193	209	219	223	223	218	120	138	153	166	178	187	201	209	211	208	200
3000	1050	80	90	98	105	111	114	117	115	109	98	83	78	88	96	102	106	109	110	105	95	82	64
	1200	93	106	117	126	133	139	146	147	143	135	122	92	104	114	122	128	133	138	136	130	118	103
	1350	107	122	135	146	155	163	174	178	177	171	162	105	120	132	142	151	157	166	168	164	155	142
9.2	1500	120	138	153	166	178	187	202	209	211	208	201	119	136	150	163	173	182	194	199	198	192	182
	1800	147	170	189	207	223	236	258	272	280	282	279	146	167	187	203	218	231	250	262	267	266	260
4000	1400	109	124	137	147	156	163	171	173	169	160	146	107	121	133	143	151	156	162	161	154	141	124
	1600	127	145	161	174	186	195	208	215	215	209	198	125	143	157	170	180	189	199	203	199	190	176
	1800	145	166	185	202	216	228	246	256	260	258	250	143	164	182	197	210	221	237	245	245	240	228
8.0	2000	163	188	209	229	246	260	283	298	306	307	303	161	185	206	224	240	254	274	286	291	289	281
	2400	199	230	258	283	305	325	358	382	397	406	407	198	228	255	279	300	319	349	370	382	387	385
5000	1750	138	158	175	189	201	211	225	231	230	222	209	136	155	171	184	195	204	215	217	213	202	185
	2000	161	184	205	223	239	252	272	283	287	284	275	159	182	201	218	233	245	262	270	270	263	250
	2250	184	211	235	257	276	293	319	335	344	345	340	182	208	232	252	270	285	308	322	327	324	315
7.1	2500	206	237	266	291	314	333	365	388	401	407	405	204	235	262	286	308	326	355	374	384	386	381
	3000	251	291	326	359	388	415	459	492	515	530	536	249	288	323	354	382	407	449	479	498	509	511

Table 4.4 Coefficients (c) for Compressive Reinforcement for Rectangular and T-Sections

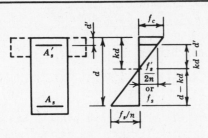

For values above heavy lines,

$$k \leq 0.333 + 0.667\frac{d'}{d}$$

and

$$c = \frac{f_s(2n-1)\left(1-\dfrac{d'}{d}\right)\left(k-\dfrac{d'}{d}\right)}{12,000n(1-k)}$$

For values below heavy lines,

$$k > 0.333 + 0.667\frac{d'}{d}$$

and

$$c = \frac{f_s\left[1-\dfrac{f_c}{f_s}\left(1-\dfrac{d'}{kd}\right)\right]\left(1-\dfrac{d'}{d}\right)}{12,000}$$

where f_s = allowable steel stress in tension, psi

Enter table with known values of $\dfrac{d'}{d}$, f_s, n and f_c; select value of c.

Compute $A'_s = \dfrac{M-KF}{cd}$.

f'_c, psi and n	f_c, psi	\multicolumn{10}{c|}{$f_s = 16{,}000$ psi, d'/d}										\multicolumn{10}{c}{$f_s = 18{,}000$ psi, d'/d}									
		.02	.04	.06	.08	.10	.12	.14	.16	.18	.20	.02	.04	.06	.08	.10	.12	.14	.16	.18	.20
2500	875	1.24	1.19	1.09	1.00	.91	.82	.73	.65	.57	.49	1.29	1.18	1.08	.98	.88	.78	.69	.60	.52	.44
	1000	1.23	1.21	1.19	1.17	1.07	.97	.88	.79	.70	.62	1.39	1.37	1.25	1.14	1.04	.94	.84	.75	.65	.57
	1125	1.22	1.20	1.18	1.16	1.14	1.11	1.03	.93	.84	.75	1.38	1.36	1.34	1.31	1.20	1.09	.99	.89	.79	.70
10.1	1250	1.21	1.19	1.17	1.15	1.13	1.11	1.09	1.06	.97	.87	1.37	1.35	1.33	1.30	1.28	1.25	1.14	1.03	.92	.82
	1500	1.19	1.17	1.15	1.13	1.11	1.09	1.07	1.05	1.03	1.01	1.35	1.33	1.31	1.29	1.26	1.24	1.22	1.19	1.17	1.08
3000	1050	1.23	1.20	1.18	1.10	1.01	.91	.82	.74	.65	.57	1.39	1.29	1.19	1.08	.98	.88	.78	.69	.60	.52
	1200	1.21	1.19	1.17	1.15	1.13	1.08	.98	.89	.80	.71	1.38	1.35	1.33	1.26	1.15	1.05	.95	.85	.75	.66
	1350	1.20	1.18	1.16	1.14	1.12	1.10	1.08	1.04	.94	.85	1.37	1.34	1.32	1.30	1.27	1.22	1.11	1.00	.90	.80
9.2	1500	1.19	1.17	1.15	1.13	1.11	1.09	1.07	1.05	1.03	.99	1.35	1.33	1.31	1.29	1.26	1.24	1.22	1.15	1.04	.94
	1800	1.17	1.15	1.13	1.11	1.09	1.07	1.05	1.03	1.01	.99	1.33	1.31	1.29	1.27	1.24	1.22	1.20	1.18	1.15	1.13
4000	1400	1.20	1.18	1.16	1.14	1.12	1.09	.99	.90	.81	.72	1.36	1.34	1.32	1.27	1.16	1.06	.96	.86	.76	.67
	1600	1.18	1.16	1.14	1.13	1.11	1.09	1.07	1.05	.98	.88	1.35	1.32	1.30	1.28	1.26	1.24	1.14	1.03	.93	.83
	1800	1.17	1.15	1.13	1.11	1.09	1.07	1.06	1.04	1.02	1.00	1.33	1.31	1.29	1.27	1.25	1.22	1.20	1.18	1.10	.99
8.0	2000	1.15	1.13	1.12	1.10	1.08	1.06	1.04	1.02	1.01	.99	1.31	1.29	1.27	1.25	1.23	1.21	1.19	1.17	1.15	1.12
	2400	1.12	1.10	1.09	1.07	1.05	1.04	1.02	1.00	.98	.97	1.28	1.26	1.24	1.22	1.20	1.18	1.16	1.14	1.12	1.10
5000	1750	1.17	1.15	1.14	1.12	1.10	1.08	1.06	1.03	.93	.84	1.33	1.31	1.29	1.27	1.25	1.20	1.09	.98	.88	.79
	2000	1.15	1.13	1.12	1.10	1.08	1.06	1.05	1.03	1.01	.99	1.31	1.29	1.27	1.25	1.23	1.21	1.19	1.17	1.07	.96
	2250	1.13	1.11	1.10	1.08	1.07	1.05	1.03	1.01	.99	.98	1.29	1.28	1.26	1.24	1.22	1.20	1.18	1.16	1.14	1.11
7.1	2500	1.11	1.10	1.08	1.06	1.05	1.03	1.02	1.00	.98	.96	1.27	1.26	1.24	1.22	1.20	1.18	1.16	1.14	1.12	1.10
	3000	1.07	1.06	1.04	1.03	1.01	1.00	.98	.97	.95	.94	1.23	1.22	1.20	1.18	1.17	1.15	1.13	1.11	1.09	1.07

f'_c, psi and n	f_c, psi	\multicolumn{10}{c|}{$f_s = 20{,}000$ psi, d'/d}										\multicolumn{10}{c}{$f_s = 22{,}000$ psi, d'/d}									
		.02	.04	.06	.08	.10	.12	.14	.16	.18	.20	.02	.04	.06	.08	.10	.12	.14	.16	.18	.20
2500	875	1.28	1.17	1.06	.95	.85	.75	.65	.56	.47	.39	1.28	1.16	1.04	.93	.82	.72	.62	.52	.43	.34
	1000	1.47	1.35	1.24	1.12	1.01	.90	.80	.70	.61	.52	1.47	1.34	1.22	1.10	.98	.87	.76	.66	.56	.47
	1125	1.55	1.52	1.41	1.29	1.17	1.06	.95	.84	.74	.65	1.66	1.53	1.39	1.27	1.14	1.03	.91	.80	.70	.59
10.1	1250	1.54	1.51	1.48	1.46	1.33	1.21	1.10	.99	.88	.77	1.70	1.67	1.57	1.44	1.31	1.18	1.06	.94	.83	.72
	1500	1.52	1.49	1.47	1.44	1.41	1.39	1.36	1.27	1.15	1.03	1.68	1.65	1.62	1.59	1.57	1.49	1.36	1.23	1.10	.98
3000	1050	1.40	1.28	1.17	1.06	.95	.85	.75	.65	.56	.47	1.39	1.27	1.15	1.03	.92	.81	.71	.61	.51	.42
	1200	1.54	1.48	1.36	1.24	1.13	1.01	.91	.80	.70	.61	1.60	1.47	1.34	1.22	1.10	.98	.87	.76	.66	.56
	1350	1.53	1.50	1.48	1.42	1.30	1.18	1.07	.96	.85	.75	1.69	1.66	1.53	1.40	1.27	1.15	1.03	.92	.80	.70
9.2	1500	1.52	1.49	1.47	1.44	1.42	1.35	1.23	1.11	1.00	.89	1.68	1.65	1.62	1.59	1.45	1.32	1.19	1.07	.95	.84
	1800	1.49	1.47	1.44	1.42	1.39	1.37	1.34	1.32	1.29	1.17	1.66	1.63	1.60	1.57	1.55	1.52	1.49	1.38	1.24	1.12
4000	1400	1.53	1.49	1.37	1.25	1.14	1.03	.92	.81	.72	.62	1.61	1.48	1.35	1.23	1.11	.99	.88	.77	.67	.57
	1600	1.51	1.49	1.46	1.44	1.34	1.22	1.10	.99	.88	.78	1.67	1.65	1.57	1.44	1.31	1.19	1.07	.95	.84	.73
	1800	1.49	1.47	1.45	1.42	1.40	1.37	1.29	1.17	1.05	.94	1.66	1.63	1.60	1.58	1.51	1.38	1.25	1.13	1.01	.89
8.0	2000	1.48	1.45	1.43	1.41	1.38	1.36	1.34	1.31	1.22	1.10	1.64	1.62	1.59	1.56	1.54	1.51	1.44	1.30	1.17	1.05
	2400	1.45	1.42	1.40	1.38	1.36	1.33	1.31	1.29	1.26	1.24	1.61	1.58	1.56	1.53	1.51	1.48	1.46	1.43	1.40	1.37
5000	1750	1.50	1.47	1.45	1.40	1.28	1.16	1.05	.94	.84	.74	1.66	1.64	1.51	1.38	1.25	1.13	1.01	.90	.79	.69
	2000	1.48	1.46	1.43	1.41	1.39	1.36	1.25	1.14	1.02	.91	1.64	1.62	1.59	1.56	1.48	1.34	1.22	1.09	.98	.86
	2250	1.46	1.44	1.41	1.39	1.37	1.35	1.32	1.30	1.21	1.09	1.62	1.60	1.57	1.55	1.52	1.50	1.42	1.29	1.16	1.04
7.1	2500	1.44	1.42	1.40	1.37	1.35	1.33	1.31	1.28	1.26	1.24	1.60	1.58	1.55	1.53	1.50	1.48	1.45	1.43	1.35	1.21
	3000	1.40	1.38	1.36	1.34	1.32	1.30	1.28	1.26	1.23	1.21	1.56	1.54	1.52	1.49	1.47	1.45	1.42	1.40	1.37	1.35

Table 4.4 *(cont.)*

f'_c, psi and n	f_c, psi	\.02	\.04	\.06	\.08	\.10	\.12	\.14	\.16	\.18	\.20	\.02	\.04	\.06	\.08	\.10	\.12	\.14	\.16	\.18	\.20
		$f_s = 24{,}000$ psi (d'/d)										$f_s = 27{,}000$ psi (d'/d)									
2500	875	1.27	1.14	1.02	.91	.79	.68	.58	.48	.38	.29	1.26	1.13	1.00	.87	.75	.63	.52	.41	.31	.21
	1000	1.46	1.33	1.20	1.07	.95	.84	.73	.62	.51	.42	1.45	1.31	1.17	1.04	.91	.79	.67	.55	.44	.34
	1125	1.65	1.51	1.38	1.24	1.12	.99	.87	.76	.65	.54	1.64	1.49	1.35	1.21	1.07	.94	.82	.70	.58	.47
10.1	1250	1.85	1.70	1.55	1.41	1.28	1.15	1.02	.90	.78	.67	1.84	1.68	1.53	1.38	1.24	1.10	.96	.84	.71	.60
	1500	1.84	1.81	1.78	1.75	1.60	1.46	1.32	1.18	1.05	.93	2.09	2.05	1.88	1.72	1.56	1.41	1.26	1.12	.98	.85
3000	1050	1.39	1.26	1.13	1.01	.89	.78	.67	.57	.47	.37	1.38	1.24	1.11	.98	.85	.73	.61	.50	.40	.29
	1200	1.60	1.46	1.32	1.19	1.07	.95	.83	.72	.61	.51	1.59	1.44	1.30	1.16	1.03	.90	.77	.66	.54	.43
	1350	1.81	1.66	1.52	1.38	1.25	1.12	.99	.87	.76	.65	1.80	1.64	1.49	1.34	1.20	1.07	.94	.81	.69	.57
9.2	1500	1.84	1.81	1.71	1.56	1.42	1.28	1.15	1.03	.90	.79	2.01	1.84	1.68	1.53	1.38	1.23	1.10	.96	.83	.71
	1800	1.82	1.79	1.76	1.73	1.70	1.62	1.47	1.33	1.20	1.07	2.07	2.03	2.00	1.90	1.73	1.57	1.42	1.27	1.13	.99
4000	1400	1.61	1.47	1.33	1.21	1.08	.96	.84	.73	.62	.52	1.60	1.45	1.31	1.17	1.04	.91	.79	.67	.55	.45
	1600	1.84	1.70	1.56	1.42	1.28	1.15	1.03	.91	.79	.68	1.84	1.68	1.53	1.38	1.24	1.10	.97	.84	.72	.61
	1800	1.82	1.79	1.76	1.63	1.49	1.35	1.21	1.08	.96	.84	2.07	1.91	1.75	1.59	1.44	1.30	1.16	1.02	.89	.77
8.0	2000	1.80	1.78	1.75	1.72	1.69	1.54	1.40	1.26	1.13	1.00	2.05	2.02	1.97	1.81	1.65	1.49	1.34	1.20	1.06	.93
	2400	1.77	1.75	1.72	1.69	1.66	1.63	1.60	1.57	1.46	1.32	2.02	1.99	1.95	1.92	1.89	1.85	1.71	1.55	1.39	1.25
5000	1750	1.78	1.63	1.49	1.36	1.22	1.10	.98	.86	.75	.64	1.77	1.61	1.47	1.32	1.18	1.05	.92	.80	.68	.56
	2000	1.81	1.78	1.73	1.59	1.45	1.31	1.18	1.05	.93	.81	2.03	1.87	1.71	1.55	1.41	1.26	1.12	.99	.86	.74
	2250	1.79	1.76	1.73	1.70	1.67	1.52	1.38	1.25	1.12	.99	2.03	2.00	1.95	1.79	1.63	1.47	1.33	1.18	1.05	.91
7.1	2500	1.77	1.74	1.71	1.68	1.66	1.63	1.59	1.44	1.30	1.17	2.01	1.98	1.95	1.92	1.85	1.69	1.53	1.38	1.23	1.09
	3000	1.73	1.70	1.68	1.65	1.62	1.60	1.57	1.54	1.51	1.49	1.97	1.94	1.91	1.88	1.85	1.82	1.79	1.76	1.60	1.44

f'_c, psi and n	f_c, psi	\.02	\.04	\.06	\.08	\.10	\.12	\.14	\.16	\.18	\.20	\.02	\.04	\.06	\.08	\.10	\.12	\.14	\.16	\.18	\.20
		$f_s = 30{,}000$ psi (d'/d)										$f_s = 33{,}000$ psi (d'/d)									
2500	875	1.25	1.11	.97	.84	.71	.58	.46	.35	.24	.14	1.24	1.09	.94	.80	.66	.53	.41	.29	.17	.06
	1000	1.44	1.29	1.15	1.00	.87	.74	.61	.49	.37	.26	1.43	1.27	1.12	.97	.83	.69	.55	.43	.30	.19
	1125	1.64	1.48	1.32	1.17	1.03	.89	.76	.63	.51	.39	1.63	1.46	1.30	1.14	.99	.84	.70	.57	.44	.32
10.1	1250	1.83	1.66	1.50	1.34	1.19	1.05	.91	.77	.64	.52	1.82	1.64	1.47	1.31	1.15	1.00	.85	.71	.57	.44
	1500	2.21	2.03	1.85	1.68	1.52	1.36	1.20	1.05	.91	.78	2.20	2.01	1.83	1.65	1.47	1.31	1.15	.99	.84	.70
3000	1050	1.37	1.22	1.08	.94	.81	.68	.56	.44	.33	.22	1.36	1.20	1.05	.91	.77	.63	.50	.38	.26	.14
	1200	1.58	1.42	1.27	1.12	.98	.85	.72	.59	.47	.36	1.57	1.40	1.24	1.09	.94	.80	.66	.53	.40	.28
	1350	1.79	1.62	1.46	1.31	1.16	1.02	.88	.75	.62	.50	1.78	1.60	1.44	1.27	1.12	.97	.82	.68	.55	.42
9.2	1500	2.00	1.82	1.66	1.49	1.34	1.19	1.04	.90	.76	.64	1.99	1.80	1.63	1.46	1.29	1.14	.98	.84	.69	.56
	1800	2.31	2.22	2.04	1.86	1.69	1.52	1.36	1.21	1.06	.91	2.40	2.21	2.01	1.83	1.65	1.47	1.30	1.14	.99	.84
4000	1400	1.59	1.43	1.28	1.14	1.00	.86	.73	.60	.48	.37	1.58	1.41	1.26	1.10	.95	.81	.67	.54	.42	.30
	1600	1.83	1.66	1.50	1.35	1.20	1.05	.91	.78	.65	.53	1.82	1.65	1.48	1.31	1.16	1.00	.86	.72	.58	.46
	1800	2.07	1.89	1.72	1.56	1.40	1.25	1.10	.96	.82	.69	2.06	1.88	1.70	1.52	1.36	1.20	1.04	.89	.75	.62
8.0	2000	2.30	2.12	1.94	1.77	1.60	1.44	1.28	1.13	.99	.85	2.30	2.11	1.92	1.74	1.56	1.39	1.23	1.07	.92	.78
	2400	2.26	2.23	2.19	2.15	2.01	1.83	1.65	1.49	1.33	1.17	2.51	2.47	2.36	2.16	1.97	1.78	1.60	1.42	1.26	1.10
5000	1750	1.76	1.60	1.44	1.29	1.14	1.00	.86	.73	.61	.49	1.75	1.58	1.41	1.25	1.10	.95	.81	.67	.54	.41
	2000	2.02	1.85	1.68	1.52	1.36	1.21	1.07	.93	.79	.66	2.01	1.83	1.66	1.49	1.32	1.16	1.01	.87	.72	.59
	2250	2.28	2.10	1.92	1.75	1.59	1.43	1.27	1.12	.98	.84	2.28	2.08	1.90	1.72	1.54	1.38	1.21	1.06	.91	.77
7.1	2500	2.26	2.22	2.17	1.99	1.81	1.64	1.47	1.32	1.16	1.02	2.50	2.34	2.14	1.95	1.77	1.59	1.42	1.25	1.09	.94
	3000	2.22	2.18	2.15	2.11	2.08	2.04	1.88	1.70	1.53	1.37	2.46	2.42	2.39	2.35	2.21	2.02	1.83	1.64	1.46	1.29

gives these coefficients for various values of h_f/d. Table 4.3 is such a table. The body of the table contains values for R, with the values for a at the top based on average values for j.

These tables neglect the compressive stress in the stem. If the stress in the stem is included, then Table 4.1 must be used along with Table 4.3. Problem 4.19 will demonstrate this procedure.

Knowing f'_c, f_c, f_s and h_f/d, the values for R and a may be determined from Table 4.3. After R is known, then d may be found by equation (4.16). If the effective flange width b, effective depth d, and the required moment capacity are known, then R can be determined by equation (4.15) and the value for h_f/d is selected from Table 4.3. Table 4.3 is reproduced with permission of the ACI.

If the expression for f'_s given in equation (4.20) is substituted in equation (4.27), the expression for the area of compression reinforcement is

$$A'_s = \frac{M - Rbd^2}{cd} \qquad\qquad (4.30)$$

If f'_s is equal to or less than f_s, then

$$c = \frac{f_s(2n-1)(1 - d'/d)(k - d'/d)}{12,000n(1-k)}$$

If f'_s is restricted to f_s, then

$$c = \frac{f_s}{12,000}\left[1 - \frac{f_c}{f_s}\frac{(kd - d')}{kd}\right](1 - d'/d)$$

In the above expressions for c, it is assumed that the effective modular ratio in compression is equal to $2n$.

Table 4.4 contains coefficients used in the design of doubly reinforced concrete beams and is reproduced with permission of the ACI.

If a doubly reinforced beam is to be designed by the tables, equation (4.26) is solved using Table 4.1 and equation (4.30) is solved using Table 4.4. Table 4.4 gives values of c for various ratios of d'/d.

If the required resisting moment M is known and if the resistance of the concrete Rbd^2 is determined by Table 4.1 and equation (4.23), the area of compressive reinforcement is then determined by equation (4.30).

The primary use for these tables is for the design of flexural members. However, these tables may also be used to investigate stresses in beams. Many times in an investigation it is convenient to use the transformed section technique rather than the tables.

Tables analogous to Tables 4.1–4.4 have not been developed for metric (SI) units because the alternate design method is not used for design purposes in countries which have adopted the metric system of units exclusively. Further, since the alternate design method is contained in the Appendix to the 1983 ACI Code, that method will eventually be eliminated from the ACI Code insofar as proportioning of reinforced concrete members is concerned. For these reasons, problems involving metric (SI) units are not included in this chapter.

In the solved problems that follow, the primary intent is to demonstrate the principles derived and discussed in this chapter. In the design of beams, factors other than flexural computations must be considered. Before a design is complete, such items as shear stresses, development length, deflections, minimum reinforcement, etc., must be checked. All of these considerations are discussed elsewhere in this book.

Solved Problems

4.1. Derive the principal expressions used in the alternate design method for rectangular reinforced concrete beams. Refer to Fig. 4-10.

For the section shown to be in equilibrium, the summation of horizontal forces must equal zero and the summation of moments must equal zero. Then $C = T$ or

$$\tfrac{1}{2}f_c kdb = A_s f_s \qquad\qquad (a)$$

And if the distance from the centroid of the compressive stress to the centroid of tensile stress is jd,

$$M_c = Cjd = \tfrac{1}{2}f_c kd^2 bj \qquad \text{and} \qquad M_s = A_s f_s jd$$

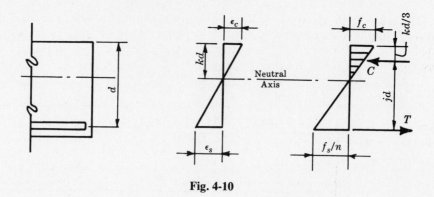

Fig. 4-10

From Fig. 4-10, $jd = d - kd/3$, or $j = 1 - k/3$. Hence to evaluate the expression for the resisting moment, the value of k must be determined.

If E = modulus of elasticity = f/ϵ, then $E_s = f_s/\epsilon_s$ and $E_c = f_c/\epsilon_c$. From the strain diagram,

$$\frac{\epsilon_c}{\epsilon_s} = \frac{kd}{d - kd} \qquad \text{or} \qquad \frac{f_c/E_c}{f_s/E_s} = \frac{kd}{d - kd} = \frac{k}{1 - k}$$

And if $n = E_s/E_c$, where n is called the modular ratio,

$$\frac{nf_c}{f_s} = \frac{k}{1 - k} \qquad\qquad\qquad (b)$$

from which

$$f_c = \frac{f_s k}{n(1 - k)}, \qquad f_s = \frac{nf_c(1 - k)}{k}$$

and

$$nf_c = f_s k + nf_c k = k(nf_c + f_s) \qquad \text{or} \qquad k = \frac{nf_c}{nf_c + f_s} = \frac{1}{1 + f_s/nf_c}$$

This is the usual form of the expression for k.

If the ratio of reinforcing steel is $p = A_s/bd$, then equation (a) becomes $\frac{1}{2} f_c kdb = pbdf_s$, or

$$f_c/f_s = 2p/k \qquad\qquad\qquad (c)$$

Substituting equation (b) into equation (c), we get

$$k^2 + 2pnk = 2pn \qquad \text{and} \qquad k = \sqrt{2pn + (pn)^2} - pn$$

Letting $R = \frac{1}{2} f_c jk$, we obtain

$$M_c = M_s = Rbd^2 \qquad \text{and} \qquad d = \sqrt{M/Rb}$$

These derived equations are developed using the straight-line theory and are applicable to the proportioning of rectangular reinforced concrete beams with tensile reinforcement only.

4.2. Derive the expressions used in the transformed section method of analysis.

Within the elastic range, stress is proportional to strain; hence $f_c = \epsilon_c E_c$ and $f_s = \epsilon_s E_s$. If the strain in concrete and steel are the same, or $\epsilon_c = \epsilon_s$, then

$$f_c/E_c = f_s/E_s \qquad \text{or} \qquad f_c = E_c f_s/E_s$$

If the modular ratio $n = E_s/E_c$, then

$$f_c = f_s/n \qquad \text{or} \qquad f_s = nf_c$$

In Fig. 4-3 the deformations or strains are assumed to be proportional to the distance from the neutral axis. Therefore at the centroid of the tensile force the strain in the concrete is the same as that in the reinforcing steel. (This assumes the concrete is capable of resisting tension.) Consequently the stress in the concrete is $f_c = f_s/n$.

If it is desired to substitute a quantity of concrete to act as the reinforcing steel, and if the stress in the concrete is less than the stress in the steel that it is replacing, then a larger area of concrete is required in order to develop the same total force. Or if the tensile force in the concrete is equal to the tensile force in the steel, then $T_c = T_s$ and $f_c A_c = f_s A_s$. Now since $f_c = f_s/n$, we have

$$f_s A_c/n = f_s A_s \qquad \text{or} \qquad A_c = nA_s$$

4.3. Derive the principal expressions used in the alternate design method for reinforced concrete T-beams. Refer to Fig. 4-11.

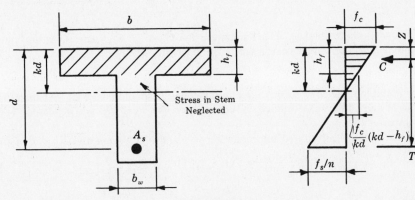

Fig. 4-11

If the compressive stress in the stem is neglected, the summation of horizontal forces is $C = T$. Since the compressive force is bounded by a trapezoid, $\frac{1}{2}[f_c + (f_c/kd)(kd - h_f)]bh_f = A_s f_s$ or

$$\frac{f_c(2kd - h_f)bh_f}{2kd} = A_s f_s \tag{a}$$

Rearranging (a),

$$\frac{f_c(2kd - h_f)bh_f}{f_s 2kd} = A_s = pbd \qquad \text{or} \qquad \frac{f_c}{f_s} = \frac{2kpd^2}{(2kd - h_f)h_f} \tag{b}$$

Substituting for f_c in (b), we obtain

$$\frac{1}{n(1-k)} = \frac{2pd^2}{(2kd - h_f)h_f} \qquad \text{from which} \qquad k = \frac{2pd^2 n + h_f^2}{2(dh_f + pd^2 n)} = \frac{pn + \frac{1}{2}(h_f/d)^2}{pn + h_f/d}$$

The distance z may be found by summing moments about the centroid of the compressive forces:

$$z = \frac{h_f(3kd - 2h_f)}{3(2kd - h_f)} \tag{c}$$

From Fig. 4-11, $jd = d - z$. Then substituting (c) into $j = 1 - z/d$, we get

$$j = \frac{6 - 6(h_f/d) + 2(h_f/d)^2 + (h_f/d)^3(\frac{1}{2}pn)}{6 - 3(h_f/d)}$$

The resisting moment of the concrete would be $M_c = C_c jd$ or $M_c = f_c(1 - h_f/2kd)bh_f jd$. If $R = f_c j(1 - h_f/2kd)(h_f/d)$, then $M_c = Rbd^2$ and $d = \sqrt{M_c/Rb}$.
These expressions were developed assuming that the compressive stress in the stem portion is negligible and that the beam has tensile reinforcement only.

4.4. Derive the principal expressions used in the alternate design method for doubly reinforced concrete beams. Refer to Fig. 4-9.

The expressions for a doubly reinforced beam will now be derived using the same basic assumptions as in Problems 4.1–4.3.

As with the section with tensile reinforcement only, $k = \dfrac{1}{1 + f_s/nf_c}$.

If the compressive stress in the compression steel is f'_s, then $\dfrac{\epsilon'_s}{\epsilon_s} = \dfrac{kd - d'}{d - kd}$; and if $E\epsilon = f$, then

$$\frac{f'_s/E_s}{f_s/E_s} = \frac{kd - d'}{d - kd} \qquad \text{or} \qquad f'_s = \frac{f_s(kd - d')}{d - kd} \tag{a}$$

Also,

$$\frac{\epsilon'_s}{\epsilon_c} = \frac{kd - d'}{kd} \qquad \text{or} \qquad \frac{f'_s/E_s}{f_c/E_c} = \frac{kd - d'}{kd}$$

Letting $n = E_s/E_c$,

$$f'_s = nf_c\left(\frac{kd - d'}{kd}\right)$$

In Fig. 4-9, if horizontal forces are summed, $T = C = C'_s + C_c$, where C'_s is the compressive force in the steel and C_c is the compressive force in the concrete. Then

$$\tfrac{1}{2} f_c k b d + A'_s f'_s = f_s A_s \tag{b}$$

It will be noted that the compressive force in the concrete should be a trifle less than that resulting from the above because the area occupied by the compressive steel was not deducted. This discrepancy is usually disregarded.

If equation (a) is substituted into equation (b), and if $A_s = pbd$ and $A'_s = p'bd$, then

$$\tfrac{1}{2} f_c k b d + p'bd f_s\left(\frac{kd - d'}{d - kd}\right) = f_s p b d \qquad \text{or} \qquad \frac{f_c}{f_s} = \frac{2}{k}\left[p - p'\left(\frac{kd - d'}{d - kd}\right)\right] \tag{c}$$

As before,

$$\frac{f_c}{f_s} = \frac{k}{n(1 - k)} \tag{d}$$

Equating f_c/f_s from (c) and (d), we obtain $k = \sqrt{2n(p + p'd'/d) + n^2(p + p')^2} - n(p + p')$.

The distance z can be determined by summing moments, $z = \dfrac{k^3 d/3 + 2np'd'(k - d'/d)}{k^2 + 2np'(k - d'/d)}$; and $jd = d - z$.

The resisting moment due to the compressive force is

$$M_c = Cjd = (C_c + C'_s)jd \qquad \text{or} \qquad M_c = jd(\tfrac{1}{2} f_c k b d + f'_s p'bd)$$

Substituting for f'_s, $M_c = \tfrac{1}{2} f_c j b d^2 [k + 2np'(1 - d'/kd)]$.

In the derivation of the above expressions for doubly reinforced beams, it was assumed that the compressive steel was n times as stiff or effective as concrete. However, because of observations of actual members in service and empirical data, the 1983 ACI Code along with most codes states that the effective modular ratio shall be taken as $2n$ for compressive steel in flexural design using the straight-line theory. But f'_s shall not exceed the allowable tensile stress of the steel. This means that in the previous expressions $2p'$ should be substituted for p'. Unless otherwise stated, a modular ratio of $2n$ shall always be used here.

The expression for k to locate the neutral axis for beams reinforced for both tension and compression should be used in calculating moments of inertia related to short-term deflections, which are not affected by creep. Hence, the compression steel is transformed to equivalent concrete using n times A'_s. Over long periods of time the effects of creep in the concrete cause the steel stresses to double; thus, the compression steel (A'_s) should be transformed to equivalent concrete using $2n$ times A'_s. However, because the compression steel (A'_s) occupies space in the compressed zone of the concrete, simply to use $2nA'_s$ would include that concrete area twice. So, it is more accurate to use $(2n - 1)A'_s$ when transforming the compression steel. When this is done, the lower expression (4.21) is obtained for k and the lower expression (4.28) is obtained for z, the point of application of the total (concrete plus steel) compression force.

In the design of doubly reinforced beams, it is assumed that the resisting moment is divided into two parts. One is the resisting moment of the beam section assuming there is no compressive reinforcement. The second is a couple formed by the additional compressive force in the compression steel and the tensile force in a proportional amount of additional tension steel.

If the resisting moment of the singly reinforced section is $M_1 = \tfrac{1}{2} f_c k j b d^2 = Rbd^2$ and the moment resisted by the compression steel is $M_2 = A'_s f'_s (d - d')$, the total moment is $M = Rbd^2 + f'_s A'_s (d - d')$.

4.5. Fig. 4-12 shows a reinforced concrete beam that must resist a 50.0 ft-kip moment. If the concrete strength is 3000 psi, determine the flexural stresses in the concrete and steel by the transformed area method.

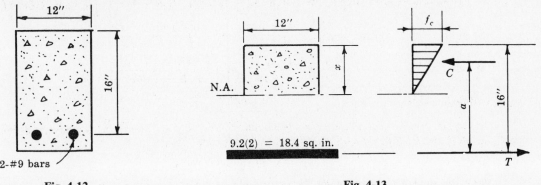

Fig. 4-12   **Fig. 4-13**

In order to determine the transformed section, the value of n must be known. Here $n = E_s/E_c$ where $E_s = 29,000,000$ psi and $E_c = 33w_c^{1.5}\sqrt{f_c'}$. If $w_c = 145$, then $E_c = 33(145)^{1.5}\sqrt{3000} = 3,160,000$ psi and $n = 29/3.16 = 9.2$.

The transformed section is as shown in Fig. 4-13.

Summing moments of the areas about the neutral axis, $12(x)(x/2) = 18.4(16 - x)$ or $x = 5.65$ in. The centroid of the compressive force would be $x/3$ below the top of the beam. Hence, $a = 16 - 5.65/3 = 14.12$ in.

Then summing moments and equating the internal resisting moment to the external moment, $Ca = Ta = M = 50.0$ ft-kips or $C = T = 50.0(12,000)/14.12 = 42,800$ lb.

The steel stress would be $f_s = T/A_s = 42,800/2.0 = 21,400$ psi.

The compressive force in the concrete is $C = f_c(x)(12)/2$; then $42,800 = f_c(5.65)(12)/2$ or $f_c = 1260$ psi.

4.6. Solve Problem 4.5 using the classic flexure formula, $f = Mc/I$.

$$\text{Moment of inertia } I = 12(5.65)^3/3 + 18.4(16 - 5.65)^2 = 269 \text{ in.}^4$$
$$f_s = Mcn/I = 50.0(12,000)(10.35)(9.2)/2691 = 21,300 \text{ psi}$$
$$f_c = Mc/I = 50.0(12,000)(5.65)/2691 = 1260 \text{ psi}$$

4.7. Solve Problem 4.5 using the formulas derived for rectangular beams with tension reinforcement only.

If $p = A_s/bd = 2.0/(12)(16)$ and $n = 9.2$, using equation (4.3), $k = \sqrt{2pn + (pn)^2} - pn = 0.352$. And from equation (4.4), $j = 1 - k/3 = 1 - 0.352/3 = 0.883$. If $M = 50.0$ ft-kips the stress in the steel and concrete can be determined from equations (4.9) and (4.1) respectively:

$$f_s = M/A_s jd = 21,300 \text{ psi} \qquad \text{and} \qquad f_c = f_s k/n(1 - k) = 1260 \text{ psi}$$

4.8. Fig. 4-14 shows a reinforced concrete T-beam that must resist a 50.0 ft-kip moment. If $f_c' = 3000$ psi, determine the flexural stresses in the concrete and steel by the transformed area method.

(a) If $n = E_s/E_c = 9.2$ for $f_c' = 3000$ psi, the transformed area of steel = 18.4 in.^{2}

If the neutral axis of the section is less than 3″ below the compression face, the section acts as a rectangular beam. If it is more than 3″, the section acts as a T-beam.

To determine whether the neutral axis is above or below the slab, moments of the transformed areas are taken about the bottom of the flange. For the concrete flange, $3(30)(1.5) = 135$ in.^{3}; for the reinforcement, $18.4(13) = 239$ in.^{3} Hence the neutral axis is below the flange.

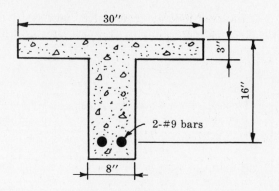

Fig. 4-14

(b) The sum of moments about the neutral axis is $8(x)(x/2) + (30-8)(3)(x-1.5) = 18.4(16-x)$ or $x = 3.92$ in.

(c) If the neutral axis is 0.92″ below the flange, the stress at the bottom of the flange is $0.92f_c/3.92 = 0.235f_c$.

The location of the centroid of the compressive force is not as readily determined here as in a rectangular beam. From Fig. 4-15, summing moments of the compressive forces about the top of the flange,

$$zC = [30.0(3.92)f_c/2](3.92/3) - [22.0(0.92)0.235f_c/2](3.0 + 0.92/3)$$
$$= 58.7f_c(1.31) - 2.38f_c(3.31) = 76.9f_c - 7.88f_c$$
$$z = 69.0f_c/(58.7f_c - 2.38f_c) = 1.22''$$

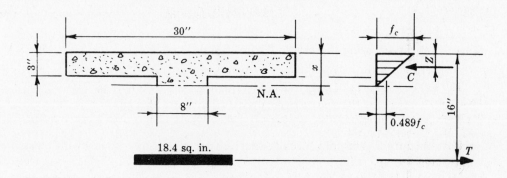

Fig. 4-15

(d) The distance from the centroid of the compressive force to the tensile force is

$$jd = 16.0 - 1.22 = 14.78''$$

Summing internal resisting moments, $M = Cjd = Tjd$. Also, $Tjd = A_s f_s jd = 50.0(12,000)$. $f_s = 20,300$ psi and $C = 40,600$ lb. Then $(58.7 - 2.38)f_c = 40,600$ or $f_c = 720$ psi.

4.9. Solve Problem 4.8 using the classic flexure formula $f = Mc/I$.

Moment of inertia $I = 30(3)^3/12 + 3(30)(2.42)^2 + 8(0.92)^3/3 + 18.4(16 - 3.92)^2 = 3276$ in.⁴
$$f_s = Mcn/I = 50.0(12,000)(12.08)(9.2)/3276 = 20,400 \text{ psi}$$
$$f_c = Mc/I = 50.0(12,000)(3.92)/3276 = 720 \text{ psi}$$

4.10. Solve Problem 4.8 using the formulas derived for T-sections.

If $p = A_s/bd = 2.0/(30.0)(16) = 0.00417$, $n = 9.2$, $h_f = 3$ and $d = 16$, equation (4.12) gives

$$k = \frac{pn + \frac{1}{2}(h_f/d)^2}{pn + (h_f/d)} = 0.250$$

Then
$$z = \frac{h_f(3kd - 2h_f)}{3(2kd - h_f)} = 1.20 \quad\text{and}\quad j = 1 - z/d = 0.922$$

If $M = 50$ ft-kips, we can determine the steel and concrete stresses by equations (4.17) and (4.10) respectively:

$$f_s = M/A_s jd = 20{,}300 \text{ psi} \quad\text{and}\quad f_c = f_s k/n(1 - k) = 736 \text{ psi}$$

4.11. The beam shown in Fig. 4-16 resists a 50.0 ft-kip moment. If $f'_c = 3000$ psi, determine the flexural stresses in the concrete and steel by the transformed area method.

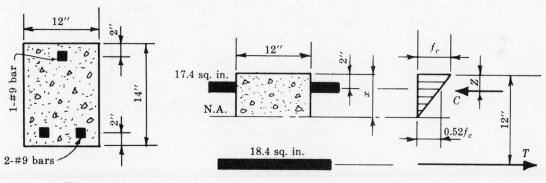

Fig. 4-16   **Fig. 4-17**

The transformed area of the compression reinforcement is $(2n - 1)A'_s$. This is in accordance with the assumption that the stress in the compression reinforcement is twice that in the concrete, owing to long-term creep. The expression also takes into account the amount of concrete displaced by the compression reinforcement.

If $n = 9.2$ for $f'_c = 3000$ psi, the transformed section is shown in Fig. 4-17.

Summing moments of the forces about the neutral axis, $12(x)(x/2) + 17.4(x - 2) = 18.4(12 - x)$ or $x = 4.17$ in. The centroid of the compressive force is located by taking moments about the top of the beam.

$$zC = (f_c/2)(12)(4.17/3) + (0.52)(f_c)(18.4)(2) = 25.02f_c(1.39) + 9.56f_c(2)$$
$$z = (34.8f_c + 19.12f_c)/(25.02f_c + 9.56f_c) = 1.56''$$

Summing internal resisting moments, $M = Cjd = Tjd$. Also, $Tjd = A_s f_s jd = 50.0(12{,}000)$; and $f_s = 28{,}800$ psi and $C = 57{,}500$ lb.

Now, $(25.02 + 9.56)f_c = 57{,}500$ or $f_c = 1670$ psi. Then $f'_s = 2n(0.52)f_c = 16{,}000$ psi.

4.12. Solve Problem 4.11 using the classic flexure formula $f = Mc/I$.

Moment of inertia $I = 12(4.17)^3/3 + 17.4(4.17 - 2.0)^2 + 18.4(12.0 - 4.17)^2 = 1502$ in.^{4}

$$f_s = Mcn/I = 50.0(12{,}000)(7.83)(9.2)/1502 = 28{,}700 \text{ psi}$$
$$f_c = Mc/I = 50.0(12{,}000)(4.17)/1502 = 1670 \text{ psi}$$
$$f'_s = 2nMy/I = 2(9.2)(50.0)(12{,}000)(2.17)/1502 = 16{,}000 \text{ psi}$$

4.13. Solve Problem 4.11 using the formulas derived for doubly reinforced beams.

If $p = A_s/bd = 2.0/(12)(12) = 0.0139$, $p' = A_s'/bd = 1.0/(12)(12) = 0.00695$, and $n = 9.2$, using equation (4.21),

$$k = \sqrt{2n(p + 2p'd'/d) + n^2(p + 2p')^2} - n(p + 2p') = 0.348$$

and
$$z = \frac{k^3 d/3 + 4np'd'(k - d'/d)}{k^2 + 4np'(k - d'/d)} = 1.56''$$

It should be noted that $2p'$ is substituted for p' in the above expressions in recognition of the doubled effectiveness of compression reinforcement. (The area of concrete replaced by the steel has not been deducted; i.e. the steel has been transformed using $2n$ instead of $(2n - 1)$. The difference is negligible.)

From equation (4.22), $j = 1 - z/d = 0.87$. If $M = 50.0$ ft-kips, the stress in the steel and concrete can be determined by use of equations (4.26), (4.18) and (4.20):

$$f_s = \frac{M}{A_s jd} = 28{,}700 \text{ psi}, \qquad f_c = \frac{f_s k}{n(1 - k)} = 1670 \text{ psi}, \qquad f_s' = 2nf_c \frac{k - d'/d}{k} = 16{,}000 \text{ psi}$$

4.14. Proportion a one-way continuous slab to support a maximum positive moment at midspan of 18.5 ft-kips/ft and a maximum negative moment at the support of 55 ft-kips/ft. Assume $f_c' = 3000$ psi, $f_s = 20{,}000$ psi and a minimum depth permitted with no compressive reinforcement.

Assume an allowable concrete stress $f_c = 0.45f_c' = 1350$ psi and $n = 9.2$. From (4.1),

$$k/(1 - k) = 9.2(1350)/20{,}000 = 0.621 \qquad \text{or} \qquad k = 0.383$$

From equation (4.4), $j = 1 - k/3 = 0.872$; and from equation (4.5), $R = f_c jk/2 = 226$.

For a unit width of slab, $b = 12''$; and from equation (4.7) the minimum depth may be determined. The negative moment at the support is maximum. Hence

$$d = \sqrt{M/Rb} = \sqrt{55(12{,}000)/[226(12)]} = 15.6''. \qquad \text{Use } d = 16.0''.$$

From equation (4.9),

$$-A_s = \frac{M}{f_s jd} = \frac{55(12{,}000)}{20{,}000(0.872)(16.0)} = 2.36 \text{ in.}^2$$

and
$$+A_s = \frac{18.5(12{,}000)}{20{,}000(0.872)(16.0)} = 0.80 \text{ in.}^2$$

In the determination of the area of reinforcement at midspan, a value of $j = 0.872$ was used. This value for j was determined for a balanced section using the larger moment at the support. Hence the concrete stress at midspan will be less than $0.45f_c'$ and the section is underreinforced. Because j varies as the concrete stress, the value for $+A_s$ is not precise. If

$$M = Rbd^2 = f_c jkbd^2/2 \qquad \text{and} \qquad f_c = f_s k/n(1 - k), \qquad j = 1 - k/3$$

then
$$M = \frac{f_s k^2}{2n(1 - k)} (1 - k/3)bd^2 \qquad \text{or} \qquad \frac{k^2(1 - k/3)}{(1 - k)} = \frac{2nM}{f_s bd^2}$$

And if the values in the problem are substituted into this expression,

$$\frac{k^2(1 - k/3)}{(1 - k)} = \frac{2(9.2)(18.5)(12{,}000)}{20{,}000(12)(16)^2} \qquad \text{or} \qquad k = 0.283$$

Then $j = 1 - k/3 = 1 - 0.094 = 0.906$. Substituting this value for j into the expression for the area of reinforcement at midspan,

$$+A_s = \frac{18.5(12{,}000)}{20{,}000(0.906)(16.0)} = 0.766 \text{ in.}^2$$

Hence with the more precise value of j, a reduction of reinforcement required of approximately 5% was obtained. Normally, such a refinement is not justified in the usual design of reinforced concrete. In the design of underreinforced sections the generally accepted practice is to use the value of j at balanced design conditions.

In underreinforced sections the value of j is greater than in balanced sections. Therefore the use of a lower value will yield a larger value for A_s and be conservative.

At midspan, $f_c = f_s k/n(1 - k) = 858$ psi.

4.15. Solve Problem 4.14 using Tables 4.1 and 4.2.

Table 4.2 for $f_s = 20,000$, $f'_c = 3000$ and $f_c = 1350$ does not contain values as high as 55 ft-kips. Hence from Table 4.1, $R = 226$ and substituting into equation (4.7), $d = \sqrt{M/Rb} = 15.6''$. Use $d = 16.0''$. From Table 4.1, $a = 1.44$; substituting in equation (4.29),

$$-A_s = M/ad = 2.39 \text{ in.}^2 \quad \text{and} \quad +A_s = 18.5/1.44(16) = 0.80 \text{ in.}^2$$

4.16. Proportion a balanced reinforced rectangular beam with tension reinforcement only to withstand a 250 ft-kip moment. Assume $f'_c = 3000$ psi and $f_y = 40,000$ psi. The length of the beam is 30'-0".

At balanced design $f_c = 0.45(3000) = 1350$ psi; and from equations (4.1) and (4.2),

$$k = \frac{1}{1 + f_s/nf_c} = \frac{1}{1 + 20,000/9.2(1350)} = 0.383$$

From equations (4.5) and (4.6),

$$R = \tfrac{1}{2} f_c jk = \tfrac{1}{2}(1350)(1 - 0.383/3)(0.383) = 226 \qquad bd^2 = M/R = 250(12,000)/226 = 13,300 \text{ in.}^3$$

Try a ratio of $d/b = 1.5$. Then $b(1.5b)^2 = 13,300$ and $b = 18.1$; use $b = 18.5''$. Thus $d = 1.5(18.5) = 27.7$; use $d = 28.0''$. From equation (4.9), $A_s = M/f_s jd = 6.14$ in.^{2}

Try a ratio of $d/b = 1.0$. then $b^3 = 13,300$ and $b = 23.7$; use $b = 24.0''$. Thus $d = 24.0''$ and $A_s = 7.17$ in.^{2}

Try a ratio of $d/b = 0.67$. Then $b(0.67b)^2 = 13,300$; $b^3 = 29,600$ and $b = 30.9$; use $b = 31.0''$. Thus $d = 0.67(31.0) = 20.8$; use $d = 21.0''$ and $A_s = 8.20$ in.^{2}

The 1983 ACI Code requires the following minimum amount of reinforcement in beams. It specifies that for flexural members, except constant thickness slabs, if positive reinforcement is required, the minimum ratio p shall not be less than $200/f_y$ unless the area of reinforcement at every section, positive or negative, is at least one-third greater than that required. If $f_y = 40,000$ psi, $200/f_y = 200/40,000 = 0.005 < 6.14/28.0(18.5)$.

The deflections of beams must be computed in accordance with the 1983 ACI Code if the depth is less than a specified value. This will be discussed in Chapter 8, which deals with serviceability.

This problem could be solved using Table 4.1 by finding the value of $R = 226$ and a value of $a = 1.44$.

4.17. Determine the area of reinforcing steel required for the section shown in Fig. 4-18. Assume $f'_c = 4000$ psi, $f_s = 24,000$ psi and $M = 300$ ft-kips.

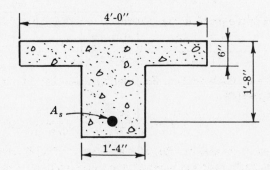

Fig. 4-18

(a) The stress in the stem will be neglected, which will yield a slightly conservative solution (typical in design offices). Because f_c is not known, the value of k must be determined by substitution into equations (4.10) through (4.17). This would be quite cumbersome due to the complex expression for j. Hence a trial and error procedure will be used.

A value $j = 0.9$ will be tried first. Substituting into equation (4.17),

$$A_s = M/f_s jd = 300(12,000)/24,000(0.9)(20) = 8.33 \text{ in.}^2$$

Now $p = 8.33/48(20) = 0.00866$, $n = 8.0$, and $h_f/d = 6.0/20.0 = 0.30$.
From equation (4.12),

$$k = \frac{pn + \frac{1}{2}(h_f/d)^2}{pn + (h_f/d)} = 0.310$$

As previously derived,

$$z = \frac{h_f(3kd - 2h_f)}{3(2kd - h_f)}$$

$$j = 1 - \frac{h_f(3k - 2h_f/d)}{3(2kd - h_f)} = 0.897$$

This value for j is near enough to the original value assumed that another trial is not necessary. For T-beams of normal proportions with the h_f/d ratio varying between 0.10 and 0.40, the value of j will vary from approximately 0.95 to 0.84. Regardless of what value of j is assumed for the initial trial, if it is within this range it will result in a design that would be close to that resulting from the "exact" value.

(b) The area of reinforcement is $A_s = 8.33 \text{ in.}^2$ Therefore if $f_y = 50,000$ psi, $200/f_y = 0.004 < 8.33/16(20)$. The minimum positive reinforcement requirement, as stated in Problem 4.16, is met.

(c) It must be determined whether or not the section depth is large enough for deflection computations to be omitted. Refer to Chapter 8.

4.18. Solve Problem 4.17 using Table 4.3.

From Table 4.3, for $f_s = 24,000$ psi, $f'_c = 4000$ psi, $f_c = 1800$ psi and $h_f/d = 0.30$: $a = 1.75$ and $R = 285$.

The area of reinforcement is $A_s = M/ad = 300/1.75(20) = 8.58 \text{ in.}^2$

As a check, the resisting moment is $M = Rbd^2 = 456$ ft-kips > 300.

4.19. Determine the area of reinforcement required for the section shown in Fig. 4-19 if the compressive stress in the stem is included. Assume $f'_c = 3000$ psi, $f_s = 20,000$ psi, and $M = 150$ ft-kips. Use Tables 4.1 and 4.3.

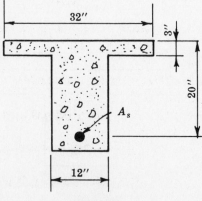

Fig. 4-19

(a) Table 4.3 neglects the effect of the stress in the stem. Hence the effect of the web must be determined using Table 4.1 and then added to the effect of the flange which has a net width equal to $32'' - 12'' = 20''$.

From Table 4.1, for $f_s = 20,000$ psi, $f'_c = 3000$ psi, $f_c = 1350$ psi: $R_w = 226$ (Balanced Design) and $a_w = 1.44$. Table 4.3, for $h_f/d = 0.15$: $R_f = 150$ (Balanced Design) and $a_f = 1.54$.

(b) The resisting moments are $M_w = R_w bd^2 = 90.3$ ft-kips and $M_f = R_f bd^2 = 100.0$ ft-kips. $M_w + M_f = 190$ ft-kips > 150. So the section will be underreinforced.

(c) The area of reinforcement required to develop the web is $A_{sw} = M_w/a_w d = 3.13$ in.^{2}

The area of reinforcement required to develop the flange is $A_{sf} = (M - M_w)/a_f d = 1.94$ in.^{2}

The total area of tension reinforcement is $A_s = A_{sw} + A_{sf} = 5.07$ in.^{2}

(d) $p = 5.07/20(12) = 0.0211 > 200/f_y$. Hence the minimum positive reinforcement requirement is met.

4.20. Compare the area of reinforcement required in Problem 4.19 to that required if the compressive stress in the stem is neglected.

From Table 4.3, for $h_f/d = 0.15$: $R = 150$ and $a = 1.54$. The resisting moment is $M = Rbd^2 = 160$ ft-kips > 150.

The area of reinforcement is $A_s = M/ad = 4.87$ in.^{2}

Because the stress in the stem is neglected, the centroid of the compressive force is located higher than in Problem 4.19. Therefore the internal moment arm jd is increased and the required area of reinforcement is decreased. The resisting moment of the concrete is decreased. The difference in reinforcement required is $(5.07 - 4.87)(100)/5.07 = 3.94\%$ on the unconservative side.

4.21. Determine the area of reinforcing steel required for the section shown in Fig. 4-20. Assume $f'_c = 3000$ psi, $f_s = 20,000$ psi, and $M = 200$ ft-kips.

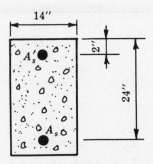

Fig. 4-20

A check should be made to determine if compressive reinforcement is required. If $f_c = 1350$ psi, $f_s = 20,000$ psi and $n = 9.2$, $k = 1/(1 + f_s/nf_c) = 0.383$; then $j = 1 - k/3 = 0.872$.

From equation (4.5), $R = \frac{1}{2} f_c jk = 226$. The resisting moment is

$$M = Rbd^2 = 226(14)(24)^2/12,000 = 152 \text{ ft-kips} < 200$$

Hence the section requires compressive reinforcement.

The resisting moment of a balanced design is 152 ft-kips. The compressive reinforcement must resist a moment $M' = 200 - 152 = 48$ ft-kips.

From equation (4.27), $A'_s = M'/f'_s(d - d')$. The value of f'_s is a function of f_c and the ratio d'/d. If $d'/d = 0.0833$, from (4.20),

$$f'_s = 2nf_c(k - d'/d)/k = 19,400 \text{ psi} < 20,000$$

Substituting into equation (4.27), $A'_s = 48(12,000)/(19,400)(22) = 1.35$ in.^{2}

From equation (*4.26*), the total area of tension reinforcement is

$$A_s = M/f_s jd = 5.73 \text{ in.}^2 \quad \text{and} \quad p = 5.73/14(24) = 0.0170 > 200/f_y$$

Hence the minimum positive reinforcement requirement is met.

4.22. Solve Problem 4.21 using Tables 4.1 and 4.4.

From Table 4.1 for $f'_c = 3000$ psi, $f_c = 1350$ psi, and $f_s = 20,000$ psi: $R = 226$, $a = 1.44$, $j = 0.872$, and $k = 0.383$. The resisting moment is $M = Rbd^2 = 152$ ft-kips < 200. Hence $M' = 200 - 152 = 48$ ft-kips.

From Table 4.4 for $f'_c = 3000$ psi, $f_c = 1350$ psi, $f_s = 20,000$ psi and $d'/d = 0.0833$; $c = 1.41$. From equations (*4.30*) and (*4.29*),

$$A'_s = M'/cd = 1.41 \text{ in.}^2 \quad \text{and} \quad A_s = M/ad = 5.79 \text{ in.}^2$$

The other design checks would be the same as in Problem 4.21.

4.23. Determine the area of tension and compression reinforcement required in the section shown in Fig. 4-21 if $M = 167$ ft-kips. Assume $f'_c = 3000$ psi and $f_s = 20,000$ psi.

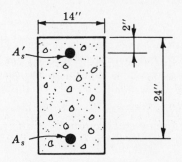

Fig. 4-21

This is the same section as in Problem 4.18. Hence $R = 226$, $a = 1.44$, $j = 0.872$, $k = 0.383$, and $c = 1.41$. Then the resisting moment is $M = 152$ ft-kips < 167. Hence $M' = 15$ ft-kips.

$$A'_s = M'/cd = 0.444 \text{ in.}^2 \quad \text{and} \quad A_s = M/ad = 4.84 \text{ in.}^2$$

4.24. Determine the area of tension reinforcement required for the section in Problem 4.23 if there is no compression reinforcement and $f_c = 1350$ psi.

If the compressive force as shown in Fig. 4-3 is made larger, the section can resist greater moments. If f_c must remain fixed, the only manner in which to accomplish this is by increasing k. By increasing the area of tension reinforcement, the neutral axis will be lowered. From equations (*4.5*) and (*4.6*), $M = Rbd^2 = \frac{1}{2}f_c jkbd^2$; and from equation (*4.4*), $j = 1 - k/3$. Hence

$$M = \tfrac{1}{2}f_c(1 - k/3)kbd^2 \quad \text{or} \quad (1 - k/3)k = 2M/f_c bd^2 = 0.368, \quad \text{and} \quad k = 0.429$$

From equations (*4.2*) and (*4.9*),

$$f_s = nf_c\left(\frac{1 - k}{k}\right) = 16,500 \text{ psi} \quad \text{and} \quad A_s = \frac{M}{f_s jd} = 5.90 \text{ in.}^2$$

In Problem 4.23 the total area of reinforcing steel is $4.84 + 0.44 = 5.28$. Comparing the answers,

$$(5.90 - 5.28)(100)/5.28 = 12\% \text{ increase}$$

in reinforcement required when tension steel only is used. This is required for a section that must resist a moment that is merely 10% greater than the balanced resisting moment.

4.25. Determine the area of reinforcement required for the edge or spandrel beam shown in Fig. 4-22. Assume $f'_c = 4000$ psi and $f_s = 24,000$ psi. The beam has a span of 10'-0" and the 5" slab has a clear span of 16'-0". The beam is continuous and must resist a negative moment at the support of 170 ft-kips and a positive moment at midspan of 55 ft-kips.

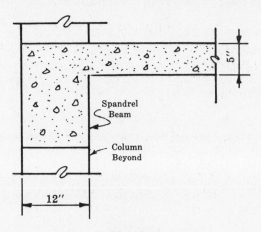

Fig. 4-22

(*a*) The negative moment at the support is the largest absolute bending moment and controls the size of the member. If the section is to be designed using balanced reinforcement, $f_c = 1800$ psi, $f_s = 24,000$ psi, and $n = 8.0$.

If no compression reinforcement is to be used at the support, then $Rbd^2 \geq 170$ ft-kips. From Table 4.1, for $f_c = 1800$ psi and $f_s = 24,000$ psi: $R = 295$ and $a = 1.76$. Hence $bd^2 = M/R = 6900$ in.^{3} Because the width of the beam has been set equal to the width of the column, 12", $d^2 = 6900/12 = 575$. Use $d = 24''$.

The area of negative reinforcement at the support is $-A_s = M/ad = 4.02$ in.^{2}

(*b*) At the supports, the beam acts as a rectangular section. However, in the positive moment region the beam acts as an unsymmetrical T-section (inverted L-section).

The 1983 ACI Code specifies that in beams having a flange on one side only the effective flange shall not be greater than 1/12 the beam span length and that the effective overhanging flange shall not be greater than six times the slab thickness nor greater than one-half the slab span. Applying these rules,

$$b = (1/12)(10)(12) + 12 = 22'', \qquad b = 12 + 6(5) = 42'', \qquad \text{or} \qquad b = 12 + \tfrac{1}{2}(16)(12) = 108''$$

Hence the assumed effective flange width is 22".

(*c*) From Table 4.3, for $f_c = 1800$ psi, $f_s = 24,000$ psi and $h_f/d = 5/24 = 0.208$: $R = 246$ and $a = 1.82$. Then $Rbd^2 = 260$ ft-kips > 55 and the area of positive reinforcement is $+A_s = M/ad = 1.26$ in.^{2}

Checking for the minimum area of reinforcement, $200/f_y = 0.0033 < 1.26/12(24)$.

Supplementary Problems

4.26. A 10" wide rectangular concrete beam has an effective depth of 14" and is reinforced with 1.32 in.^{2} of steel. If $n = 15$, determine the moment of inertia of the transformed section.
Ans. $I = 1980$ in.^{4}

4.27. A 12" wide rectangular concrete beam is reinforced with 4-#8 bars. If $n = 12$, determine the location of the neutral axis.   *Ans.* $kd = 8.9''$

4.28. If $b = 12''$, $d = 16''$, $f_c = 1350$ psi, and $f_s = 20,000$ psi, determine the required reinforcement for a rectangular beam to resist a moment of 60 ft-kips. Use the transformed section method.
Ans. 2.59 in.^{2}

4.29. Repeat Problem 4.28 using the flexure formula and Table 4.1.

4.30. Given a T-beam with $b = 50''$, $b_w = 10''$, $h_f = 6''$, $d = 20''$, $A_s = 2.50$ in.^{2}, and an applied moment of 65 ft-kips, determine the concrete and steel stresses by the transformed section method if $n = 10$.
Ans. $f_c = 460$ psi, $f_s = 16,800$ psi

4.31. Repeat Problem 4.30 using the flexure formula and Table 4.3.

4.32. Given a rectangular doubly reinforced beam with $b = 14''$, $d = 16''$, $f_c = 1350$ psi, $f_s = 20,000$ psi, and an applied moment of 140 ft-kips, determine the area of reinforcement required. Assume $d' = 2''$ and $n = 9.2$.   *Ans.* $A_s = 6.08$ in.^{2}, $A'_s = 2.91$ in.^{2}

4.33. Proportion a one-way slab with simple supports to resist an applied moment of 12.0 ft-kips/ft. Assume balanced design, $f'_c = 3000$ psi, and $f_s = 24,000$ psi. Use equations (*4.1*) through (*4.9*).
Ans. $d = 8''$, $A_s = 0.85$ in.^{2}/ft

4.34. Repeat Problem 4.33 but use Tables 4.1 and 4.2.

4.35. Proportion by use of equations (*4.1*) through (*4.9*) a balanced reinforced rectangular beam to resist a moment of 300 ft-kips. Assume $f'_c = 5000$ psi and $f_s = 24,000$ psi.
Ans. $b = 15''$, $d = 25''$, $A_s = 6.82$ in.^{2}

4.36. Repeat Problem 4.35 but use Table 4.1.

4.37. If $b = 35''$, $b_w = 12''$, $d = 20''$, $h_f = 3''$, $f'_c = 3000$ psi, $f_s = 20,000$ psi and $M = 65$ ft-kips, determine the reinforcement required in a T-beam. Use equations (*4.10*) through (*4.17*). Assume balanced design.   *Ans.* $A_s = 2.14$ in.^{2}

4.38. Using equations (*4.18*) through (*4.27*), proportion a doubly reinforced beam if $b = 13''$, $d = 26''$, $d' = 2.5''$, $f'_c = 2500$ psi, $f_s = 20,000$ psi, and $M = 200$ ft-kips.   *Ans.* $A_s = 5.30$ in.^{2}, $A'_s = 2.16$ in.^{2}

4.39. Repeat Problem 4.38 but use Table 4.4.

Chapter 5

Strength Design

NOTATION

A_s = area of tension reinforcement

A'_s = area of compression reinforcement

A_{sf} = area of reinforcement to develop compressive strength of overhanging flanges in I- and T-sections

a = depth of equivalent rectangular stress block = $\beta_1 c$

b = width of compression face of flexural member

b_w = width of web in I- and T-sections

c = distance from extreme compression fiber to neutral axis at ultimate strength

D = dead load

d = distance from extreme compression fiber to centroid of tension reinforcement

d' = distance from extreme compression fiber to centroid of compression reinforcement

E = earthquake load

f'_c = compressive strength of concrete

f_y = yield strength of reinforcement

h_f = flange thickness in I- and T-sections

L = specified live load plus impact

M_n = nominal (or design) resisting moment of a cross section = M_u/ϕ

M_u = ultimate resisting moment of a cross section, from frame analysis

ρ = A_s/bd

ρ' = A'_s/bd

ρ_b = reinforcement ratio producing balanced conditions at ultimate strength

ρ_f = $A_{sf}/b_w d$

ρ_w = $A_s/b_w d$

ω = $A_s f_y/bdf'_c$

U = required ultimate load capacity of section

W = wind load

ϕ = capacity reduction factor

(The notation used here is the same as in the 1983 ACI Building Code.)

INTRODUCTION

Strength design procedures differ from the alternate design procedures of Chapter 4. In the former it is recognized that at high stress levels in concrete, stress is not proportional to strain and, second, in strength design procedures design loads are multiples of anticipated service loads. In the alternate design method, stress is assumed to be proportional to strain and design loads are equal to service loads.

There are several advantages or reasons for using strength design:

(1) Strength design better predicts the strength of a section because of the recognition of the non-linearity of the stress-strain diagram at high stress levels.

(2) Because the dead loads to which a structure is subjected are more certainly determined than the live loads, it is unreasonable to apply the same factor of safety to both.

(3) Elastic column design is a modification of strength design and therefore is not compatible with the alternate design method for flexural members. Hence a consistent design technique is desirable.

(4) A more certain evaluation of the critical moment-thrust ratio for columns is possible with strength design than with the alternate design method.

(5) Strength design must be used when determining the ultimate capacity of prestressed concrete.

Before derivation of the basic equations and relationships it is necessary to delineate the basic assumptions, which are:

(1) Plane sections before bending remain plane after bending.

(2) At ultimate capacity, strain and stress are not proportional.

(3) Strain in the concrete is proportional to the distance from the neutral axis.

(4) Tensile strength of concrete is neglected in flexural computations.

(5) The ultimate concrete strain is 0.003.

(6) The modulus of elasticity of the reinforcing steel is 29,000,000 psi (200 000 MPa).

(7) The average compressive stress in the concrete is $0.85f_c'$.

(8) The average tensile stress in the reinforcement does not exceed f_y.

In addition to these eight assumptions, the assumed compressive stress distribution in the concrete is most important (Fig. 5-1). Any distribution such as a rectangle, trapezoid, parabola, sine wave or any other shape is permitted if the predicted ultimate capacity is in close agreement with test data.

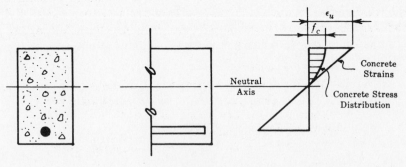

Fig. 5-1

Many stress distributions have been proposed, the three most common being the parabola, trapezoid and rectangle, Fig. 5-2. All yield reasonable results.

The mechanics and derivations using the equivalent rectangular stress distribution, as assumed in the 1983 ACI Code, are somewhat simpler. It is further assumed that $a = \beta_1 c$ and that β_1 be taken as 0.85 for concrete strengths of 4000 psi (30 MPa) or less; whereas, for greater strength, β_1 shall be reduced at the rate of 0.05 for each 1000 psi of strength in excess of 4000 psi (at the rate of 0.008 for each 1.0 MPa of strength in excess of 30 MPa), provided β_1 does not fall below 0.65.

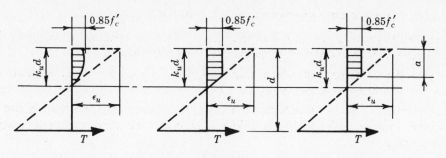

Fig. 5-2

LOAD FACTORS AND UNDERSTRENGTH FACTORS

In the ASCE-ACI Report and in the 1956 edition of the ACI Code, all of the factor of safety in strength design was provided for ostensibly in the load factor. That is, the service loads were increased by some multiple and the idealized capacity, or strength, of the member had to be equal to or greater than this assumed loading. However, in the 1983 edition of the ACI Code, the factor of safety is applied both to the load and to the idealized capacity sides of the equation. The 1956 ACI Code required that

$$\text{idealized capacity } (U) = \text{ultimate load } (C_1 D + C_2 L)$$

Later (1963, 1971, 1977, 1983) Codes substitute

$$\phi \times \text{idealized capacity } (\phi U) = \text{ultimate load } (C_3 D + C_4 L) \tag{5.1}$$

which provides for possible overloads (right hand side of equation) and possible under-capacity (left hand side of equation). The 1983 ACI Code requires that designs be made for the *nominal capacity* of a cross section, which is the frame-analysis capacity divided by the understrength factor, ϕ. For bending moment, the nominal moment is $M_u/\phi = M_n$. Similarly, $V_n = V_u/\phi$, etc.

This idealized capacity reduction factor provides for the possibility of the concrete or reinforcing steel being of less strength than required and for the possibility of members being understrength due to inaccuracies or mistakes in construction. This reduction factor ϕ does vary with the importance of the member and with the mode of anticipated failure. In design the following are required by the Code:

$$\phi = 0.90 \text{ for flexure}$$
$$= 0.85 \text{ for diagonal tension, bond and anchorage}$$
$$= 0.75 \text{ for spirally reinforced compression members*}$$
$$= 0.70 \text{ for tied compression members*}$$
$$= 0.70 \text{ for bearing on concrete}$$
$$= 0.65 \text{ for flexure in plain concrete (not reinforced)}$$

It is noted that for flexural members such as beams and girders, the capacity reduction factor is the highest (0.90). As will be noted later, the amount of reinforcing steel in flexure is limited so that the steel will yield in tension before the concrete fails in compression. Therefore the failure in flexure will normally be due to a gradual yielding of the reinforcing. In columns, however, the mode of failure could be an explosive compression failure of the concrete. Too, in most structures, the action of columns is more important than that of beams. Consequently the idealized capacity of columns is reduced more than that for beams, 0.75 or 0.70 vs. 0.90.

---

* For members subjected to axial load and bending moment, these values are increased linearly to 0.90 as the axial load decreases from $0.1 f_c' A_g$ to zero, or from "balanced P_u" to zero. The lesser of P_b and $0.1 f_c' A_g$ is used.

The 1983 ACI Code provides three basic equations for "factoring" service-level dead loads (D), service-level live loads (L) and service-level wind (W) or earthquake (E) loads. Wind and earthquake are interchangeable in equation (5.3) since there is little probability of high winds and an earthquake occurring simultaneously. The equations are:

$$U = 1.4D + 1.7L \qquad\qquad (5.2)$$
$$U = 0.75(1.4D + 1.7L + 1.7W) \qquad\qquad (5.3)$$
$$U = 0.9D + 1.3W \qquad\qquad (5.4)$$

All three conditions must be investigated and the most critical case is used in the design. Equation (5.2) usually governs the design; however, for tall structures, equation (5.3) or (5.4) might be critical. The factor 0.75 in equation (5.3) is used because of the probability that full live load would not be present during high winds or an earthquake. Equation (5.4) applies to tall structures subjected to lateral loads. For example, on the windward side of a building, column loads will be tension forces due to the wind; the dead load (weight of the structure) will cause compressive forces which will counteract these tension forces. The factor 0.9 is applied to the dead load because of the probability of overestimating the dead loads due to the weights of the materials.

If the structure resists lateral liquid pressure (H), as in a water or sewage treatment tank, then (H) is included as

$$U = 1.4D + 1.7L + 1.7H \qquad\qquad (5.5)$$

However, if dead load (D) or live load (L) reduces the effect of earth pressure (H), $0.9D$ is used in lieu of $1.4D$, and live load is excluded.

If the structure resists lateral liquid pressure (H), as in a water or sewage treatment tank, then

$$U = 1.4D + 1.7L + 1.4H \qquad\qquad (5.6)$$

The liquid pressure is regarded as dead load insofar as depth is concerned. The lateral pressure is a function of liquid depth as is the vertical pressure. The same is true for granular materials such as corn and wheat stored in silos.

If "impact forces" are involved, such as elevators in elevator shafts or vehicles in a parking building, such forces are included as live load (L). Local building codes provide requirements concerning such forces.

Effects of differential settlement of foundations, creep, shrinkage and temperature are included as (T). Combinations of such effects can exist simultaneously. The design strength must be

$$U = 0.75(1.4D + 1.7L + 1.4T) \qquad\qquad (5.7)$$

but the required strength shall not be less than

$$U = 1.4(D + T) \qquad\qquad (5.8)$$

It is highly probable that the effects of differential settlement, temperature, creep and shrinkage will be present during high winds or earthquake. Thus, it would be reasonable to consider the strength requirement:

$$U = 0.75(1.4D + 1.7L + 1.7W + 1.4T) \qquad\qquad (5.9)$$

However, the 1983 ACI Code does not list equation (5.9) as a requirement for investigation.

In all cases, equation (5.2) must be satisfied.

A joint committee of code writing organizations was established to study the load factors and understrength (resistance) factors (ϕ) as related to structures composed of several different structural materials. An example would be a structure that utilizes structural steel beams that are connected with shear connectors to a concrete slab so that both the steel beam and the concrete slab will act as a unit. Previously, load factors and understrength factors varied for various structural materials. The intent of the committee was to establish common load factors for all structural materials and to adjust the understrength factors to reflect the behavior of the different materials. The advantage of common load factors is that ultimate loads will be material independent, and that adjustments can be made during the design stage using appropriate understrength (resistance) factors.

FLEXURAL COMPUTATIONS BY STRENGTH DESIGN

Test data show that the assumed ultimate concrete strain of 0.003 is conservative yet reasonable. Further, the maximum allowable concrete stress of $0.85f'_c$ is compatible with the ultimate strain. If the free-body diagram given in the ASCE-ACI Report of October 1955 is used (Fig. 5-3), the summation of horizontal forces shows that

$$C = T \qquad \text{or} \qquad 0.85f'_c k_1 k_u bd = A_s f_s \qquad (a)$$

k_1 is the average compressive stress divided by $0.85f'_c$. The summation of moments about T yields

$$M = 0.85f'_c k_1 k_u (1 - k_2 k_u) bd^2 \qquad (b)$$

where $k_2 k_u d$ = distance from compression face to centroid of compressive force.

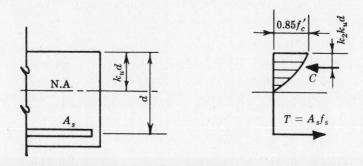

Fig. 5-3

If $f_s = f_y$, $\rho = A_s/bd$ and $\mu = f_y/0.85f'_c$; combining (a) and (b) above, we obtain

$$M_n = A_s f_y d\left(1 - \frac{k_2}{k_1}\right)\rho\mu \qquad (5.10)$$

which is the expression for the nominal resisting moment for an underreinforced beam regardless of the assumed shape of the compressive force diagram.

If the second degree parabola of the 1955 ASCE-ACI Report is used, it is found that $k_2/k_1 = 0.554$ and $k_1 = 0.925 - 0.438f'_c/E_c\epsilon_u$. If $\epsilon_u = 0.0035$ and $E_c = 1000f'_c$ (as in the 1956 ACI Code, page 935), then $k_1 = 0.80$. Substituting for k_1 and k_2 in the expression for the nominal resisting moment,

$$M_n = f'_c bd^2 \omega(1 - 0.65\omega) \qquad (5.11)$$

in which $\omega = \rho f_y/f'_c$.

Likewise, if the trapezoidal distribution contained in the ASCE-ACI Report is assumed, $k_2/k_1 = 0.50$ and $k_1 = 0.879$. Substituting values of k_1 and k_2, the nominal resisting moment is

$$M_n = f'_c bd^2 \omega(1 - 0.59\omega) \qquad (5.12)$$

If the beam is underreinforced so that the failure in flexure is due to a yielding of the steel, the expression for the nominal resisting moment assuming a rectangular stress distribution is

$$M_n = f'_c bd^2 \omega(1 - 0.59\omega) \qquad (5.13)$$

See Problem 5.1.

If the condition is such that the concrete and reinforcing steel are stressed to the allowable ultimate values at the same time, then the section or beam has balanced reinforcement. Tests show that when $\omega = 0.456$, the condition of balanced reinforcement exists. Because of assumed underreinforcement in the derivations and because of the desirability of tension control of failure, it is necessary to limit the value of ρ or ω in order to assure this condition.

In the ASCE-ACI Report and in the 1956 Building Code, ω was limited to a maximum value of 0.40. However, in the 1983 Code an expression for the *balanced steel ratio* was presented:

$$\rho_b = \frac{0.85 f'_c \beta_1}{f_y} \frac{87,000}{87,000 + f_y} \quad \left(\frac{0.85 f'_c \beta_1}{f_y} \frac{600}{600 + f_y} \right) \tag{5.14}$$

Equation (5.14) was derived for beams reinforced for tension only. Similar equations have been derived for T-beams and beams reinforced for both tension and compression. Rather than derive separate equations for different shapes of beams and different conditions of reinforcement, it is easier to set the maximum steel ratio as $0.75\rho_b$ and locate the neutral axis as 75 percent of that which corresponds to *balanced strain conditions*. In this way, the balanced conditions and the maximum steel area permitted are independent of member shape and whether or not the member contains compression side reinforcement.

Fig. 5-4 shows a series of curves obtained by plotting values of $\rho_{\max}$ versus f_y for various concrete strengths for rectangular beams with tension reinforcement only.

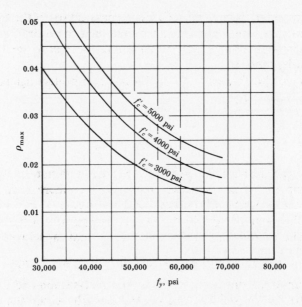

Fig. 5-4

COMPRESSION REINFORCEMENT

Compression reinforcement in flexural members designed by strength design procedures will very seldom be required. Because of the much increased value given to the concrete in compression due to the shape and increased depth of the compression zone, it will be practically unsound to place enough reinforcement in a beam or girder to attain the balanced condition. If $f'_c = 3000$ psi and $f_y = 40,000$ psi, then $\rho = 0.18 f'_c / f_y$ for a beam of balanced design using the alternate design method. For the same conditions in strength design, $\rho_b = 0.37 f'_c / f_y$. It is generally impractical, if not impossible, to use twice as much reinforcement in a beam as that which would be required by balanced conditions in the alternate design method. Hence the discussion and derivation of flexural formulas for members requiring compression reinforcement is somewhat academic rather than practical.

In strength design of doubly reinforced beams, beams requiring compression reinforcement, it is assumed that the resisting moment is composed of two separate moments. First, there is the resisting capacity of the concrete and the maximum tension reinforcement. Second, there is the resisting capacity of the compression steel and a balancing amount of additional tension steel.

If the expression for the nominal resisting moment of a beam with tensile reinforcement is rearranged,

$$M_n = A_s f_y (d - a/2) \qquad (5.12a)$$

The moment of the additional compression steel would be $M'_n = A'_s f_y (d - d')$ if the compression steel is stressed to the yielding. Adding these two moments and deducting the effect of the concrete in compression occupied by the compressive reinforcement,

$$M_n = (A_s - A'_s) f_y (d - a/2) + A'_s f_y (d - d') \qquad (5.15)$$

For the expression to be valid, it is necessary that the compressive reinforcement reach its yield strength at the ultimate strength of the member. This is satisfied if

$$\rho - \rho' \geqq \frac{0.85 f'_c \beta_1 d'}{f_y d} \frac{87,000}{87,000 - f_y} \qquad \text{(U.S. customary units)} \qquad (5.16)$$

This is the expression presented in the 1983 ACI Code commentary. If the total tensile reinforcement is greater than this minimum value, there is more than enough tensile force, $A_s f_y$, to develop the concrete and any compression reinforcement. Therefore the tensile steel is not at yield and the failure is not controlled by tension.

The quantity $\rho - \rho'$ is also limited to $0.75 \rho_b$ in order to protect against brittle or compression failures of the concrete.

I- AND T-SECTIONS

In flanged sections such as an I- or T-section in strength design, it is again assumed that the resisting moment is composed of two separate moments. First, there is the moment capacity of the rectangular portion of the concrete, b_w times d, and a corresponding amount of tensile reinforcement. Second, there is the moment capacity of the overhang portion of the concrete, $(b - b_w) h_f$, and another amount of tensile reinforcement. The nominal flexural capacity of an I- or T-section is

$$M_n = (A_s - A_{sf}) f_y (d - 0.5a) + A_{sf} f_y (d - 0.5 h_f) \qquad (5.17)$$

This expression assumes that the section acts as an I- or T-section and that the neutral axis of the sections falls without the flanged section. The neutral axis falls within the flange if

$$h_f \geqq 1.18 \omega d / \beta_1 \qquad (5.18)$$

Then the member acts as a rectangular beam and it can be proportioned according to requirements of rectangular beams with tension reinforcement.

In an investigation of a member, it is easy to substitute in equation (5.18). However, in a design it might be more practical to determine if the flange of the section is capable of resisting the applied moment. If not, then the neutral axis falls outside of the flange and the section does not act as a rectangular member.

If the nominal moment capacity of the flange,

$$M_{nf} = 0.85 f'_c b h_f (d - 0.5 h_f)$$

is greater than the nominal applied moment, then the member is proportioned as a rectangular beam.

In I- and T-sections, the area of tension steel available to develop the concrete of the web in compression is limited to $0.75\rho_b$, that is,

$$A_s/b_w d - A_{sf}/b_w d \leqq 0.75\rho_b \qquad \text{or} \qquad \rho_w - \rho_f \leqq 0.75\rho_b$$

This, too, is to guard against possible brittle or compressive failures of the concrete.

Design specifications require that in reinforced concrete structures proportioned by strength design techniques the analysis of the frame to determine shears, moments, and thrusts be based on elastic behavior. The 1983 ACI Code contains provisions for moment redistribution so that limit design may be considered to some extent.

SOLUTION OF EQUATIONS AND DESIGN AIDS

The solution and application of the expressions derived in this chapter will be demonstrated in the solved problems. However, it is appropriate to discuss the solution of the expression for the nominal moment capacity of a rectangular beam with tension reinforcement only. In the solution of the expression $M_n = bd^2 f'_c \omega(1 - 0.59\omega) = A_s f_y(d - a/2)$, there are the variables b, d and A_s which are not known before the design is completed. It is obvious that there are an infinite number of solutions to the equations that would yield the same value of M_n. Hence the designer must assume the value of one or more of these variables before he can determine the unique solution he is seeking.

The designer may assume the overall dimensions of the member; or may assume a value of ρ; or may assume that the member shall have minimum depth or maximum ρ permitted by specifications. As previously shown, the last assumption will seldom be used because of the extremely high quantity of reinforcing steel required, which is not an economical solution. Often the overall width or depth, or both, of a beam are established by some other structural or architectural consideration.

If the minimum depth of the overall member size is not determined or assumed, then the designer must assume a value for ρ or ω.

There are three basic techniques for the solution of singly reinforced beams:

(1)   Direct substitution into formulas

(2)   Use of tables for the formulas and constants

(3)   Use of curves or charts for the formulas and constants

All three methods will be used in the solved problems.

Table 5.1 is a solution of a dimensionless form of equation (*5.13*),

$$M_n/f'_c bd^2 = \omega(1 - 0.59\omega)$$

with the first column being the first two decimal places and the first row being the third decimal place for the value of ω; and the body of the table gives the corresponding values for $M_n/f'_c bd^2$. With Table 5.1, the design may be accomplished by assuming a value of ω or ρ and solving for $M_n/f'_c bd^2$ and subsequently for b and d; or, a value for bd^2 may be assumed and the value of ω determined.

Fig. 5-5 is a chart or series of curves that accomplish the solution similar to Table 5.1. This is the same chart contained in the ASCE-ACI Joint Committee Report and is reproduced with permission of the ACI.

The chart may be entered with a value of M_n/bd^2; then traverse horizontally to the concrete strength, then vertically to the yield strength of the steel, then horizontally to the value of ρ. If ρ is assumed, this procedure is reversed.

In Chapter 4 it was shown that the basic equations using the transformed section method could be reduced to simpler forms, and with the help of design aid tables or charts the flexural computations could be made much easier. A method similar to this can be employed in strength design procedures.

If $a = A_s f_y/0.85 f'_c b$ or $a/d = \rho\mu$, where $\mu = f_y/0.85 f'_c$, the distance from the centroid of the

Table 5.1　Nominal Resisting Moment of Rectangular Sections (Tension Reinforcement)

ω	.000	.001	.002	.003	.004	.005	.006	.007	.008	.009
					$M_n/f_c'bd^2$					
.0	0	.0010	.0020	.0030	.0040	.0050	.0060	.0070	.0080	.0090
.01	.0099	.0109	.0119	.0129	.0139	.0149	.0159	.0168	.0178	.0188
.02	.0197	.0207	.0217	.0226	.0236	.0246	.0256	.0266	.0275	.0285
.03	.0295	.0304	.0314	.0324	.0333	.0343	.0352	.0362	.0372	.0381
.04	.0391	.0400	.0410	.0420	.0429	.0438	.0448	.0457	.0467	.0476
.05	.0485	.0495	.0504	.0513	.0523	.0532	.0541	.0551	.0560	.0569
.06	.0579	.0588	.0597	.0607	.0616	.0625	.0634	.0643	.0653	.0662
.07	.0671	.0680	.0689	.0699	.0708	.0717	.0726	.0735	.0744	.0753
.08	.0762	.0771	.0780	.0789	.0798	.0807	.0816	.0825	.0834	.0843
.09	.0852	.0861	.0870	.0879	.0888	.0897	.0906	.0915	.0923	.0932
.10	.0941	.0950	.0959	.0967	.0976	.0985	.0994	.1002	.1011	.1020
.11	.1029	.1037	.1046	.1055	.1063	.1072	.1081	.1089	.1098	.1106
.12	.1115	.1124	.1133	.1141	.1149	.1158	.1166	.1175	.1183	.1192
.13	.1200	.1209	.1217	.1226	.1234	.1243	.1251	.1259	.1268	.1276
.14	.1284	.1293	.1301	.1309	.1318	.1326	.1334	.1342	.1351	.1359
.15	.1367	.1375	.1384	.1392	.1400	.1408	.1416	.1425	.1433	.1441
.16	.1449	.1457	.1465	.1473	.1481	.1489	.1497	.1506	.1514	.1522
.17	.1529	.1537	.1545	.1553	.1561	.1569	.1577	.1585	.1593	.1601
.18	.1609	.1617	.1624	.1632	.1640	.1648	.1656	.1664	.1671	.1679
.19	.1687	.1695	.1703	.1710	.1718	.1726	.1733	.1741	.1749	.1756
.20	.1764	.1772	.1779	.1787	.1794	.1802	.1810	.1817	.1825	.1832
.21	.1840	.1847	.1855	.1862	.1870	.1877	.1885	.1892	.1900	.1907
.22	.1914	.1922	.1929	.1937	.1944	.1951	.1959	.1966	.1973	.1981
.23	.1988	.1995	.2002	.2010	.2017	.2024	.2031	.2039	.2046	.2053
.24	.2060	.2067	.2075	.2082	.2089	.2096	.2103	.2110	.2117	.2124
.25	.2131	.2138	.2145	.2152	.2159	.2166	.2173	.2180	.2187	.2194
.26	.2201	.2208	.2215	.2222	.2229	.2236	.2243	.2249	.2256	.2263
.27	.2270	.2277	.2284	.2290	.2297	.2304	.2311	.2317	.2324	.2331
.28	.2337	.2344	.2351	.2357	.2364	.2371	.2377	.2384	.2391	.2397
.29	.2404	.2410	.2417	.2423	.2430	.2437	.2443	.2450	.2456	.2463
.30	.2469	.2475	.2482	.2488	.2495	.2501	.2508	.2514	.2520	.2527
.31	.2533	.2539	.2546	.2552	.2558	.2565	.2571	.2577	.2583	.2590
.32	.2596	.2602	.2608	.2614	.2621	.2627	.2633	.2639	.2645	.2651
.33	.2657	.2664	.2670	.2676	.2682	.2688	.2694	.2700	.2706	.2712
.34	.2718	.2724	.2730	.2736	.2742	.2748	.2754	.2760	.2766	.2771
.35	.2777	.2783	.2789	.2795	.2801	.2807	.2812	.2818	.2824	.2830
.36	.2835	.2841	.2847	.2853	.2858	.2864	.2870	.2875	.2881	.2887
.37	.2892	.2898	.2904	.2909	.2915	.2920	.2926	.2931	.2937	.2943
.38	.2948	.2954	.2959	.2965	.2970	.2975	.2981	.2986	.2992	.2997
.39	.3003	.3008	.3013	.3019	.3024	.3029	.3035	.3040	.3045	.3051
.40	.3056									

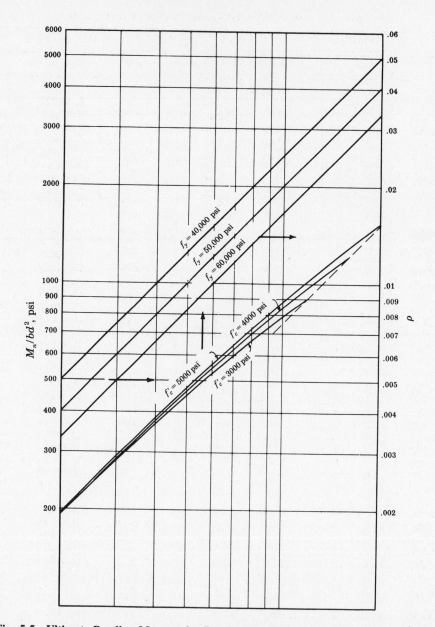

Fig. 5-5 Ultimate Bending Moment for Rectangular Beams (Tension Reinforcement)

compression force in the concrete to the tensile reinforcement $j_u d = d - a/2$ or $j_u = 1 - a/2d$. The resisting moment $M_u/\phi = T j_u d = A_s f_y j_u d$. In the alternate design method, $R = M/bd^2 = f_c/2kj$. In strength design if the member is underreinforced, then the analogous R is $R_u = f'_c \omega (1 - 0.59 \omega) = \rho f_y (1 - 0.59 \rho f_y / f'_c)$ and $M_u = \phi R_u bd^2$. In the alternate design method, for a given steel stress, the values of j were averaged and a value was computed so that $a = f_s j/12{,}000$ and hence $A_s = M/ad$.

In the alternate design method, for a given steel stress, typical values of j would not vary more than 5 to 10 percent regardless of the concrete stress. However, in strength design, because of the wider range of possible reinforcing steel percentages, there obviously would be a greater variation in the values of j as the concrete stress varies. Therefore, if it is considered that a_u (this is not the depth of the stress block) $= f_y j_u$, the area of steel must be $A_s = M_u / \phi f_y j_u d$.

Table 5.2 (in U.S. customary units) and Table 5.3 (in SI units) list the various constants for strength design analogous to those developed for the alternate design method. The values of the

Table 5.2 Strength Design Factors (U.S. Customary Units)

$f'_c = 3000.0$ psi		$f_y = 40,000.0$ psi		$\rho_{min} = 200/f_y = 0.0050$		$\phi = 0.9$

ρ	ϕa_u	ϕj_u	ϕR_u	c/d	a/d	ω
0.00200	2.953	0.8858	71	0.0369	0.0314	0.0267
0.00250	2.941	0.8823	88	0.0461	0.0392	0.0333
0.00300	2.929	0.8788	105	0.0554	0.0471	0.0400
0.00350	2.917	0.8752	123	0.0646	0.0549	0.0467
0.00400	2.906	0.8717	139	0.0738	0.0627	0.0533
0.00450	2.894	0.8681	156	0.0830	0.0706	0.0600
0.00500	2.882	0.8646	173	0.0923	0.0784	0.0667
0.00550	2.870	0.8611	189	0.1015	0.0863	0.0733
0.00600	2.858	0.8575	206	0.1107	0.0941	0.0800
0.00650	2.847	0.8540	222	0.1200	0.1020	0.0867
0.00700	2.835	0.8504	238	0.1292	0.1098	0.0933
0.00750	2.823	0.8469	254	0.1384	0.1176	0.1000
0.00800	2.811	0.8434	270	0.1476	0.1255	0.1067
0.00850	2.799	0.8398	286	0.1569	0.1333	0.1133
0.00900	2.788	0.8363	301	0.1661	0.1412	0.1200
0.00950	2.776	0.8327	316	0.1753	0.1490	0.1267
0.01000	2.764	0.8292	332	0.1845	0.1569	0.1333
0.01050	2.752	0.8257	347	0.1938	0.1647	0.1400
0.01100	2.740	0.8221	362	0.2030	0.1725	0.1467
0.01150	2.729	0.8186	377	0.2122	0.1804	0.1533
0.01200	2.717	0.8150	391	0.2215	0.1882	0.1600
0.01250	2.705	0.8115	406	0.2307	0.1961	0.1667
0.01300	2.693	0.8080	420	0.2399	0.2039	0.1733
0.01350	2.681	0.8044	434	0.2491	0.2118	0.1800
0.01400	2.670	0.8009	448	0.2584	0.2196	0.1867
0.01450	2.658	0.7973	462	0.2676	0.2275	0.1933
0.01500	2.646	0.7938	476	0.2768	0.2353	0.2000
0.01550	2.634	0.7903	490	0.2860	0.2431	0.2067
0.01600	2.622	0.7867	503	0.2953	0.2510	0.2133
0.01650	2.611	0.7832	517	0.3045	0.2588	0.2200
0.01700	2.599	0.7796	530	0.3137	0.2667	0.2267
0.01750	2.587	0.7761	543	0.3230	0.2745	0.2333
0.01800	2.575	0.7726	556	0.3322	0.2824	0.2400
0.01850	2.563	0.7690	569	0.3414	0.2902	0.2467
0.01900	2.552	0.7655	582	0.3506	0.2980	0.2533
0.01950	2.540	0.7619	594	0.3599	0.3059	0.2600
0.02000	2.528	0.7584	607	0.3691	0.3137	0.2667
0.02050	2.516	0.7549	619	0.3783	0.3216	0.2733
0.02100	2.504	0.7513	631	0.3875	0.3294	0.2800
0.02150	2.493	0.7478	643	0.3968	0.3373	0.2867
0.02200	2.481	0.7442	655	0.4060	0.3451	0.2933
0.02250	2.469	0.7407	667	0.4152	0.3529	0.3000
0.02300	2.457	0.7372	678	0.4245	0.3608	0.3067
0.02350	2.445	0.7336	690	0.4337	0.3686	0.3133
0.02400	2.434	0.7301	701	0.4429	0.3765	0.3200
0.02450	2.422	0.7265	712	0.4521	0.3843	0.3267
0.02500	2.410	0.7230	723	0.4614	0.3922	0.3333
0.02550	2.398	0.7195	734	0.4706	0.4000	0.3400
0.02600	2.386	0.7159	745	0.4798	0.4078	0.3467
0.02650	2.375	0.7124	755	0.4890	0.4157	0.3533
0.02700	2.363	0.7088	766	0.4983	0.4235	0.3600
0.02750	2.351	0.7053	776	0.5075	0.4314	0.3667
3/4 X BALANCED STEEL RATIO						
0.02784	2.343	0.7029	783	0.5138	0.4367	0.3712

Table 5.2   (*cont.*)

$f'_c = 3000.0$ psi		$f_y = 50,000.0$ psi		$\rho_{min} = 200/f_y = 0.0040$	$\phi = 0.9$	
ρ	ϕa_u	ϕj_u	ϕR_u	c/d	a/d	ω
0.00200	3.676	0.8823	88	0.0461	0.0392	0.0333
0.00250	3.658	0.8779	110	0.0577	0.0490	0.0417
0.00300	3.639	0.8734	131	0.0692	0.0588	0.0500
0.00350	3.621	0.8690	152	0.0807	0.0686	0.0583
0.00400	3.602	0.8646	173	0.0923	0.0784	0.0667
0.00450	3.584	0.8602	194	0.1038	0.0882	0.0750
0.00500	3.566	0.8557	214	0.1153	0.0980	0.0833
0.00550	3.547	0.8513	234	0.1269	0.1078	0.0917
0.00600	3.529	0.8469	254	0.1384	0.1176	0.1000
0.00650	3.510	0.8425	274	0.1499	0.1275	0.1083
0.00700	3.492	0.8381	293	0.1615	0.1373	0.1167
0.00750	3.473	0.8336	313	0.1730	0.1471	0.1250
0.00800	3.455	0.8292	332	0.1845	0.1569	0.1333
0.00850	3.437	0.8248	351	0.1961	0.1667	0.1417
0.00900	3.418	0.8203	369	0.2076	0.1765	0.1500
0.00950	3.400	0.8159	388	0.2191	0.1863	0.1583
0.01000	3.381	0.8115	406	0.2307	0.1961	0.1667
0.01050	3.363	0.8071	424	0.2422	0.2059	0.1750
0.01100	3.344	0.8027	441	0.2537	0.2157	0.1833
0.01150	3.326	0.7982	459	0.2653	0.2255	0.1917
0.01200	3.308	0.7938	476	0.2768	0.2353	0.2000
0.01250	3.289	0.7894	493	0.2883	0.2451	0.2083
0.01300	3.271	0.7850	510	0.2999	0.2549	0.2167
0.01350	3.252	0.7805	527	0.3114	0.2647	0.2250
0.01400	3.234	0.7761	543	0.3230	0.2745	0.2333
0.01450	3.215	0.7717	559	0.3345	0.2843	0.2417
0.01500	3.197	0.7673	575	0.3460	0.2941	0.2500
0.01550	3.178	0.7628	591	0.3576	0.3039	0.2583
0.01600	3.160	0.7584	607	0.3691	0.3137	0.2667
0.01650	3.142	0.7540	622	0.3806	0.3235	0.2750
0.01700	3.123	0.7496	637	0.3922	0.3333	0.2833
0.01750	3.105	0.7451	652	0.4037	0.3431	0.2917
0.01800	3.086	0.7407	667	0.4152	0.3529	0.3000
0.01850	3.068	0.7363	681	0.4268	0.3627	0.3083
0.01900	3.049	0.7319	695	0.4383	0.3725	0.3167
0.01950	3.031	0.7274	709	0.4498	0.3824	0.3250
0.02000	3.013	0.7230	723	0.4614	0.3922	0.3333
0.02050	2.994	0.7186	737	0.4729	0.4020	0.3417
3/4 X BALANCED STEEL RATIO						
0.02065	2.989	0.7173	740	0.4763	0.4048	0.3441

STRENGTH DESIGN

Table 5.2 (*cont.*)

	$f'_c = 3000.0$ psi	$f_y = 60,000.0$ psi		$\rho_{min} = 200/f_y = 0.0033$	$\phi = 0.9$	
ρ	ϕa_u	ϕj_u	ϕR_u	c/d	a/d	ω
0.00200	4.394	0.8788	105	0.0554	0.0471	0.0400
0.00250	4.367	0.8734	131	0.0692	0.0588	0.0500
0.00300	4.341	0.8681	156	0.0830	0.0706	0.0600
0.00350	4.314	0.8628	181	0.0969	0.0824	0.0700
0.00400	4.288	0.8575	206	0.1107	0.0941	0.0800
0.00450	4.261	0.8522	230	0.1246	0.1059	0.0900
0.00500	4.234	0.8469	254	0.1384	0.1176	0.1000
0.00550	4.208	0.8416	278	0.1522	0.1294	0.1100
0.00600	4.181	0.8363	301	0.1661	0.1412	0.1200
0.00650	4.155	0.8310	324	0.1799	0.1529	0.1300
0.00700	4.128	0.8257	347	0.1938	0.1647	0.1400
0.00750	4.102	0.8204	369	0.2076	0.1765	0.1500
0.00800	4.075	0.8150	391	0.2215	0.1882	0.1600
0.00850	4.049	0.8097	413	0.2353	0.2000	0.1700
0.00900	4.022	0.8044	434	0.2491	0.2118	0.1800
0.00950	3.996	0.7991	455	0.2630	0.2235	0.1900
0.01000	3.969	0.7938	476	0.2768	0.2353	0.2000
0.01050	3.942	0.7885	497	0.2907	0.2471	0.2100
0.01100	3.916	0.7832	517	0.3045	0.2588	0.2200
0.01150	3.889	0.7779	537	0.3183	0.2706	0.2300
0.01200	3.863	0.7726	556	0.3322	0.2824	0.2400
0.01250	3.836	0.7673	575	0.3460	0.2941	0.2500
0.01300	3.810	0.7619	594	0.3599	0.3059	0.2600
0.01350	3.783	0.7566	613	0.3737	0.3176	0.2700
0.01400	3.757	0.7513	631	0.3875	0.3294	0.2800
0.01450	3.730	0.7460	649	0.4014	0.3412	0.2900
0.01500	3.704	0.7407	667	0.4152	0.3529	0.3000
0.01550	3.677	0.7354	684	0.4291	0.3647	0.3100
0.01600	3.650	0.7301	701	0.4429	0.3765	0.3200
3/4 X BALANCED STEEL RATIO						
0.01604	3.649	0.7297	702	0.4439	0.3773	0.3207

Table 5.3 Strength Design Factors (SI Units)

$f_c' = 20.0$ MPa		$f_y = 280.0$ MPa		$\rho_{min} = 1.4/f_y = 0.0050$	$\phi = 0.9$	
ρ	$10^3 \phi a_u$	ϕj_u	ϕR_u	c/d	a/d	ω
0.00200	0.248	0.8851	0.4957	0.0388	0.0329	0.0280
0.00250	0.247	0.8814	0.6170	0.0484	0.0412	0.0350
0.00300	0.246	0.8777	0.7373	0.0581	0.0494	0.0420
0.00350	0.245	0.8740	0.8566	0.0678	0.0576	0.0490
0.00400	0.244	0.8703	0.9747	0.0775	0.0659	0.0560
0.00450	0.243	0.8665	1.0919	0.0872	0.0741	0.0630
0.00500	0.242	0.8628	1.2080	0.0969	0.0824	0.0700
0.00550	0.241	0.8591	1.3231	0.1066	0.0906	0.0770
0.00600	0.240	0.8554	1.4371	0.1163	0.0988	0.0840
0.00650	0.238	0.8517	1.5501	0.1260	0.1071	0.0910
0.00700	0.237	0.8480	1.6620	0.1356	0.1153	0.0980
0.00750	0.236	0.8442	1.7730	0.1453	0.1235	0.1050
0.00800	0.235	0.8405	1.8828	0.1550	0.1318	0.1120
0.00850	0.234	0.8368	1.9917	0.1647	0.1400	0.1190
0.00900	0.233	0.8331	2.0994	0.1744	0.1482	0.1260
0.00950	0.232	0.8294	2.2062	0.1841	0.1565	0.1330
0.01000	0.231	0.8257	2.3119	0.1938	0.1647	0.1400
0.01050	0.230	0.8219	2.4166	0.2035	0.1729	0.1470
0.01100	0.229	0.8182	2.5202	0.2131	0.1812	0.1540
0.01150	0.228	0.8145	2.6228	0.2228	0.1894	0.1610
0.01200	0.227	0.8108	2.7243	0.2325	0.1976	0.1680
0.01250	0.226	0.8071	2.8248	0.2422	0.2059	0.1750
0.01300	0.225	0.8034	2.9243	0.2519	0.2141	0.1820
0.01350	0.224	0.7996	3.0227	0.2616	0.2224	0.1890
0.01400	0.223	0.7959	3.1201	0.2713	0.2306	0.1960
0.01450	0.222	0.7922	3.2164	0.2810	0.2388	0.2030
0.01500	0.221	0.7885	3.3117	0.2907	0.2471	0.2100
0.01550	0.220	0.7848	3.4060	0.3003	0.2553	0.2170
0.01600	0.219	0.7811	3.4992	0.3100	0.2635	0.2240
0.01650	0.218	0.7773	3.5913	0.3197	0.2718	0.2310
0.01700	0.217	0.7736	3.6825	0.3294	0.2800	0.2380
0.01750	0.216	0.7699	3.7726	0.3391	0.2882	0.2450
0.01800	0.215	0.7662	3.8616	0.3488	0.2965	0.2520
0.01850	0.213	0.7625	3.9496	0.3585	0.3047	0.2590
0.01900	0.212	0.7588	4.0366	0.3682	0.3129	0.2660
0.01950	0.211	0.7550	4.1225	0.3779	0.3212	0.2730
0.02000	0.210	0.7513	4.2074	0.3875	0.3294	0.2800
0.02050	0.209	0.7476	4.2913	0.3972	0.3376	0.2870
0.02100	0.208	0.7439	4.3741	0.4069	0.3459	0.2940
0.02150	0.207	0.7402	4.4558	0.4166	0.3541	0.3010
0.02200	0.206	0.7365	4.5366	0.4263	0.3624	0.3080
0.02250	0.205	0.7327	4.6163	0.4360	0.3706	0.3150
0.02300	0.204	0.7290	4.6949	0.4457	0.3788	0.3220
0.02350	0.203	0.7253	4.7725	0.4554	0.3871	0.3290
0.02400	0.202	0.7216	4.8491	0.4650	0.3953	0.3360
0.02450	0.201	0.7179	4.9246	0.4747	0.4035	0.3430
0.02500	0.200	0.7142	4.9991	0.4844	0.4118	0.3500
0.02550	0.199	0.7104	5.0725	0.4941	0.4200	0.3570
0.02600	0.198	0.7067	5.1449	0.5038	0.4282	0.3640
3/4 X BALANCED STEEL RATIO						
0.02639	0.197	0.7038	5.2007	0.5114	0.4347	0.3695

Table 5.3 (*cont.*)

ρ	$10^3\phi a_u$	ϕj_u	ϕR_u	c/d	a/d	ω
\multicolumn{7}{c}{$f'_c = 20.0$ MPa $f_y = 350.0$ MPa $\rho_{min} = 1.4/f_y = 0.0040$ $\phi = 0.9$}

ρ	$10^3\phi a_u$	ϕj_u	ϕR_u	c/d	a/d	ω
0.00200	0.308	0.8814	0.6170	0.0484	0.0412	0.0350
0.00250	0.307	0.8768	0.7672	0.0606	0.0515	0.0438
0.00300	0.305	0.8721	0.9158	0.0727	0.0618	0.0525
0.00350	0.304	0.8675	1.0627	0.0848	0.0721	0.0613
0.00400	0.302	0.8628	1.2080	0.0969	0.0824	0.0700
0.00450	0.300	0.8582	1.3517	0.1090	0.0926	0.0787
0.00500	0.299	0.8535	1.4937	0.1211	0.1029	0.0875
0.00550	0.297	0.8489	1.6342	0.1332	0.1132	0.0962
0.00600	0.295	0.8442	1.7730	0.1453	0.1235	0.1050
0.00650	0.294	0.8396	1.9101	0.1574	0.1338	0.1137
0.00700	0.292	0.8350	2.0457	0.1695	0.1441	0.1225
0.00750	0.291	0.8303	2.1796	0.1817	0.1544	0.1312
0.00800	0.289	0.8257	2.3119	0.1938	0.1647	0.1400
0.00850	0.287	0.8210	2.4426	0.2059	0.1750	0.1487
0.00900	0.286	0.8164	2.5716	0.2180	0.1853	0.1575
0.00950	0.284	0.8117	2.6990	0.2301	0.1956	0.1662
0.01000	0.282	0.8071	2.8248	0.2422	0.2059	0.1750
0.01050	0.281	0.8024	2.9490	0.2543	0.2162	0.1837
0.01100	0.279	0.7978	3.0715	0.2664	0.2265	0.1925
0.01150	0.278	0.7931	3.1924	0.2785	0.2368	0.2012
0.01200	0.276	0.7885	3.3117	0.2907	0.2471	0.2100
0.01250	0.274	0.7838	3.4294	0.3028	0.2574	0.2187
0.01300	0.273	0.7792	3.5454	0.3149	0.2676	0.2275
0.01350	0.271	0.7746	3.6598	0.3270	0.2779	0.2362
0.01400	0.269	0.7699	3.7726	0.3391	0.2882	0.2450
0.01450	0.268	0.7653	3.8837	0.3512	0.2985	0.2537
0.01500	0.266	0.7606	3.9933	0.3633	0.3088	0.2625
0.01550	0.265	0.7560	4.1012	0.3754	0.3191	0.2712
0.01600	0.263	0.7513	4.2074	0.3875	0.3294	0.2800
0.01650	0.261	0.7467	4.3121	0.3997	0.3397	0.2887
0.01700	0.260	0.7420	4.4151	0.4118	0.3500	0.2975
0.01750	0.258	0.7374	4.5165	0.4239	0.3603	0.3062
0.01800	0.256	0.7327	4.6163	0.4360	0.3706	0.3150
0.01850	0.255	0.7281	4.7144	0.4481	0.3809	0.3237
0.01900	0.253	0.7234	4.8109	0.4602	0.3912	0.3325
0.01950	0.252	0.7188	4.9058	0.4723	0.4015	0.3412
\multicolumn{7}{c}{3/4 X BALANCED STEEL RATIO}
0.01956	0.251	0.7183	4.9164	0.4737	0.4026	0.3422

Table 5.3 (*cont.*)

ρ	$10^3\phi a_u$	ϕj_u	ϕR_u	c/d	a/d	ω
	$f'_c = 20.0$ MPa		$f_y = 400.0$ MPa	$\rho_{min} = 1.4/f_y = 0.0035$	$\phi = 0.9$	
0.00200	0.352	0.8788	0.7031	0.0554	0.0471	0.0400
0.00250	0.349	0.8734	0.8735	0.0692	0.0588	0.0500
0.00300	0.347	0.8681	1.0418	0.0830	0.0706	0.0600
0.00350	0.345	0.8628	1.2080	0.0969	0.0824	0.0700
0.00400	0.343	0.8575	1.3721	0.1107	0.0941	0.0800
0.00450	0.341	0.8522	1.5340	0.1246	0.1059	0.0900
0.00500	0.339	0.8469	1.6938	0.1384	0.1176	0.1000
0.00550	0.337	0.8416	1.8515	0.1522	0.1294	0.1100
0.00600	0.335	0.8363	2.0071	0.1661	0.1412	0.1200
0.00650	0.332	0.8310	2.1606	0.1799	0.1529	0.1300
0.00700	0.330	0.8257	2.3119	0.1938	0.1647	0.1400
0.00750	0.328	0.8204	2.4611	0.2076	0.1765	0.1500
0.00800	0.326	0.8150	2.6082	0.2215	0.1882	0.1600
0.00850	0.324	0.8097	2.7531	0.2353	0.2000	0.1700
0.00900	0.322	0.8044	2.8960	0.2491	0.2118	0.1800
0.00950	0.320	0.7991	3.0367	0.2630	0.2235	0.1900
0.01000	0.318	0.7938	3.1752	0.2768	0.2353	0.2000
0.01050	0.315	0.7885	3.3117	0.2907	0.2471	0.2100
0.01100	0.313	0.7832	3.4460	0.3045	0.2588	0.2200
0.01150	0.311	0.7779	3.5782	0.3183	0.2706	0.2300
0.01200	0.309	0.7726	3.7083	0.3322	0.2824	0.2400
0.01250	0.307	0.7673	3.8363	0.3460	0.2941	0.2500
0.01300	0.305	0.7619	3.9621	0.3599	0.3059	0.2600
0.01350	0.303	0.7566	4.0858	0.3737	0.3176	0.2700
0.01400	0.301	0.7513	4.2074	0.3875	0.3294	0.2800
0.01450	0.298	0.7460	4.3269	0.4014	0.3412	0.2900
0.01500	0.296	0.7407	4.4442	0.4152	0.3529	0.3000
0.01550	0.294	0.7354	4.5595	0.4291	0.3647	0.3100
0.01600	0.292	0.7301	4.6725	0.4429	0.3765	0.3200
3/4 X BALANCED STEEL RATIO						
0.01626	0.291	0.7274	4.7297	0.4500	0.3825	0.3251

parameters f'_c and f_y in Table 5.3 are the *rounded* equivalents to those of Table 5.2. Hence, even the nondimensional columns in the two tables (e.g. ω) differ somewhat. In Table 5.2, the values in the second column have been divided by 12,000 in order that the formula $A_s = M_u/\phi a_u d$ shall give A_s in in^2 when M_u is in ft-kips (instead of in-lbs). In like manner, the second column of Table 5.3 has been normalized to put A_s in mm^2 when M_u is in kN·m. Note that both tables include the effect of ϕ (0.90); some design handbooks include the ϕ-factor and some do not.

The 1963 ACI Code required deflection checks when ω exceeded 0.18. This value is still in common use to control deflections, even though it is not mentioned in the 1983 ACI Code.

Solved Problems

5.1. Derive the expression given in the 1983 ACI Code for the nominal flexural capacity of an underreinforced rectangular beam assuming a rectangular stress distribution. Refer to Fig. 5-6.

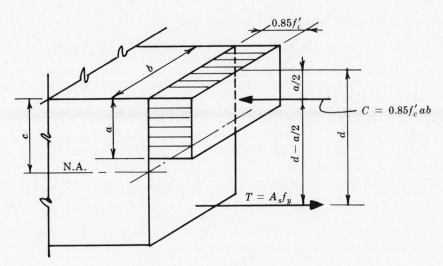

Fig. 5-6

Summing horizontal forces, $T = C$. Then

$$A_s f_y = 0.85 f'_c ab \quad \text{and} \quad a = A_s f_y/0.85 f'_c b$$

Summing moments about the centroid of the compressive force, $M_n = A_s f_y (d - a/2)$. Substituting for a,

$$M_n = A_s f_y d(1 - 0.59 \rho f_y/f'_c) \quad \text{or} \quad M_n = f'_c b d^2 \omega (1 - 0.59\omega)$$

5.2. Derive the expression for balanced reinforcement of a beam for tension only, as given by equation (5.14).

Referring to Fig. 5-7, by geometry,

$$\epsilon_y c = \epsilon_u(d - c) \quad \text{or} \quad c = \epsilon_u d/(\epsilon_u + \epsilon_y)$$

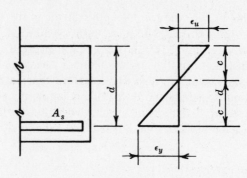

Fig. 5-7

But, $c = a/\beta_1 = A_s f_y/0.85 f'_c b \beta_1$. Then

$$\epsilon_u d/(\epsilon_u + \epsilon_y) = A_s f_y/0.85 f'_c b \beta_1$$

or

$$\frac{A_s}{bd} = \frac{0.85 f'_c \beta_1}{f_y} \frac{\epsilon_u}{\epsilon_u + \epsilon_y} = \rho$$

If ρ is defined as the balanced reinforcement, ρ_b, and if we let $\epsilon_u = 0.003$, then

$$\rho_b = \frac{0.85 f'_c \beta_1}{f_y} \frac{0.003 E_s}{(0.003 + \epsilon_y) E_s}$$

If $E_s = 29,000,000$ psi $(200\,000$ MPa$)$ and $f_y = \epsilon_y E_s$, then

$$\rho_b = \frac{0.85 f'_c \beta_1}{f_y} \frac{87,000}{87,000 + f_y} \qquad \left(\frac{0.85 f'_c \beta_1}{f_y} \frac{600}{600 + f_y} \right)$$

5.3 Derive equation *(5.16)*.

Referring to Fig. 5-8, by geometry,

$$\epsilon'_s c = \epsilon_u (c - d') \qquad \text{or} \qquad c = \epsilon_u d'/(\epsilon_u - \epsilon'_s)$$

But $c = a/\beta_1 = A_s f_y/0.85 f'_c b \beta_1$. Then

$$\epsilon_u d'/(\epsilon_u - \epsilon'_s) = A_s f_y/0.85 f'_c b \beta_1$$

or

$$\frac{A_s}{bd} = \frac{0.85 f'_c \beta_1 d'}{f_y d} \frac{\epsilon_u}{\epsilon_u - \epsilon'_s}$$

At ultimate strength $\epsilon'_s = \epsilon_y$, and substituting for $f_y = E_s \epsilon_y$,

$$\rho - \rho' = \frac{0.85 f'_c \beta_1 d'}{f_y d} \frac{87,000}{87,000 - f_y}$$

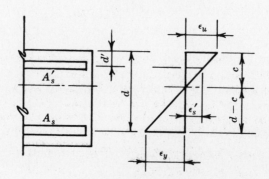

Fig. 5-8

5.4. Derive equation (*5.17*).

Referring to Fig. 5-9, it is assumed that the effect of the overhanging portion of the flange is the same as the effect of compression reinforcement in a double reinforced beam. This "additional" reinforcement is A_{sf}.

The compressive strength of the overhang would be $C' = 0.85f'_c(b - b_w)h_f$, and transforming to reinforcing steel, $A_{sf} = C'/f_y = 0.85(b - b_w)h_f f'_c/f_y$. It is further assumed that this compressive force acts at the centroid of the flange. Therefore, the moment resistance due to the flange would be $M_{nf} = A_{sf}f_y(d - 0.5h_f)$. Substituting these values in the expression for a double reinforced beam,

$$M_n = (A_s - A_{sf})f_y(d - 0.5a) + A_s f_y(d - 0.5h_f)$$

where $a = (A_s - A_{sf})f_y/0.85f'_c b_w$.

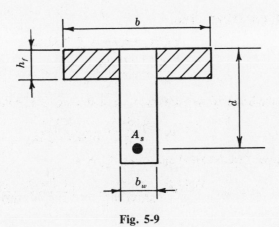

Fig. 5-9

5.5. The three span beam shown in Fig. 5-10 is subjected to a dead load of 1.0 kip/ft and a live load of 2.0 kips/ft. Determine the ultimate positive and negative moments and shears. Use the load factors of the 1983 ACI Code.

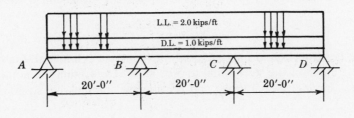

Fig. 5-10

Since the L.L. could be on any number of the spans, it will be necessary to determine the critical loading pattern. The critical loading condition for maximum shear at A and for maximum positive moment in span AB would be live load in spans AB and CD only. The critical loading for maximum positive moment in span BC would be live load in span BC only. The critical loading condition for maximum shear and maximum negative moment at B would be live load in spans AB and BC only.

An elastic analysis of the structure yields the following shears and moments. For dead load:

$$V_{AB} = 0.4wl = 0.4(1.0)(20) = 8.0 \text{ kips} \qquad +M_{AB} = 0.08wl^2 = 0.08(1.0)(20)^2 = 32.0 \text{ ft-kips}$$

$$V_{BA} = 0.6wl = 0.6(1.0)(20) = 12.0 \text{ kips} \qquad -M_B = 0.10wl^2 = 0.10(1.0)(20)^2 = 40.0 \text{ ft-kips}$$

$$V_{BC} = 0.5wl = 0.5(1.0)(20) = 10.0 \text{ kips} \qquad +M_{BC} = 0.025wl^2 = 0.025(1.0)(20)^2 = 10.0 \text{ ft-kips}$$

For live load on spans AB and CD only:

$$V_{AB} = 0.45wl = 0.45(2)(20) = 18.0 \text{ kips}$$
$$+M_{AB} = 0.101wl^2 = 0.101(2)(20)^2 = 80.8 \text{ ft-kips}$$

For live load on span BC only:

$$+M_{BC} = 0.075wl^2 = 0.075(2)(20)^2 = 60.0 \text{ ft-kips}$$

For live load on spans AB and BC only:

$$V_{BA} = 0.617wl = 0.617(2)(20) = 24.7 \text{ kips}$$
$$V_{BC} = 0.583wl = 0.583(2)(20) = 23.3 \text{ kips}$$
$$-M_B = 0.117wl^2 = 0.117(2)(20)^2 = 93.6 \text{ ft-kips}$$

Summarizing,

	D.L.	Max. L.L.
$V_{AB} =$	8.0 kips	18.0 kips
$V_{BA} =$	12.0	24.7
$V_{BC} =$	10.0	23.3
$+M_{AB} =$	32.0 ft-kips	80.0 ft-kips
$-M_B =$	40.0	93.6
$+M_{BC} =$	10.0	60.0

The appropriate load factor equations are

$$U = 1.4D + 1.7L \qquad \text{and} \qquad U = 0.75(1.4D + 1.7L + 1.7W)$$

It is apparent that the first equation governs, since there are no wind load effects. Therefore:

$$\text{Ult.} \quad V_{AB} = 1.4(8.0) + 1.7(18) = 41.8 \text{ kips}$$
$$\text{Ult.} \quad V_{BA} = 1.4(12.0) + 1.7(24.7) = 58.8 \text{ kips}$$
$$\text{Ult.} \quad V_{BC} = 1.4(10.0) + 1.7(23.3) = 53.6 \text{ kips}$$
$$\text{Ult.} \ +M_{AB} = 1.4(32.0) + 1.7(80.8) = 182.2 \text{ ft-kips}$$
$$\text{Ult.} \ -M_B = 1.4(40.0) + 1.7(93.6) = 215.1 \text{ ft-kips}$$
$$\text{Ult.} \ +M_{BC} = 1.4(10.0) + 1.7(60.0) = 116.0 \text{ ft-kips}$$

It will also be necessary to investigate the load factor equation $U = 0.9D + 1.3W$ when wind is involved.

5.6. Given the beam section in Fig. 5-11 with the concrete cylinder strength equal to 3500 psi and the yield strength of the steel equal to 50,000 psi. Determine the nominal flexural capacity of the beam assuming the parabolic and rectangular compressive stress distributions.

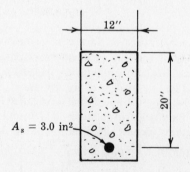

Fig. 5-11

By (5.10),

$$M_n = A_s f_y d\left(1 - \frac{k_2}{k_1}\, \rho\mu\right)$$

It was shown that for the parabolic distribution, $k_2/k_1 = 0.554$ when $f_c = 0.85 f_c'$ and $f_s = f_y$. Then

$$M_n = 3.0(50{,}000)(20)\left[1 - \frac{0.554(3.0)(50)}{12(20)(0.85)(3.5)}\right] = 2{,}650{,}000 \text{ in-lb}$$

If the rectangular distribution is assumed, $k_2/k_1 = 0.50$ and $M_n = 2{,}690{,}000$ in-lb. A comparison of the results yields $2{,}690{,}000/2{,}650{,}000 = 1.016$, or less than 2% difference.

5.7. Design a one way continuous slab to support the moments tabulated. Assume $f_c' = 3000$ psi, $f_y = 40{,}000$ psi, and design for minimum depth permitted with no compressive reinforcement.

$$\text{Dead Load Moment} = +9.5 \text{ ft-kips/ft at midspan}$$
$$= -15.0 \text{ ft-kips/ft at supports}$$
$$\text{Live Load Moment} = +9.0 \text{ ft-kips/ft at midspan}$$
$$= -10.0 \text{ ft-kips/ft at supports}$$
$$\text{Wind Load Moment} = \pm 30.0 \text{ ft-kips/ft at the supports}$$

(Wind load produces no moment at midspan.)

(a) The first step is to determine the ultimate moments, both positive and negative, by use of load factor equations:

$$U = 1.4D + 1.7L\,, \qquad U = 0.75(1.4D + 1.7L + 1.7W)\,, \qquad U = 0.9D + 1.3W$$

For positive moments,

$$U = 1.4(9.5) + 1.7(9.0) = 28.6 \text{ ft-kips/ft}$$

This will be the ultimate positive moment because wind has no effect at midspan and the last load factor equation will be critical only at the supports.
For negative moment,

$$U = 1.4(-15.0) + 1.7(-10.0) = -38.0 \text{ ft-kips/ft}$$
$$U = 0.75[(1.4)(-15) + (1.7)(-10) + (1.7)(-30)] = -66.75 \text{ ft-kips/ft}$$
$$U = 0.9(15.0) \pm 1.3(30.0) = -25.5 \text{ ft-kips/ft} \qquad \text{or} \qquad +52.5 \text{ ft-kips/ft}$$

(b) The maximum ultimate negative moment of -66.75 ft-kips/ft is the largest absolute value and governs the thickness of the slab. It is necessary to determine the maximum percentage of reinforcement permitted. Then

$$\rho_{max} = 0.75\left[\frac{0.85\beta_1 f_c'}{f_y}\, \frac{87{,}000}{87{,}000 + f_y}\right] = 0.75\left[\frac{0.85(0.85)(3000)}{40{,}000}\, \frac{87{,}000}{127{,}000}\right] = 0.0279$$

(c) With ρ_{max} determined, the depth of the section is determined by substitution into the expression for the ultimate moment capacity; assuming the rectangular distribution,

$$M_u = \phi[bd^2 f_c' \omega(1 - 0.59\omega)] \qquad (1)$$

where $\phi = 0.90$ for flexure and $\omega = \rho(f_y/f_c') = 0.0279(40{,}000/3000) = 0.372$. Substituting into equation (1):

$$\frac{M_u}{\phi bd^2 f_c'} = \omega(1 - 0.59\omega) = 0.372[1 - (0.59)(0.372)] = 0.2904$$

so

$$\frac{(66.75)(1200)}{(0.9)(12)(0.2904)(3000)} = d^2 = 85.13 \text{ in}^2$$

and $d = 9.23''$. Assuming that No. 4 bars will be used, with the required $\frac{3}{4}''$ cover, the total depth of the slab is $9.23 + 0.25 + 0.75 = 10.23''$. Use a total depth of $10.5''$, so the furnished effective depth is $10.5 - 0.25 - 0.75 = 9.5''$. Use an effective depth of $9.5''$.

The reinforcement at all sections can be determined by use of equation (1), wherein every term is known but ω. But substituting the numerical values, a quadratic equation develops in terms of ω. Once ω is known, the steel ratio can be obtained using $\rho = \omega f_c'/f_y$. Then, the steel area required can be obtained as $A_s = \rho bd$. The negative moment reinforcement and the positive moment reinforcement can thus be determined.

5.8. Solve Problem 5.7 by using charts or tables for proportioning the member.

(a)   From Fig. 5-4, $\rho_{max} = 0.028$; then $\omega_{max} = \rho_{max}(f_y/f_c') = 0.372$.

(b)   Solving for $M_n/bd^2f_c' = M_u/\phi bd^2 f_c'$ by Table 5.1 for $\omega = 0.37$,

$$\frac{M_u}{\phi f_c' bd^2} = 0.2892 \quad \text{or} \quad d^2 = \frac{66.75(12,000)}{0.9(3000)(12)(0.2892)} = 85.48 \text{ in}^2$$

from which $d = 9.25''$. Considering No. 4 bars and $\frac{3}{4}''$ cover, the total depth is

$$9.25 + \tfrac{1}{2}(0.5) + 0.75 = 10.25''$$

Use total depth $= 10.5''$, so the effective depth $d = 10.5 - 0.75 - \tfrac{1}{2}(0.5) = 9.5''$.

(c)   Negative reinforcement at the support is $-A_s = 0.028(12)(9.5) = 3.19 \text{ in}^2/\text{ft}$.

(d)   Positive reinforcement at support is determined by solving for ω. We have

$$\frac{M_u}{\phi f_c' bd^2} = \frac{52.5(12,000)}{(0.9)(3000)(12)(9.5)^2} = 0.215$$

From Table 5.1, $\omega = 0.253$. Then $+A_s = [0.253(3000/40,000)](12)(9.5) = 2.16 \text{ in}^2/\text{ft}$.

(e)   Similarly, the positive reinforcement at midspan is determined.

$$\frac{M_u}{\phi f_c' bd^2} = \frac{28.6(12,000)}{(0.9)(3000)(12)(9.5)^2} = 0.117$$

From Table 5.1, $\omega = 0.1265$. Then $A_s = 1.075 \text{ in}^2/\text{ft}$.

5.9. Solve Problem 5.8 using Fig. 5-5.

After determining maximum ρ and minimum d as before, enter Fig. 5-5 with values of $M_n/bd^2 = M_u/\phi bd^2$:

$$\frac{M_u}{\phi bd^2} = \frac{52.5(12,000)}{(0.9)(12)(9.5)^2} = 646.4$$

From Fig. 5-5, $\rho = 0.019$. Then $+A_s = 0.019(12)(9.5) = 2.167 \text{ in}^2/\text{ft}$.

$$\frac{M_u}{\phi bd^2} = \frac{28.6(12,000)}{(0.9)(12)(9.5)^2} = 352$$

and $\rho = 0.010$. Then $A_s = 0.010(12)(9.5) = 1.14 \text{ in}^2/\text{ft}$.

5.10. Solve Problem 5.7 using Table 5.2.

(a)   Using Table 5.2, the maximum resisting moment at three-fourths balanced conditions is $M_u = \phi R_u bd^2$. From the table, for $f_c' = 3000$ psi and $f_y = 40,000$ psi, $\phi R_u = 783$. Then

$$d^2 = \frac{66.75(12,000)}{(12)(783)} = 85.24 \quad \text{and} \quad d = 9.23$$

Use $d = 9.25''$.

(b)   From Table 5.2, $\rho_{max} = 0.02784$. Then $-A_s = 0.02784(12)(9.25) = 3.09\ \text{in}^2/\text{ft}$.

(c)   Using the average value $\phi j_u = 0.80$ selected from Table 5.2,

$$+A_s = \frac{M_u}{(\phi j_u)f_y d} = \frac{52.5(12,000)}{0.80(40,000)(9.25)} = 2.13\ \text{in}^2/\text{ft}$$

and $\rho = 2.13/(9.25)(12) = 0.019$. This value is very close to the value for ρ given in the table. If greater refinement is required, the next trial may be made using $\rho = 0.019$ and the corresponding $\phi a_u = 2.552$. The value of ϕa_u is determined so that the moment is in foot-kips.

$$+A_s = \frac{M}{\phi a_u d} = \frac{52.5}{(2.552)(9.25)} = 2.22\ \text{in}^2/\text{ft}$$

In normal design procedures, the first trial yielding an area of 2.13 in²/ft would be sufficiently accurate. This is comparable to the accuracy attained in the alternate design method.

(d)   Because the value of the midspan moment is less than that in (c), a value for ϕa_u somewhat greater than that given above will be used. Hence try $\phi a_u = 2.847$:

$$A_s = \frac{M}{\phi a_u d} = \frac{28.6}{(2.847)(9.25)} = 1.086\ \text{in}^2/\text{ft} \quad \text{and} \quad \rho = \frac{1.086}{9.25(12)} = 0.0098$$

or, say, $\rho = 0.01$. In Table 5.2, the corresponding value of ϕa_u is 2.764, so that further iteration yields

$$A_s = \frac{M_u}{\phi a_u d} = \frac{28.6}{2.764(9.25)} = 1.12\ \text{in}^2/\text{ft} \quad \text{and} \quad \rho = \frac{1.12}{9.25(12)} = 0.01$$

The iteration converges.

5.11.   Design a minimum depth rectangular beam with tension reinforcement only to carry a D.L. moment of 50 ft-kips and a L.L. moment of 200 ft-kips. Assume $f'_c = 3000$ psi and $f_y = 50,000$ psi.

(a)   The ultimate moment must be determined by use of the load factor equation (5.2), which gives $U = 410$ ft-kips.

(b)   A minimum depth section requires maximum percentage of steel reinforcement. Therefore, substituting in the expression for ρ_b,

$$\rho_b = \frac{0.85f'_c\beta_1}{f_y}\ \frac{87,000}{87,000 + f_y} = 0.0276 \quad \text{and} \quad \rho_{max} = 0.75(0.0276) = 0.0207$$

(c)   With ρ_{max} determined, the value for bd^2 is found by substitution in the expression for the ultimate moment capacity.

$$M_u = \phi[bd^2 f'_c \omega(1 - 0.59\omega)]$$

where $\omega = \rho(f_y/f'_c) = 0.0207(50,000/3000) = 0.345$. Substituting,

$$410(12,000) = (0.9)(bd^2)(3000)(0.345)(1 - 0.204) \quad \text{or} \quad bd^2 = 6645\ \text{in}^3$$

There would be many combinations of the beam width and depth that would satisfy the above expression. A width or a depth is often set by some other architectural or structural consideration. Try a ratio $d/b = 1.5$. Then $b(1.5b)^2 = 6645$, $b^3 = 2953$, $b = 14.35$, use $b = 14.5''$. Thus $d = 1.5(14.5) = 21.8$; use $d = 22''$. Then $A_s = \rho bd = 0.0207(14.5)(22) = 6.60\ \text{in}^2$.

(d)   Try a ratio of $d/b = 1.0$. Then $b^3 = 6645$, $b = 18.8$; use $b = 19.0''$. Use $d = 19.0''$. Then $A_s = \rho bd = 0.0207(19.0)^2 = 7.47\ \text{in}^2$.

(e)   The 1983 ACI Code requires a minimum amount of reinforcement in beams. The specification states that for flexural members, except constant thickness slabs, if tension reinforcement is required, the minimum reinforcement ratio ρ shall be not less than $200/f_y$ unless the area of reinforcement at every section, positive or negative, is at least one third greater than that required.

Of course, in a section of maximum ρ, this requirement will not control. However, for $f_y = 50,000$ psi, $200/f_y = 0.004$.

(*f*) For a beam such as this, the deflections at service loads must be checked. The 1983 ACI Code requires that the deflection of a beam must be checked if the depth is less than stated minimum depths (see Chapter 8).

5.12. Solve Problem 5.11 using Table 5.2.

If $f'_c = 3000$ psi, $f_y = 50,000$ psi, $M_u = 410$ ft-kips, from Table 5.2 the maximum values for ϕR_u and ρ are $\phi R_u = 740$, $\rho = 0.02065$. Since in strength design $\phi R_u = M_u/bd^2$, then $bd^2 = 410(12,000)/740 = 6649$ in^3.

Using the same ratios of d/b and selecting ϕa_u from Table 5.2, $\phi a_u = 2.989$.

For $d/b \approx 1.5$, use $b = 14.5''$ and $d = 22''$ and $A_s = 410/(2.989)(22) = 6.23$ in^2.

For $d/b \approx 1.0$, use $b = d = 19''$ and $A_s = 410/(2.989)(19) = 7.22$ in^2.

It will be noted that the area of reinforcement above for $d/b = 1.0$ does not check with that of Problem 5.11. This discrepancy is due to the fact that the values determined for d and b were rounded off.

In Problem 5.11, if the actual values were used for $d/b = 1.0$, $b = d = 18.8''$ and $A_s = 0.02065(18.8)^2 = 7.30$ in^2.

In this Problem 5.12, if these same values for d and b are used, $A_s = 7.22$ in^2. Thus when the "exact" values for d and b are used, both methods yield the same results.

Regardless of which method is used, the rounded-off values result in areas of reinforcement within 1.1 percent of the results determined using the more precise values for b and d. This degree of accuracy would be normally acceptable in ordinary design procedures provided the results are conservative.

5.13. Determine the area of reinforcing steel required for the section shown in Fig. 5-12. Assume $f'_c = 3500$ psi, $f_y = 50,000$ psi and $M_u = 150$ ft-kips.

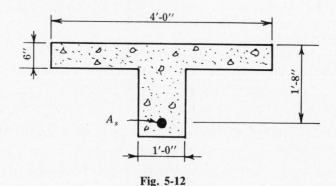

Fig. 5-12

(*a*) The first step in the solution is to determine if the neutral axis of the section falls within the stem and the section acts as a T-beam. If not, it is proportioned as a rectangular beam. If the moment capacity of the flange is greater than 150 ft-kips, then the section acts as a rectangular beam.

The moment capacity of the flange is

$$M_{nf} = 0.85f'_c bh_f(d - 0.5h_f) = 0.85(3500)(48)(6)(20 - 3)/12,000 = 1215 \text{ ft-kips} > 150/0.9 = M_n$$

Hence the section does not act as a T-beam and should be designed as a rectangular section.

(*b*) Calculate

$$\frac{M_n}{f'_c bd^2} = \frac{M_u}{\phi f'_c bd^2} = \frac{150(12,000)}{(0.90)(3500)(48)(20)^2} = 0.0298$$

Now using Table 5.1, we find $\omega = 0.030$. Then $\rho = \omega f'_c / f_y = 0.030(3500)/50,000 = 0.00210$ and $A_s = \rho bd = 0.00210(48)(20) = 2.02$ in^2.

(c)   As discussed in Problem 5.11, the 1983 ACI Code requires that the minimum area of tension reinforcement shall not be less than $200/f_y$. $200/f_y = 200/50,000 = 0.004 < 2.02/(12)(20)$.

It should be noted that here ρ is defined as the ratio of the area of tension reinforcement to the effective area of concrete in the web of a flanged member.

(d) $h_f = 6'' > 1.18\omega d/\beta_1 = 1.18(0.030)(20)/0.85 = 0.83''$ also establishes that the section is to be proportioned as a rectangular beam.

5.14.   Determine the area of reinforcement for a T-section with the following given: $h_f = 2.0''$, $b = 30''$, $b_w = 14''$, $d = 30''$, $f'_c = 3000$ psi, $f_y = 40,000$ psi, $M_u = 600$ ft-kips.

(a)   As in Problem 5.13, we must determine if the neutral axis of the section falls within the stem. The nominal moment capacity of the concrete flange is

$$M_{nf} = 0.85 f'_c b h_f (d - 0.5 h_f) = 370 \text{ ft-kips} < 600/0.90 = M_n$$

Hence the section acts as a T-beam.

(b)   In the design of T-beams, the overhanging portion of the flange is considered the same as an equivalent amount of compression steel.

$$A_{sf} = 0.85(b - b_w)h_f f'_c / f_y = 2.04 \text{ in}^2$$

The nominal moment capacity of the overhanging portion is $M'_{nf} = A_{sf} f_y (d - 0.5 h_f) = 197$ ft-kips.

(c)   The nominal moment capacity of the rectangular section, b_w times d, must be

$$M_w = M_n - M'_{nf} = 667 - 197 = 470 \text{ ft-kips}$$

And the nominal moment capacity of the web portion is $M_w = b_w d^2 f'_c \omega (1 - 0.59\omega)$. Then

$$\frac{M_w}{f'_c b_w d^2} = \frac{470(12,000)}{3000(14)(30)^2} = 0.149 = \omega(1 - 0.59\omega)$$

From Table 5.1, $\omega = 0.165$ or the reinforcement available to develop the web is $A_{sw} = \rho_{sw} b_w d$. Then $A_{sw} = 0.165(3000)(14)(30)/40,000 = 5.19$ in^2.

(d)   The nominal moment capacity of the overhanging portion of the section is 197 ft-kips, and the nominal moment capacity of the web is 470 ft-kips. Thus $197 + 470 = 667 = 600/0.9$.

(e)   The total area of the tension reinforcement is $A_s = A_{sf} + A_{sw} = 2.04 + 5.19 = 7.23$ in^2.

(f)   As a check, substituting in the expression derived for the ultimate moment capacity of a T-section,

$$M_u = \phi[(A_s - A_{sf})f_y(d - a/2) + A_{sf}f_y(d - 0.5 h_f)]$$

where $a = A_{sw}f_y/0.85 f'_c b_w = 5.80$. Then $M_u = 7,200,000$ in-lb $= 600$ ft-kips.

(g)   The 1983 ACI Code requires that in I- and T-sections the area of the tension reinforcement available to develop the web portion be limited to prevent the possibility of brittle or compression failures of the concrete. Therefore $\rho_w - \rho_f \leqq 0.75\rho_b$ or

$$\frac{A_s}{b_w d} - \frac{A_{sf}}{b_w d} \leqq 0.75\rho_b \qquad \text{or} \qquad \frac{A_{sw}}{b_w d} \leqq 0.75\rho_b$$

In this example, $\dfrac{A_{sw}}{b_w d} = \dfrac{5.19}{14(30)} = 0.0124 < 0.75\rho_b = 0.75(0.0371) = 0.0278$

(h)   Checking for minimum reinforcement requirements:

$$\rho_{\min} = 200/40,000 = 0.005 \qquad \text{and} \qquad A_{s\,\min} = 0.005(14)(30) = 2.10 \text{ in}^2 < 7.23$$

(i) $h_f = 2.0'' < 1.18\omega d/\beta_1 = 1.18(0.165)(30)/0.85 = 6.9''$

5.15. With the given data, determine the area of reinforcement required to resist an ultimate moment of 900 ft-kips: $b = 12''$, $d = 30''$, $f'_c = 3000$ psi, $f_y = 40,000$ psi.

(a) It is necessary to determine the area of tension reinforcement. Examining Table 5.2, a value for $\phi a_u = 2.4$ is selected. Then

$$A_s = \frac{M_u}{\phi a_u d} = \frac{900}{2.4(30)} = 12.5 \text{ in}^2 \qquad \rho = \frac{A_s}{12(30)} = \frac{12.5}{12(30)} = 0.0348$$

This value is greater than the maximum value of $\rho = 0.0278$ given in Table 5.2. Hence compression reinforcement will be required.

Using the maximum value of $\rho = 0.0278$, $A_s = 0.0278(12)(30) = 10.0 \text{ in}^2$. Then the nominal moment capacity of this area of reinforcement is

$$M_n = A_s(\phi a_u)d/\phi = 10.0(2.343)(30)/0.9 = 782 \text{ ft-kips}$$

(b) The second portion of the nominal resisting moment is the couple formed by the compression steel and a proportional amount of tension steel. Or, assuming $f'_s = f_y$,

$$M'_n = 900/\phi - 782 = 218 \text{ ft/kips}$$

Assuming the depth to the compression reinforcement $d' = 2.0''$, the resisting moment is

$$M'_n = A'_s f_y(d - d') ; \qquad 218(12,000) = A'_s(40,000)(28) ; \qquad A'_s = 2.34 \text{ in}^2$$

Checking to see if the compression steel is at yield,

$$\rho - \rho' \geqq \frac{0.85\beta_1 f'_c d'}{f_y d}\ \frac{87,000}{87,000 - f_y} = 0.00668$$

$$\rho - \rho' = \frac{10.0}{12(30)} = 0.0278 > 0.00668$$

Therefore the condition is satisfied.

(c) The total area of tension reinforcement is $A_s = 10.0 + 2.34 = 12.34 \text{ in}^2$. The area of compression reinforcement is $A'_s = 2.34 \text{ in}^2$.

(d) The minimum steel requirement $\rho \geqq 200/f_y$ is satisfied.

(e) The requirement that $\rho - \rho' \leqq 0.75\rho_b$ is met.

5.16. Because Table 5.2 is limited to certain combinations of f_y and f'_c, solve Problem 5.15 without the use of this table. (Use Table 5.1.)

First, to determine the area of reinforcement required,

$$\frac{M_n}{f'_c bd^2} = \frac{M_u}{\phi f'_c bd^2} = \frac{900(12,000)}{0.9(3000)(12)(30)^2} = 0.370 = \omega(1 - 0.59\omega)$$

Entering Table 5.1, it is seen that there is no value of $M_n/f'_c bd^2 = 0.370$. Therefore, it will be necessary to solve for ω,

$$\omega(1 - 0.59\omega) = 0.370 , \qquad \omega = 0.551$$

This value exceeds the maximum value for ω. Hence $\rho > \rho_b$ and compression reinforcement is required.

$$\rho_{\max} = \frac{0.75(0.85)f'_c\beta_1}{f_y}\ \frac{87,000}{87,000 + f_y} = 0.0278 \qquad \text{and} \qquad \omega_{\max} = \frac{0.0278 f_y}{f'_c} = 0.371$$

Entering Table 5.1 with $\omega = 0.371$,

$$\frac{M_n}{f'_c bd^2} = 0.2898 \qquad \text{or} \qquad M_n = \frac{0.2898(3000)(12)(30)^2}{12,000} = 782 \text{ ft-kips}$$

This is the value of M_n found in Problem 5.15, and so A'_s and A_s will be the same as in Problem 5.15.

5.17. For the basic factored load equation $U = 1.4D + 1.7L$, find the range of *composite load factors* (C_f) for various ratios of L/D (Live Load/Dead Load), where $U = C_f(D + L)$.

Multiply the right-hand side of the load factor equation by $(L + D)/(L + D)$. Then,

$$U = \left[\frac{(1.4D + 1.7L)}{L + D}\right](L + D)$$

Divide all terms within the brackets by D, giving

$$U = \left[\frac{1.4 + 1.7L/D}{L/D + 1}\right](L + D) = C_f(L + D)$$

By substituting various values of L/D, the range of C_f can be established, as in Table 5.4.

Table 5.4

L/D	C_f
1.0	1.55
2.0	1.60
3.0	1.625
4.0	1.64
5.0	1.65
6.0	1.657
7.0	1.662
10.0	1.672
20.0	1.686

In the practical range of values, L/D ranging from 1.0 to 5.0, C_f would vary from 1.55 to 1.65. For practical purposes, using $U = 1.65(L + D)$ would normally be conservative. For $L/D = 10.0$, $C_f = 1.65$ would be unconservative by $(1.672 - 1.65) \times 100/1.672 = 1.32$ percent. For $L/D = 1.0$, $C_f = 1.65$ would be conservative by $(1.65 - 1.55) \times 100/1.55 = 6.45$ percent.

5.18. Design a simply supported rectangular beam supporting a uniform dead load of 15 kN/m and a uniform live load of 20 kN/m, using a steel ratio $\rho = 0.0186$. The concrete strength is $f'_c = 30$ MPa and the steel yield strength is $f_y = 300$ MPa. Assume that the beam weight is 8.5 kN/m. The beam span is 10 m.

The ultimate uniform load is $w_u = (1.4)(15.0 + 8.5) + (1.7)(20) = 66.9$ kN/m, so

$$M_u = (66.9)(10)^2/8 = 836 \text{ kN} \cdot \text{m} = 836 \times 10^6 \text{ N} \cdot \text{mm}$$

Then, by equation (*5.13*),

$$M_u = \phi \rho f_y bd^2(1 - 0.59\rho f_y/f'_c)$$
$$836 \times 10^6 = 0.9(0.0186)(300)(bd^2)[1 - 0.59(0.0186)(300/30)]$$
$$836 \times 10^6 = bd^2(4.471)$$

or $bd^2 = 187 \times 10^6$ mm^3. Using a width $b = 400$ mm, then $d^2 = (187)(10^6)/400 = 467\,500$ mm^2. So $d = \sqrt{467\,500} = 684$ mm.

The area of steel required is equal to ρbd, or $A_s = (0.0186)(400)(684) = 5089$ mm^2. This steel area can be provided by using eight No. 30M bars, $A_{st} = 5600$ mm^2. It is necessary to determine whether or not these eight bars will fit in one row or two rows, considering the required bar cover and shear stirrup clearances (see Table 1.13).

5.19. Reconsider Problem 5.18 as being the SI translation of a problem in U.S. customary units. The given loads, beam weight, and beam span were obtained by "soft conversion" of the U.S. values; that is, they are the exact metric equivalents (to within unavoidable truncation errors). However, the given concrete strength and steel strength were obtained by "hard conversion" of $f'_c = 3000$ psi and $f_y = 40{,}000$ psi; this means that 30 MPa and 300 MPa are *convenient metric approximations*. (Perhaps these are standard values in Canada, from which the materials will be ordered.) Rework Problem 5.18, making a totally soft conversion of the data with the aid of Table 5.5.

Table 5.5

psi (U.S.)	MPa (SI)
3,000	20.7
4,000	27.6
5,000	34.5
40,000	275.8
50,000	344.8
60,000	413.7

As before, $M_u = 836 \times 10^6$ N·mm and $\rho = 0.0186$, and so the equation for bd^2 is

$$836 \times 10^6 = (0.9)(0.0186)(275.8)(bd^2)[1 - (0.59)(0.0186)(275.8/20.7)]$$

from which $bd^2 = 212\,134\,550$ mm³. Considering a beam width of 450 mm, then

$$d^2 = 212\,134\,552/450 = 471\,410 \text{ mm}^2 \quad \text{and} \quad d = \sqrt{471\,410} = 686.6 \text{ mm}$$

The area of steel required will be $(0.0186)(450)(686.6) = 5747$ mm². This steel area can be satisfied by using six No. 35M bars ($A_s = 6000$ mm²) or four No. 45M bars ($A_s = 6000$ mm²). If the metric bars are not available, the U.S. standard bars of equivalent area may be used. In this problem, six No. 11 U.S. standard bars may be used to provide 6036 mm² of steel area. Refer to Tables 1.12 and 1.13 (Chapter 1).

5.20. Using a steel ratio $\rho = A_s/bd$ of 0.0200, concrete strength $f'_c = 30$ MPa and steel yield strength $f_y = 400$ MPa, determine the required effective depth (d) from the compression face to the centroid of the tensile reinforcing steel, if the beam width is 300 mm and the nominal moment is $M_n = 80.9$ kN·m.

$$M_n = bd^2 f'_c(1 - 0.59\omega) \quad \text{and} \quad \omega = \rho f_y/f_c = (0.02)(400)/30 = 0.267$$

so
$$80.9 \times 10^6 = (300)d^2(30)0.267[1 - 0.59(0.267)]$$
$$80.9 \times 10^6 = 2024.5 d^2$$
$$d = \sqrt{39\,960.5} = 200.0 \text{ mm}$$

Supplementary Problems

5.21. A rectangular beam that has a width of 12″ and an effective depth of 13″ is reinforced at the bottom with 2-#10 bars. If $f'_c = 3000$ psi and $f_y = 40,000$ psi, determine the nominal flexural capacity of the beam assuming the parabolic stress distribution.   *Ans.*   94.5 ft-kips.

5.22. Determine the capacity of the beam in Problem 5.21 assuming the rectangular stress distribution. *Ans.*   95.8 ft-kips

5.23. A simply supported one way slab has a span of 12′-0″ and must resist service uniform dead and live loads of 750 lb/ft^2 and 1000 lb/ft^2 respectively. Design using strength design a minimum depth slab if $f'_c = 3000$ psi and $f_y = 50,000$ psi.   *Ans.* $d = 9″$, $A_s = 2.05$ in^2/ft

5.24. A simply supported rectangular beam must resist service dead load and live load moments of 60.0 ft-kips and 30.0 ft-kips respectively. If $f'_c = 4000$ psi and $f_y = 50,000$ psi, proportion a minimum depth beam with tension reinforcement only.   *Ans.* $b = 12″$, $d = 12.5″$, $A_s = 3.85$ in^2

5.25. Repeat Problem 5.24 but assume $\omega = 0.18$.   *Ans.* $b = 12″$, $d = 16″$, $A_s = 2.70$ in^2

5.26. A concrete T-beam must resist an ultimate moment of 125 ft-kips. If $b = 55″$, $b_w = 12″$, $d = 20″$, $h_f = 6″$, $f'_c = 4000$ psi, and $f_y = 50,000$ psi, determine the required area of tension reinforcement. *Ans.* $A_s = 1.66$ in^2

5.27. Repeat Problem 5.26 but let $h_f = 3″$ and the ultimate moment = 900 ft-kips. *Ans.* $A_{sw} = 4.53$ in^2, $A_{sf} = 8.77$ in^2

5.28. A rectangular beam with a width of 11″ and an effective depth of 25″ must resist an ultimate moment of 700 ft-kips. If $f'_c = 3000$ psi, $f_y = 40,000$ psi, and $d' = 2″$, determine the reinforcement required. *Ans.* $A_s = 11.30$ in^2, $A'_s = 3.65$ in^2

5.29. A *fixed-end* rectangular beam must support uniform service dead and live loads of 15.0 kips/ft and 10.0 kips/ft. If $l = 20.0′$, $f'_c = 4000$ psi and $f_y = 60,000$ psi, determine the flexural reinforcement requirements and dimensions, using $b = 24″$, so that $\omega < 0.18$. *Ans.* $d = 35″$, $+A_s = 4.15$ in^2, $-A_s = 9.60$ in^2

5.30. For $f'_c = 20.0$ MPa and $f_y = 350.0$ MPa, determine $\omega = \rho f_y / f'_c$ corresponding to the following values of the steel ratio: (*a*) 0.006, (*b*) 0.007, (*c*) 0.010, (*d*) 0.012, (*e*) 0.015, (*f*) 0.165, (*g*) 0.0195. *Ans.*   (*a*) 0.1050, (*b*) 0.1225, (*c*) 0.1750, (*d*) 0.2100, (*e*) 0.2625, (*f*) 0.2887, (*g*) 0.3412

5.31. For slab sections 300 mm wide and for $f'_c = 30$ MPa and $f_y = 400$ MPa, determine the values of M_n (in kN·m) corresponding to the following steel ratio–effective depth pairs: (*a*) $\rho = 0.0050$, $d = 190.0$ mm; (*b*) 0.0060, 170.0; (*c*) 0.0110, 160.0; (*d*) 0.0130, 180.0; (*e*) 0.0150, 180.0; (*f*) 0.0200, 200.0; (*g*) 0.0220, 200.0; (*h*) 0.0230, 210.0. *Ans.*   (*a*) 22.80, (*b*) 19.83, (*c*) 29.60, (*d*) 45.38, (*e*) 51.44, (*f*) 80.90, (*g*) 87.33, (*h*) 99.70

Chapter 6

Shear in Beams and Slabs

NOTATION

A_g = gross area of section

A_s = area of tension reinforcement

A_v = total area of web reinforcement in tension within a distance, s, measured in a direction parallel to the longitudinal reinforcement

α = angle between inclined web bars and longitudinal axis of member

b = lease transverse dimension of a column or rectangular member

b_o = perimeter of critical section for slabs and footings

b_w = width of web in I- and T-sections

β_c = aspect ratio of column = h/b

d = distance from extreme compression fiber to centroid of tension reinforcement

d' = distance from the concrete face to the centroid of the near compression reinforcement

f'_c = compressive strength of concrete

f_y = yield strength of reinforcement, including web reinforcement

h = largest transverse dimension of column or beam

M = bending moment

M_n = nominal bending moment

M_u = ultimate bending moment = ϕM_n

N = load normal to the cross section, to be taken as positive for compression, negative for tension, and to include the effects of tension due to shrinkage and creep

$\rho_w = A_s/b_w d$

s = spacing of stirrups or bent bars in a direction parallel to the longitudinal reinforcement

V_c = shear force withstandable by the concrete alone

V_n = nominal shear force

V_s = shear carried by web reinforcement = $V_n - V_c$

V_u = total ultimate shear at section = ϕV_n

ϕ = capacity reduction factor = 0.85 for shear

SHEAR STRESS IN HOMOGENEOUS BEAMS

In the study of Strength of Materials it is shown that, for *homogeneous elastic materials*, the unit horizontal shear stress at a section of a beam may be calculated using the equation

$$v = VA\bar{y}/Ib \qquad (6.1)$$

where v = unit horizontal shear stress

A = cross-sectional area above element in question

$\bar{y}$ = distance from the centroid of A to the neutral axis

I = moment of inertia of the beam cross-sectional area

b = thickness of the element

This shear stress develops due to the change in bending moment from one side of the element to the other.

Concrete is neither truly elastic nor homogeneous, and results of experiments related to shear stress in concrete cannot be correlated directly with equation (6.1). At the present time, statistical correlation of test data must be utilized in order to provide rational equations for use in designing concrete elements to successfully resist shear forces.

Shear stress does not work alone to cause failure of concrete beams. The complex mechanism of concrete and reinforcing bars provides for an equally complex resistance to stress. Although no precise theory has been developed to properly explain shear as related to failure, it is known that *diagonal tension stresses* develop and that failure is due to tension in the concrete, rather than shear stress.

In order to develop a rational method of design, the shear stress distribution is assumed as shown in Fig. 6-1. Tension is neglected in the concrete, causing a constant shear stress distribution to be assumed below the neutral axis.

Since the equations for shear design have been empirically devised using test data obtained for *first-cracking* of the concrete, the equations are, of necessity, *strength design* equations. There is in actuality no true *working stress method* for shear design.

The total *nominal shear force*, V_n is obtained by dividing the calculated ultimate shear force,

$$V_u = 1.4V_D + 1.7V_L \tag{6.2}$$

by the understrength factor, $\phi = 0.85$. Here, V_D is the service dead-load shear force and V_L is the service live-load shear force. Thus,

$$V_n = V_u/\phi \tag{6.3}$$

The 1983 ACI Code contains empirical equations which are used to calculate the shear force that can be resisted by the concrete cross section alone, without the addition of steel web reinforcement. There are two sets of equations, one for beams and one for slabs; these are given below, with each statement in U.S. customary units followed by its "hard conversion" (approximate) SI equivalent.

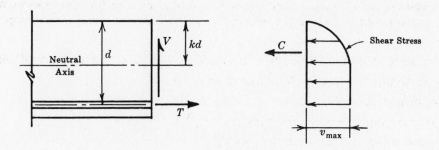

Fig. 6-1

Beams

The shear force that can be taken by a concrete beam alone (without stirrups), when subjected to shear and flexure without torsion or axial force, is

$$V_c = 2\sqrt{f'_c}\, b_w d \qquad (\tfrac{1}{6}\sqrt{f'_c}\, b_w d) \tag{6.4}$$

or, with the effect of longitudinal tension steel included,

$$V_c = (1.9\sqrt{f'_c} + 2500\rho_w V_u d/M_u)b_w d \qquad ((\tfrac{1}{6}\sqrt{f'_c} + 100\rho_w V_u d/M_u)b_w d) \tag{6.5}$$

so long as $V_c \leq 3.5\sqrt{f'_c}\, b_w d$ $(0.29\sqrt{f'_c}\, b_w d)$. In equation (6.5), $V_u d/M_u$ is *always* positive and is never greater than 1.0.

When the cross section of the beam is subjected to axial tension force (N_u), the shear force taken by the concrete is *reduced* to

$$V_c = 2(1 + N_u/500A_g)\sqrt{f'_c}\, b_w d \quad (\tfrac{1}{6}(1 + 0.3N_u/A_g)\sqrt{f'_c}\, b_w d) \qquad (6.6)$$

where N_u is negative for tension and $A_g = b_w h$ is the gross cross-sectional area.

When the beam cross section is subjected to axial compression (N_u positive), the shear capacity of the concrete is *increased* to

$$V_c = 2(1 + N_u/2000A_g)\sqrt{f'_c}\, b_w d \quad (\tfrac{1}{6}(1 + 0.07N_u/A_g)\sqrt{f'_c}\, b_w d) \qquad (6.7)$$

Also, for members subjected to axial compression, equation (6.5) is valid with $V_u d/M_u$ not limited to 1.0, and with

$$M_m = M_u - N_u(4h - d)/8$$

substituted for M_u, provided M_m is positive and

$$V_c \leqq 3.5\sqrt{f'_c}\, b_w d\, \sqrt{1 + N_u/500A_g} \quad (0.3\sqrt{f'_c}\, b_w d\, \sqrt{1 + 0.3N_u/A_g})$$

If M_m is negative, equation (6.6) is to be used instead.

In equations (6.4) through (6.7) the choice of units is as follows: V_c, V_o, and N_u in lbs (kN); f'_c in psi (MPa); M_u or M_m in in.-lbs (kN $\cdot$ m); b_w, d, b, and h in in. (mm). Observe that this implies the units psi (MPa) for the quantity N_u/A_g.

The *critical section* for shear is at a distance equal to the effective depth, d, from the face of the support when the support reaction is compressive. When the support reaction is tensile, the critical section is taken at the face of the support.

Slabs

Slabs are subjected to two-way shear, or *punching shear*. The column attempts to *punch* through the slab at a distance $d/2$ all around the column. The critical situation is shown in Fig. 6-2. The area contributing to shear around the column extends from centerline to centerline of slab between columns in both directions, and the total area (A_t) is reduced to the net area (A_{net}) by subtracting the peripheral area $(b + d)(h + d)$:

$$A_{net} = A_t - (b + d)(h + d) \qquad (6.8)$$

The ultimate uniform load, w_u, is given by

$$w_u = 1.4w_D + 1.7w_L \qquad (6.9)$$

and

$$V_u = A_{net}w_u \qquad (6.10)$$

The shear force that can be taken by the concrete is

$$V_c = (2 + 4/\beta_c)\sqrt{f'_c}\, b_o d \quad (\tfrac{1}{6}(1 + 2/\beta_c)\sqrt{f'_c}\, b_o d) \qquad (6.11)$$

so long as

$$V_c \leqq 4\sqrt{f'_c}\, b_o d \quad (\tfrac{1}{3}\sqrt{f'_c}\, b_o d) \qquad (6.12)$$

In other words, (6.11) is valid if the column aspect ratio $\beta_c \geqq 2$. In (6.11), the perimeter b_o is given by

$$b_o = 2(h + d) + 2(b + d)$$

If the shear force exceeds V_c, web reinforcement may be used for the excess shear, but V_c is then limited to

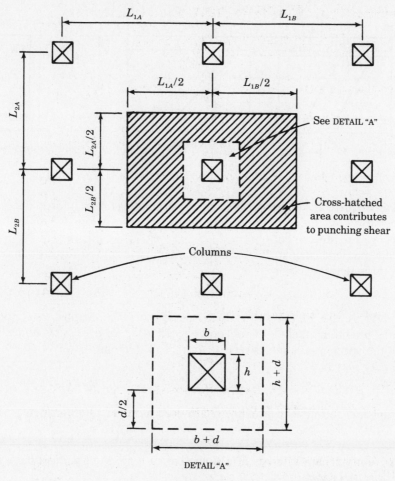

Fig. 6-2

$$V_{c\,\mathrm{max}} = 2\sqrt{f'_c}\,b_o d \qquad \left(\tfrac{1}{6}\sqrt{f'_c}\,b_o d\right) \tag{6.13}$$

and the nominal shear force $V_n = V_u/\phi$, including that taken by steel stirrups, may not exceed

$$V_{n\,\mathrm{max}} = 6\sqrt{f'_c}\,b_o d \qquad \left(\tfrac{1}{2}\sqrt{f'_c}\,b_o d\right) \tag{6.14}$$

It should be noted that in equations (6.8)–(6.10), linear dimensions are measured in ft (m) and loads in kips/ft^2 (kN/m^2). In equations (6.11)–(6.14), all quantities (including b, d and h) bear the same units as in the preceding beam equations. The perimeter b_o is in in. (mm).

When steel shear reinforcement in the form of stirrups is used in slabs, the critical section at $d/2$ from the face of the column must be investigated for shear. In addition, similar peripheral sections more distant than $d/2$ from the column face must also be investigated for two-way shear.

WEB REINFORCEMENT

When the concrete cross section has insufficient area to maintain shear forces below the permissible values, additional resistance to shear must be provided. One form of shear reinforcement consists of hoops or *stirrups*, which may be placed vertically or at some angle with the horizontal. The stirrups may be of the closed type, or else U-shaped, with appropriate *hooks* on the ends. Another form of web reinforcement may consist of flexural reinforcement which can be bent

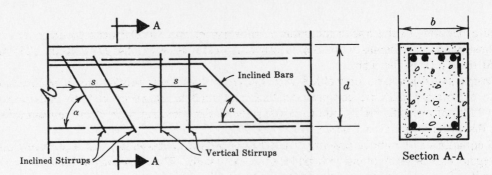

Fig. 6-3

diagonally upward (where no longer needed to resist bending) to reinforce the web. (Bent-up, or *trussed*, bars being costly, they are used only in special cases.) Fig. 6-3 shows a combination of various types of web reinforcement.

When stirrups perpendicular to the axis of the member are used to resist shear, the theoretical stirrup spacing (s) is determined as

$$s = A_v f_y d / V_s \qquad\qquad (6.15)$$

where $V_s = V_n - V_c$ is the excess shear. The spacing of the stirrups may not exceed $d/2$ or 24 in. (600 mm), providing that V_s does not exceed $4\sqrt{f_c'}\, b_w d$ ($\frac{1}{6}\sqrt{f_c'}\, b_w d$). If that value is exceeded, then the spacing may not exceed $d/4$ or 12 in. (300 mm).

In no case may s exceed

$$s_{max} = A_v f_y / 50 b_w \qquad (3 A_v f_y / b_w) \qquad\qquad (6.16)$$

When stirrups inclined upward at angle α are used as shear reinforcement,

$$s = A_v f_y (\sin \alpha + \cos \alpha) d / V_s \qquad\qquad (6.17)$$

When shear reinforcement consists of a single bar or a group of parallel bars, all bent up at the same distance from the support,

$$V_s = A_v f_y \sin \alpha \qquad\qquad (6.18)$$

so long as this does not exceed $3\sqrt{f_c'}\, b_w d$ ($\frac{1}{4}\sqrt{f_c'}\, b_w d$). However, when shear reinforcement consists of parallel bent-up bars or groups of parallel bent-up bars at different distances from the support, shear strength V_s shall be computed using equation (6.17).

When more than one type of shear reinforcement is used in the same portion of a member, shear strength shall be computed as the sum of the V_s values for the various types.

Shear strength V_s shall not be taken greater than

$$V_{s\,max} = 8\sqrt{f_c'}\, b_w d \qquad (\tfrac{2}{3}\sqrt{f_c'}\, b_w d) \qquad\qquad (6.19)$$

When this limit is exceeded, the dimensions of the cross section must be increased, or the concrete strength f_c' must be increased, or both.

Design strength of shear reinforcement may not exceed 60,000 psi (400 MPa).

OTHER PROVISIONS FOR SHEAR

Although web reinforcement in the form of steel stirrups is theoretically not needed when the shear at a point in a beam is equal to or less than V_c, the 1983 ACI Code requires additional shear reinforcement. Minimum shear reinforcement must be provided from the support to a point along the member where $V_n = V_c/2$. This minimum reinforcement is provided by using the maximum spacing for stirrups as stipulated above.

The exceptions to the above minimum reinforcement are: (*a*) slabs and footings; (*b*) concrete joist construction; (*c*) beams with total depth not greater than the larger of 10 in. (250 mm) and one-half the width of the web.

Stirrups and other bars or welded wire fabric used as shear reinforcement shall extend to a distance d from the extreme compression fiber and shall be anchored at both ends according to Section 12.14 of the 1983 ACI Code to develop the design yield strength of the reinforcement. This will be discussed in detail in Chapter 7.

All equations given above properly apply to *normal-density concrete*; that is, concrete consisting of aggregate such as sand, along with gravel or crushed stone. The equations will hold for *low-density concrete* if $\sqrt{f'_c}$ is in every occurrence multiplied by 0.85 (assuming low-density coarse aggregate and ordinary sand) or by 0.75 (assuming low-density coarse aggregate and low-density fine aggregate). An exception is equation (*1.3*) for E_c, to which the multipliers are not applied.

Solved Problems

6.1. Use Fig. 6-4, with $L = 32$ ft, $d = 24$ in., $b_w = 14$ in. and $w_u = 5.6$ kips/ft, to determine whether or not the beam requires stirrups. If required, find the stirrup spacing. Use $f'_c = 4000$ psi, $f_y = 40,000$ psi, and No. 4 U-stirrups.

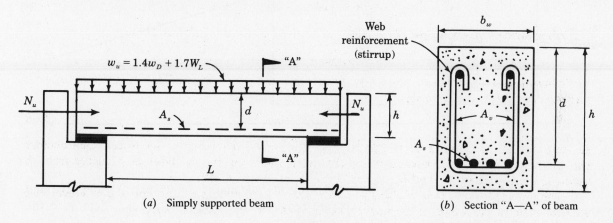

$w_u = 1.4w_D + 1.7W_L$

(*a*)   Simply supported beam   (*b*)   Section "A—A" of beam

Fig. 6-4

It is best to use $w_n = w_u/\phi$ from the beginning, and not use ϕ thereafter. Hence,

$$w_n = 5.6/0.85 = 6.59 \text{ kips/ft}$$

The end reaction at the face of the support is $R_n = w_n L/2 = (6.59)(32/2) = 105.4$ kips. By (*6.19*), the maximum shear permitted to be taken by steel is

$$8\sqrt{f'_c}\, b_w d = 8\sqrt{4000}\,(14)(24)/1000 = 170.0 \text{ kips}$$

Also, maximum spacings $d/2$ or 24 in. are multiplied by $1/2$ if V_s exceeds

$$4\sqrt{f'_c}\, b_w d = 85.0 \text{ kips}$$

The concrete can take $V_c = 2\sqrt{f'_c}\, b_w d = 42.5$ kips. The design shear at the *critical section* (at d from the face of the support) is

$$V_n = R_n - w_n d/12 = 105.4 - (6.59)(24/12) = 92.2 \text{ kips}$$

Because $V_s = V_n - V_c = 92.2 - 42.5 = 49.7$ kips, which is less than 85.0 kips, maximum stirrup spacings are $d/2$ or 24 in., whichever is lesser. Further, $s_{max} = A_v f_y / 50 b_w$. For two No. 4 stirrup legs, $A_v = 2(0.2) = 0.4$ in.² Hence,

$$s_{max} = (0.4)(40,000)/(50)(16) = 20.0 \text{ in.}$$

Since $d/2 = 24.0/2 = 12.0$ in., this spacing is the maximum.

At the critical section, $V_s = 49.7$ kips and $s = A_v f_y d/V_s = (0.4)(40)(24)/49.7 = 7.72$ in. This spacing is used from the critical section back to the face of the support.

The 1983 ACI Code requires stirrups to be used over a distance from the face of the support to a point where $V_n = V_c/2$. From similar triangles in the shear diagram, Fig. 6-5,

$$x_m/84.15 = 16/105.4 \qquad \text{or} \qquad x_m = (16)(84.15)/105.4 = 12.77 \text{ ft}$$

Using the equation $s = A_v f_y d/V_n$, spacings are found to be 8.57 in. at $x = 33$ in., and 9.91 in. at $x = 44$ in., from the face of the support. Thus, spacings from the face of the support would be one of 3.5 in., five of 7.5 in., two of 8.5 in., one of 11.0 in. and the remaining eight of the maximum at 12.0 in.

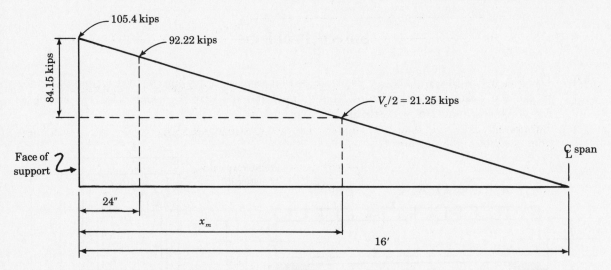

Fig. 6-5

6.2. Use Fig. 6-4, with $L = 10.0$ m, $w_u = 90$ kN/m, $d = 600$ mm, $b_w = 400$ mm, $f'_c = 20$ MPa and $f_y = 300$ MPa, to determine if stirrups are required. If stirrups are needed, determine the required spacing of No. 10M bar U-stirrups.

The nominal uniform load is given by $w_n = w_u/\phi = 90/0.85 = 105.88$ kN/m and $R_n = w_n L/2 = (105.88)(10/2) = 529.4$ kN. At distance d from the support, $V_n = 529.4 - (105.88)(0.600) = 465.9$ kN.

$$V_c = \sqrt{f'_c}\, b_w d/6 = \sqrt{20}\,(400)(600)/[6(1000)] = 178.9 \text{ kN}$$

So $V_s = V_n - V_c = 465.9 - 178.9 = 287.0$ kN.

Maximum stirrup spacings are the lesser of $d/2 = 600/2 = 300$ mm and 600 mm, i.e. 300 mm, unless V_s exceeds $\sqrt{f'_c}\, b_w d/3 = 357.8$ kN. Also, by (6.16), maximum spacing of stirrups may not exceed

$$3A_v f_y/b_w = 3(2 \times 100)(300)/400 = 450 \text{ mm}$$

When V_s exceeds 357.8 kN, the maximum spacing of stirrups becomes $\frac{1}{2}(300) = 150$ mm. Since the shear at the critical section is 465.9 kN, these maximum spacings apply. The total shear V_n may not exceed

$$2\sqrt{f'_c}\, b_w d/3 = 2\sqrt{20}\,(400)(6600)/[3(1000)] = 715.5 \text{ kN}$$

which is greater than $V_n = 465.9$ kN. The spacing at the critical section is

$$s = A_v f_y d/V_s = (2 \times 100)(300)(600)/[(287.0)(1000)] = 125.4 \text{ mm}$$

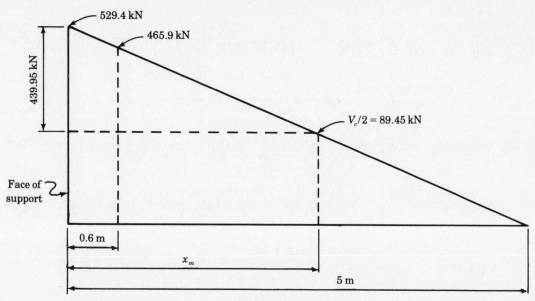

Fig. 6-6

By similar triangles in the shear diagram, Fig. 6-6, stirrups are required over a distance

$$x_m = (439.95)(5.0)/529.4 = 4.155 \text{ m}$$

Subsequent calculations show that the theoretical (and maximum) spacings $s = A_v f_y d / V_s$ are as follows, from the face of the support:

Distance x, m	Spacings, mm
0 to 0.6	125
0.66	128
0.99	147
1.324	171
1.655	205
1.986	257
1.986 to 4.155	300

A practical way to space stirrups is to plot s (vertically) versus x (horizontally) and to select spacings such as 125 mm, 150 mm or 175 mm, accordingly.

6.3. Refer to Fig. 6-4. Use span length $L = 22.3$ ft, $w_u = 2.0$ kips/ft, $d = 13.75$ in., $b_w = 12$ in., and a compression axial load $N_u = 10,000$ lb. Determine the required spacing of No. 3 U-stirrups, and the distance over which stirrups are required. The overall depth of the beam is 16 in., $f_c' = 3000$ psi, and $f_y = 40,000$ psi.

The nominal uniformly distributed load is $w_n = w_u/\phi = 2.0/0.85 = 2.353$ kips/ft, giving the end reaction

$$R_n = w_n L/2 = 2.353(22.3)/2 = 26.235 \text{ kips}$$

At the critical section, $d = 13.75$ in. from the face of the support,

$$V_n = 26.235 - (2.353)(13.75/12) = 23.539 \text{ kips}$$

For a compressive axial load, with $A_g = (12)(16) = 192$ in.^{2},

$$V_c = 2[1 + N_u/(2000 A_g)]\sqrt{f_c'} b_w d$$
$$= 2\{1 + (10,000)/[(2000)(192)]\}\sqrt{3000}\,(12)(13.75)/1000 = 18.546 \text{ kips}$$

But V_c may not exceed

$$3.5\sqrt{f'_c}\,b_w d\,\sqrt{1 + N_u/500A_g} = 3.5\sqrt{3000}\,(12)(13.75)\,\sqrt{1 + (10{,}000)/(500)(192)}\,/1000$$
$$= 33.238 \text{ kips}$$

Thus, $V_c = 18.546$ kips governs.

Stirrups are required to a point where $V_n = V_c/2 = 18.546/2 = 9.273$ kips. This distance is

$$x_m = (R_n - V_c/2)(L/2)/R_n = (26.235 - 9.273)(22.3/2)/26.235 = 7.209 \text{ ft} = 86.51 \text{ in.}$$

At the critical section, $V_s = V_n - V_c = 23.539 - 18.546 = 4.993$ kips, so the spacing there is

$$s = A_v f_y d/V_s = (0.22)(40)(13.75)/4.993 = 24.23 \text{ in.}$$

If V_s exceeds $4\sqrt{f'_c}\,b_w d = 4\sqrt{3000}(12)(13.75/1000) = 36.15$ kips, the spacings in (a) and (b) below are multiplied by 0.5. But actual $V_s = 4.993$ kips, so that maximum spacing is the least of

(a) $d/2 = 13.75/2 = 6.875$ in.
(b) 24 in.
(c) $A_v f_y/50b_w = (0.11)(40{,}000)/50(12) = 7.33$ in.

The maximum permissible spacing is therefore 6.875 in. Use one stirrup at 3 in. from the support, and stirrups at 6.5 in. spacing over the distance $x_m = 86.51$ in.

6.4. Consider the beam shown in Fig. 6-4, with $L = 20$ ft, $w_u = 3.32$ kips/ft, $d = 16$ in., $b_w = 10.5$ in., $f'_c = 3600$ psi, $f_y = 40{,}000$ psi and with a tension axial load $N_u = -28{,}600$ lb and a total depth 18 in. Determine the spacing for No. 3 U-stirrups and the distance over which stirrups are required by the 1983 ACI Code.

The end reaction at the face of the support is $R_n = w_n L/2$ where $w_n = w_u/\phi = 3.32/0.85 = 3.906$ kips/ft. So, $R_n = 3.906(20/2) = 39.06$ kips. For a tension axial load, the shear is critical at the face of the support, so $V_n = 39.06$ kips.

With $A_g = (10.5)(18) = 189$ in.^{2}, the allowable ultimate shear is

$$V_c = 2(1 + N_u/500A_g)\sqrt{f'_c}\,b_w d = 2[1 - (28{,}600)/(500)(189)]/\sqrt{3600}\,(10.5)(16)/1000$$
$$= 14.06 \text{ kips}$$

and $V_s = V_n - V_c = 39.06 - 14.06 = 25.0$ kips. At the critical section (face of support),

$$s = A_v f_y d/V_s = (0.22)(40)(16)/25.0 = 5.632 \text{ in.}$$

The maximum stirrup spacing is the smallest of

(a) $d/2 = 16.0/2 = 8.0$ in.
(b) 24 in.
(c) $A_v f_y/50b_w = (0.22)(40{,}000)/50(10.5) = 16.76$ in.

Thus, $s_{max} = 8.0$ in. (V_s does not exceed $4\sqrt{f'_c}\,b_w d = 40.32$ kips, so limits (a) and (b) are not halved). The distance over which stirrups are required is

$$x_m = (R_n - V_c/2)(L/2)/R_n = (39.06 - 7.03)(10)/39.06 = 8.20 \text{ ft} = 98.4 \text{ in.}$$

Subsequent calculations show the following theoretical and maximum spacings of No. 3 U-stirrups from the face of the support:

x, in.	s, in.
0 to 16	5.632
21	7.751
28 to 98.4	8.0

Spacings at practical distances are selected accordingly.

The reader is reminded that, for shear with or without axial load, V_s may never exceed $8\sqrt{f'_c}\,b_w d$, which here is $(2)(40.32) = 80.64$ kips   (O.K.).

Supplementary Problems

6.5. With reference to Fig. 6-4, verify the solutions of subproblems (a) through (l) as listed in the last five columns of Table 6.1. The first seven columns give the data. Required are:

s_c = stirrup spacing at the critical section

s_{max} = maximum stirrup spacing

x_m = distance over which stirrups are required

V_n = nominal shear force acting on the critical section

V_c = allowable ultimate shear force

In all subproblems, f_y = 40,000 psi and U-stirrups using No. 3 bars are assumed.

Table 6.1

	Span L, ft	w_u, kips/ft	f'_c, psi	Axial Load N_u, lb	b_w, in.	d, in.	h, in.	s_c, in.	s_{max}, in.	x_m, in.	V_n, kips	V_c, kips
(a)	20.0	3.32	3000	−28,600	10.5	16.0	18.0	5.81	8.00	97.24	39.06	14.82
(b)	19.5	3.50	2000	11,000	12.0	12.0	16.0	5.33	6.00	93.36	36.03	16.23
(c)	20.0	3.00	4000	−21,500	11.0	15.0	17.0	6.87	7.50	92.68	35.29	16.07
(d)	24.5	2.90	3000	10,000	12.0	13.0	16.0	5.56	6.50	116.16	38.10	17.53
(e)	20.5	4.00	3000	10,000	12.0	11.5	13.5	3.60	5.75	103.13	43.73	15.58
(f)	20.0	4.50	4000	−28,600	13.5	16.0	17.0	4.34	8.00	96.75	52.94	20.51
(g)	24.0	3.40	4000	−16,800	10.5	16.0	18.0	4.61	8.00	117.79	48.00	17.47
(h)	23.0	4.20	3000	10,000	12.0	13.75	16.0	3.71	6.88	115.48	51.16	18.55
(i)	20.0	4.40	5000	−15,400	10.5	16.0	17.0	4.39	8.00	97.21	51.77	19.65
(j)	21.5	3.80	3000	10,000	13.0	13.75	16.0	5.29	6.88	102.09	42.94	20.05
(k)	20.0	3.50	5000	−28,600	10.5	18.0	20.0	7.29	9.00	91.66	41.18	19.45
(l)	28.0	3.50	5000	−21,200	12.5	15.0	17.0	3.62	7.50	137.07	57.65	21.23

6.6. The situation is as in Problem 6.5, except that no axial load is applied. In all subproblems of Table 6.2, f'_c = 20 MPa, f_y = 300 MPa, and No. 10M U-stirrups are assumed.

Table 6.2

	Span L, m	w_u, kN/m	b_w, mm	d, mm	V_n, kN	V_c, kN	s_c, mm	s_{max}, mm	x_m, m
(a)	10.0	90.0	400.0	600.0	465.9	178.9	125.4	300.0	4.16
(b)	12.0	100.0	380.0	700.0	623.5	198.3	98.8	175.0	5.16
(c)	7.0	80.0	410.0	500.0	282.4	152.8	231.6	250.0	2.69
(d)	6.5	85.0	350.0	480.0	277.0	125.2	189.7	240.0	2.62
(e)	8.5	90.0	320.0	520.0	394.9	124.0	115.2	130.0	3.66
(f)	7.5	100.0	380.0	600.0	370.6	169.9	179.4	300.0	3.03

6.7. The situation is as in Problem 6.6 (no axial load). In all subproblems of Table 6.3, $f_y = 40,000$ psi and No. 4 bars are used for U-stirrups.

Table 6.3

	GIVEN					SOLUTIONS				
	Span L, ft	w_u, kips/ft	f'_c, psi	b_w, in.	d, in.	V_n, kips	V_c, kips	s_c, in.	s_{max}, in.	x_m, in.
(a)	35.0	6.5	4000	14.0	24.0	118.53	42.50	5.05	12.00	176.65
(b)	22.0	5.4	3000	12.0	19.5	59.56	25.64	9.20	9.75	107.79
(c)	28.0	5.1	4000	13.0	21.0	73.50	34.53	8.62	10.50	133.47
(d)	32.0	5.6	4000	14.0	24.0	92.24	42.50	7.72	12.00	153.29
(e)	25.0	4.0	3000	12.0	14.5	53.14	19.06	6.81	7.25	125.70
(f)	36.0	6.3	4000	13.0	22.0	119.80	36.18	4.21	5.50	186.71
(g)	25.0	3.8	3000	10.0	15.0	50.29	16.43	7.09	3.75	127.95
(h)	26.0	4.6	4000	12.5	18.0	62.24	28.46	8.53	9.00	124.45
(i)	26.0	4.4	3000	12.0	18.5	59.31	24.32	8.45	9.25	127.81
(j)	26.0	4.8	4000	10.0	21.0	63.53	26.56	9.09	10.50	127.78
(k)	23.0	3.8	3000	10.0	15.0	45.82	16.43	8.17	7.50	115.95

6.8. The first nine columns of Table 6.4 list data for slabs (see Fig. 6-2). Required are actual shear V_n and allowable punching shear V_c.

Table 6.4

	GIVEN									SOLUTIONS	
	L_{1A}, ft	L_{1B}, ft	L_{2A}, ft	L_{2B}, ft	d, in.	Col. b, in.	Col. h, in.	f'_c, psi	w_u, kips/ft^2	V_n, kips	V_c, kips
(a)	15.0	16.0	15.0	16.0	7.00	20.0	20.0	4000	0.45	124.4	191.3
(b)	16.0	18.0	18.0	18.0	6.75	18.0	18.0	4000	0.35	124.3	169.1
(c)	14.0	15.0	14.0	15.0	5.75	16.0	16.0	4000	0.40	97.3	126.6
(d)	12.0	12.0	13.0	13.0	5.25	12.0	18.0	3500	0.35	63.2	100.6
(e)	15.0	16.0	16.0	15.0	6.00	14.0	20.0	3500	0.55	*153.1	130.6
(f)	20.0	16.0	20.0	16.0	6.25	12.0	16.0	5000	0.29	108.2	143.2

*Increase d, since $V_n > V_c$.

6.9. For each subproblem of Table 6.5, $f'_c = 30$ MPa. Solve like Problem 6.8.

Table 6.5

	GIVEN								SOLUTIONS	
	L_{1A}, m	L_{1B}, m	L_{2A}, m	L_{2B}, m	d, mm	Col. b, mm	Col. h, mm	w_u, kN/m^2	V_n, kN	V_c, kN
(a)	5.5	5.5	6.0	5.0	147.0	375.0	375.0	12.1	465.9	560.3
(b)	6.0	6.0	6.0	6.0	185.0	400.0	400.0	15.0	629.3	790.3
(c)	5.75	5.5	5.75	5.5	200.0	380.0	380.0	12.5	460.1	847.1
(d)	6.5	6.5	6.5	6.5	210.0	410.0	410.0	14.0	689.6	950.8
(e)	7.0	7.0	6.8	6.8	190.0	400.0	420.0	13.0	722.6	832.5
(f)	6.2	6.4	6.5	6.2	210.0	415.0	415.0	10.0	466.1	958.4
(g)	6.4	6.2	6.3	6.4	170.0	375.0	375.0	8.0	373.7	676.6

Chapter 7

Development of Reinforcement and Splices

NOTATION

a = depth of equivalent rectangular stress block

A_b = area of an individual bar

A_s = area of tension reinforcement

A_v = area of shear reinforcement within a distance s

A_w = area of an individual wire to be developed or spliced

b_w = web width, or diameter of circular section

d = distance from extreme compression fiber to centroid of tension reinforcement

d_b = nominal diameter of bar or wire

f'_c = specified compressive strength of concrete

f_y = specified yield strength of reinforcement

h = overall thickness of member

L_s = additional embedment length at support or at point of inflection

L_d = development length

   = L_{db} × applicable modification factors

L_{db} = basic development length

L_{dh} = development length of standard hook in tension, measured from critical section to outside end of hook (straight embedment length between critical section and start of hook [point of tangency] plus radius of bend and one bar diameter)

   = L_{hb} × applicable modification factors

L_{hb} = basic development length of standard hook in tension

M_n = nominal moment strength at section

   = $A_s f_y (d - a/2) = M_u / \phi$

s = spacing of stirrups or ties

s_w = spacing of wire to be developed or spliced

V_u = factored shear force at section = ϕV_n

β_b = ratio of area of reinforcement cut of to total area of tension reinforcement at section

DEVELOPMENT OF TENSION REINFORCEMENT

To satisfy strength and serviceability requirements of the 1983 ACI Code, it is necessary that the *bond* between the reinforcing steel and the concrete be maintained. If the forces in the steel should fracture or split the concrete, failure may occur at load levels even less than the service loads. In earlier ACI Codes, *bond stresses* were calculated and compared with permissible bond stresses determined from laboratory tests, with appropriate safety factors applied. The 1983 ACI Code provides equations for computing *development lengths* and *splice lengths* that are based on the *empirical* bond stresses. The design philosophy involves the transfer of forces from the concrete to the reinforcement (or vice versa) with regard to development length of the reinforcement (Fig. 7-1(a)), and transfer of forces from one reinforcing bar to another where lapped splices (Fig. 7-1(b)) or butt splices (Fig. 7-1(c)) are required.

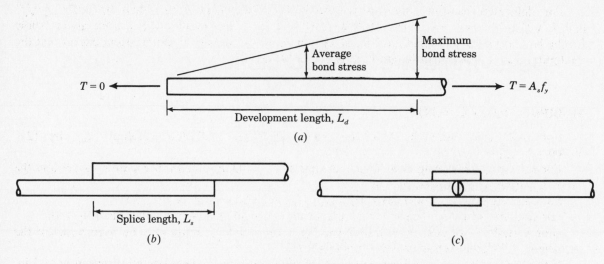

Fig. 7-1

The 1983 ACI Code provides development lengths for *deformed* reinforcing bars, but does not provide information concerning *plain* or *smooth* (undeformed) reinforcing bars. This is the case because plain bars are to be used only for spirals or lap spliced spiral cages in reinforced concrete columns. However, in many countries plain bars are the only type available. Research has shown that plain bars are approximately 50 percent as effective as deformed bars in bond with the concrete. Thus, when smooth bars are used, the development lengths and splice lengths provided herein should be multiplied by 2.

The development length of reinforcing bars depends on the concrete strength (f'_c), the steel yield strength (f_y), the cross-sectional area of the reinforcing bars (A_b) and the diameter of the reinforcing bars (d_b), as well as on whether or not the reinforcing bars are stressed in compression or tension. In addition, development length is dependent on whether or not the reinforcing bars are *top bars* or *non-top* bars. The reason for this difference is that, as the concrete is placed, it is usually *vibrated* to insure that it will completely surround and bond with the reinforcing bars. This process causes the water to rise in the concrete form, so that the concrete at the top of a member has a higher water/cement ratio than does the concrete at the bottom. Thus, *top* concrete has slightly less strength than *non-top* concrete.

The Code provisions regarding splices take into account the quantity of reinforcement that is spliced at a given point. The more steel area that is spliced at a particular location, the longer the splice lengths must be. Further, when low-density concrete or concrete with part low-density coarse aggregate and sand is used, the development lengths and splice lengths must normally be increased. The equations and modification factors related to development lengths and splice lengths are as follows.

BASIC DEVELOPMENT LENGTHS

The basic development lengths, L_{db}, shall be as follows for bars in tension:

For No. 11 (35M) bars and smaller bars, $0.04A_b f_y/\sqrt{f'_c}$ $(0.02A_b f_y/\sqrt{f'_c})$, but not less than $0.0004d_b f_y$ $(0.06d_b f_y)$.

For No. 14 (45M) bars, $0.085 f_y/\sqrt{f'_c}$ $(25 f_y/\sqrt{f'_c})$.

For No. 18 (55M) bars, $0.11 f_y/\sqrt{f'_c}$ $(35 f_y/\sqrt{f'_c})$.

For deformed wire, $0.03d_b f_y/\sqrt{f'_c}$ $(\frac{3}{8}d_b f_y/\sqrt{f'_c})$.

For deformed bars, not less than 12 in. (300 mm), except in the case of lapped splices.

For deformed bars in compression, the basic development length shall be $0.02d_b f_y/\sqrt{f_c'}$ ($\frac{1}{4}d_b f_y/\sqrt{f_c'}$), but not less than $0.0003d_b f_y$ ($0.04d_b f_y$). The modification factors given below for the basic development length of bars in tension apply also to bars in compression, except that the minimum basic development length here is 8 in. (200 mm).

MODIFICATION FACTORS

For steel yield strengths f_y greater than 60,000 psi (400 MPa), multiply L_{db} by ($2 - 60,000/f_y$) (($2 - 400/f_y$)).

For *top bars* (reinforcing bars having at least 12 in. (300 mm) of fresh concrete cast below the bars), L_{db} shall be multiplied by 1.4.

For concrete with all low-density aggregate, L_{db} is to be multiplied by 1.33.

For concrete with low-density coarse aggregate and sand, L_{db} is to be multiplied by 1.18.

(For partial use of sand with low-density aggregates, interpolation may be used between the coefficients 1.18 and 1.33 as provided above.)

If the reinforcing bars are spaced at least 6 in. (150 mm) on centers laterally, with at least 3 in. (70 mm) from the edge of a bar to the nearest face (*clear cover*), L_{db} may be multiplied by 0.8

When tensile reinforcement (A_s) is provided in excess of that required to resist the tensile forces, L_{db} may be multiplied by the fraction (A_s required/A_s provided).

If the reinforcement is enclosed within spirals not less than $\frac{1}{4}$ in. (5 mm) in diameter, with not more than 4 in. (100 mm) *pitch*, L_{db} may be multiplied by 0.75. (The *pitch* is the center-to-center distance between the spirals of the bar.)

REINFORCING BARS WITH HOOKS AT ENDS

The 1983 ACI Code contains special provisions for development lengths for bars in tension that are *hooked* at their ends. Here, *standard hooks* are involved, as defined in Fig. 7-2.

When standard hooks are provided at the ends of reinforcing bars *in tension*, they may be assumed to replace part of the required basic development length L_{db}, which is to be multiplied by the factors previously given for variations described concerning material types and strength. Hooks may not be considered to be effective for partial development length for bars *in compression*.

The hook development length, L_{dh}, must exceed both $8d_b$ and 6 in. (150 mm). When the steel yield strength is 60,000 psi (400 MPa), $L_{hb} = 1200d_b/\sqrt{f_c'}$ ($100d_b/\sqrt{f_c'}$). At other values of f_y, L_{hb} is to be multiplied by $f_y/60,000$ ($f_y/400$). All other multipliers for development length also apply to development length additions for bars in tension that are hooked at their ends with standard hooks.

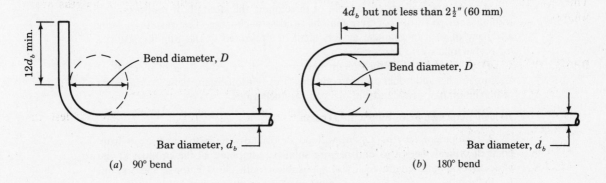

(a) 90° bend   (b) 180° bend

Fig. 7-2

Table 7.1 gives minimum bend diameters (D) and Table 7.2 gives tolerances for the placement of reinforcing bars. Exceptions to Table 7.2 are as follows: Tolerance for the clear distance to formed soffits shall be $-\frac{1}{4}$ in. (-6 mm) and tolerance for cover shall not exceed $-\frac{1}{3}$ the minimum concrete cover required in the design drawings or specifications.

Table 7.1

Bar Size	Minimum D
No. 3 (10M) through No. 8 (25M)	$6d_b$
Nos. 9, 10, 11 (30M and 35M)	$8d_b$
No. 14 (45M) and No. 18 (55M)	$10d_b$

Table 7.2

Depth	Tolerance on d	Tolerance on Minimum Concrete Cover
$d \leq 8$ in. (200 mm)	$\pm\frac{3}{8}$ in. (± 10 mm)	$-\frac{3}{8}$ in. (-10 mm)
$d > 8$ in. (200 mm)	$\pm\frac{1}{2}$ in. (± 12 mm)	$-\frac{1}{2}$ in. (-12 mm)

BUNDLED BARS

Frequently, various sizes of bars are placed together in a group of three or four to provide a given area of reinforcement. The bars are called *bundled bars* because they are normally tied together with bailing wire. Use of bundled bars also facilitates staggering of bar lapped splices. However, when bundled bars are used, the development length of each individual bar must be increased by 20 percent when three bars are bundled together, and by 33 percent when four bars are bundled together. The 1983 ACI Code does not place similar restrictions on two bars bundled together, but the development length of each individual bar applies when two different sizes of bars are used.

ADDITIONAL REQUIREMENTS

The following is adapted from the 1983 ACI Code, with the permission of ACI.

Mechanical Anchorage

Any mechanical device capable of developing the strength of reinforcement without damage to the concrete may be used as anchorage. Test results showing adequacy of such mechanical devices shall be presented to the Building Official for review relative to approval.

Development of reinforcement may consist of a combination of mechanical anchorage plus additional embedment length of reinforcement between the point of maximum bar stress and the mechanical anchorage.

Welded Deformed-Wire Fabric in Tension

Development length, L_d, of welded deformed-wire fabric measured from point of critical section to end of wire shall be computed as the product of the basic development length, L_{db}, and applicable modification factor or factors; but L_d shall not be less than 8 in. (200 mm), except in computation of lap splices and development of web reinforcement.

Basic development length, L_{db}, of welded deformed-wire fabric, with at least one cross wire within the development length not less than 2 in. (50 mm) from point of critical section, shall be

$$0.03 d_b (f_y - 20,000)/\sqrt{f_c'} \qquad (\tfrac{3}{8} d_b (f_y - 140)/\sqrt{f_c'}) \qquad (7.1)$$

but not less than

$$0.20 \frac{A_w}{s_w} \frac{f_y}{\sqrt{f_c'}} \qquad \left(2.5 \frac{A_w}{s_w} \frac{f_y}{\sqrt{f_c'}}\right) \qquad (7.2)$$

Basic development length, L_{db}, of welded deformed-wire fabric, with no cross wires within the development length, shall be determined as for deformed wire.

Welded Smooth-Wire Fabric in Tension

Yield strength of welded smooth-wire fabric shall be considered developed by embedment of two cross wires with the closer cross wire not less than 2 in. (50 mm) from point of critical section. However, basic development length, L_{db}, measured from point of critical section to outermost cross wire shall not be less than

$$0.27 \frac{A_w}{s_w} \frac{f_y}{\sqrt{f_c'}} \qquad \left(3.3 \frac{A_w}{s_w} \frac{f_y}{\sqrt{f_c'}}\right) \qquad (7.3)$$

modified by the factor $(A_s \text{ required})/(A_s \text{ provided})$ for reinforcement in excess of that required by analysis, and by the factor for low-density aggregate concrete; but L_d shall not be less than 6 in. (150 mm), except in computation of lap splices.

FLEXURAL REINFORCEMENT

Tension reinforcement may be developed by bending across the web to be anchored or it may be made continuous with the reinforcement on the opposite face of the member. Critical sections for development of reinforcement in flexural members are at points of maximum stress and at points within the span where adjacent reinforcement terminates or is bent. Reinforcement shall extend beyond the point at which it is no longer required to resist flexure, for a distance equal to the effective depth of the member or $12 d_b$, whichever is greater, except at supports of simple spans and at free ends of cantilevers. Continuing reinforcement shall have an embedment length not less than the development length L_d beyond the point where bent or terminated tension reinforcement is no longer required to resist flexure.

Flexural reinforcement shall not be terminated in a tension zone unless one of the following conditions is satisfied:

(1) Shear at the cutoff point does not exceed two-thirds that permitted, including shear strength of shear reinforcement provided.

(2) Stirrup area in excess of that required for shear and torsion is provided along each terminated bar or wire over a distance from the termination point equal to three-fourths the effective depth of the member. Excess stirrup area A_v shall be not less than $60 b_w s/f_y$ $(0.4 b_w s/f_y)$. Spacing s shall not exceed $(d/8)\beta_b$, where β_b is the ratio of the area of reinforcement cut off to the total area of tension reinforcement at the station.

(3) For No. 11 (35M) bars and smaller, continuing reinforcement must provide double the area required for flexure at the cutoff point, and shear does not exceed three-fourths that permitted.

Adequate anchorage shall be provided for tension reinforcement in flexural members where reinforcement stress is not directly proportional to moment, such as: sloped, stepped or tapered footings; brackets; deep flexural members; or members in which the tension reinforcement is not parallel to the compression face.

Positive Moment Reinforcement

At least one-third of the positive moment reinforcement in simple members and one-fourth of the positive moment reinforcement in continuous members shall extend along the same face of the member into the support. In beams, such reinforcement shall extend into the support at least 6 in. (150 mm). When a flexural member is part of a primary lateral-load-resisting system, positive moment reinforcement required to be extended into the support shall be anchored to develop the specified yield strength f_y in tension at the face of support.

At simple supports and at points of inflection, positive moment tension reinforcement shall be limited to a diameter such that

$$L_d \leqq \frac{M_n}{V_u} + L_a \qquad (7.4)$$

where: M_n is nominal moment strength assuming all reinforcement at the section to be stressed to the specified yield strength f_y; V_u is factored shear force at the section; L_a at a support shall be the embedment length beyond the center of support, whereas L_a at a point of inflection shall be limited to the effective depth of the member or $12d_b$, whichever is greater.

The value of M_n/V_u in (7.4) may be increased 30 percent when the ends of reinforcement are confined by a compressive reaction. Condition (7.4) need not be satisfied for reinforcement, terminating beyond the centerline, of simple supports by a standard hook (or a mechanical anchorage at least equivalent to a standard hook).

Negative Moment Reinforcement

Negative moment reinforcement in a continuous, restrained or cantilever member, or in any member of a rigid frame, shall be anchored in or through the supporting member by embedment length, hooks or mechanical anchorage. Embedment length into the span shall equal the effective depth of the member or $12d_b$, whichever is greater.

At least one-third of the total tension reinforcement provided for negative moment at a support shall have an embedment length beyond the point of inflection not less than the effective depth of the member, $12d_b$, or one-sixteenth the clear span, whichever is greater.

Web Reinforcement

Web reinforcement shall be carried as close to the compression and tension surfaces of a member as cover requirements and proximity of other reinforcement will permit. Ends of single-leg, simple U-, or multiple U-stirrups shall be anchored by one of the following means:

A standard hook plus an embedment of $0.5L_d$. The $0.5L_d$ embedment of a stirrup leg shall be taken as the distance between the middepth of the member, $d/2$, and the start of the hook (point of tangency).

Embedment $d/2$ above or below middepth on the compression side of the member for a full development length L_d, but not less than $24d_b$ or, for deformed bars or deformed wire, 12 in. (300 mm).

For No. 5 (15M) bars and D31 wire, and smaller, bending around longitudinal reinforcement through at least 135°, plus, for stirrups with design stress exceeding 40,000 psi (300 MPa), an embedment of $0.33L_d$. The $0.33L_d$ embedment of a stirrup leg shall be taken like the above $0.5L_d$ embedment.

For each leg of welded smooth-wire fabric forming simple U-stirrups, there shall be either (i) two longitudinal wires with a 2 in. (50 mm) spacing along the member at the top of the U; or (ii) one longitudinal wire located not more than $d/4$ from the compression face and a second wire closer to the compression face and spaced not less than 2 in. (50 mm) from the first wire. (The second wire may be located on the stirrup leg beyond a bend, or on a bend with an inside diameter of bend not less than $8d_b$.)

For each end of a single-leg stirrup of welded smooth- or deformed-wire fabric, there shall be two longitudinal wires at a minimum spacing of 2 in. (50 mm) and with the inner wire at least the greater of $d/4$ and 2 in. (50 mm) from middepth of the member, $d/2$. The outer longitudinal wire at the tension face shall not be farther from the face than the portion of primary flexural reinforcement closest to the face.

Between anchored ends, each bend in the continuous portion of a simple U-stirrup or multiple U-stirrup shall enclose a longitudinal bar.

Longitudinal bars bent to act as shear reinforcement, if extended into a region of tension, shall be continuous with longitudinal reinforcement and, if extended into a region of compression, shall be anchored beyond middepth, $d/2$, as specified for development length for that part of f_y required to satisfy (6.18) and the condition attached thereto.

Pairs of U-stirrups or ties so placed as to form a closed unit shall be considered properly spliced when length of laps are $1.7L_d$. In members at least 18 in. (500 mm) deep, such splices with $A_b f_y$ not more than 9000 lb (40 kN) per leg may be considered adequate if stirrup legs extend the full available depth of the member.

SPLICES OF REINFORCEMENT

Splices of reinforcement shall be made only as required or permitted on the design drawings, or in the specifications, or as authorized by the Engineer.

Lap splices shall not be used for bars larger than No. 11 (35M), except as provided for end-bearing splices. Lap splices of bundled bars shall be based on the lap splice length required for individual bars within a bundle, increased 20 percent for a three-bar bundle and 33 percent for a four-bar bundle. Individual bar splices within a bundle shall not overlap (the splices must be staggered). Bars spliced by noncontact lap splices in flexural members shall be spaced transversely closer than one-fifth the required lap splice length and 6 in. (150 mm).

Welded splices and other mechanical connections may be used as follows:

(*a*) Except as provided in the 1983 ACI Code, all welding shall conform to "Structural Welding Code-Reinforcing Steel" (AWS D1.4).

(*b*) A full-welded splice shall have bars butted and welded to develop in tension at least 125 percent of specified yield strength f_y of the bar.

(*c*) A full mechanical connection shall develop in tension or compression, as required, at least 125 percent of specified yield strength f_y of the bar.

Deformed Bars and Deformed Wire in Tension

Minimum length of lap for tension lap splices shall be as required for a Class A $(1.0L_d)$, Class B $(1.3L_d)$ or Class C $(1.7L_d)$ splice, but shall not be less than 12 in. (300 mm). Moreover, Table 7.3 shall be conformed with.

Table 7.3

$\dfrac{A_s \text{ provided}}{A_s \text{ required}}$	Maximum percent of A_s spliced within required lap length		
	50	75	100
Equal to or greater than 2	Class A	Class A	Class B
Less than 2	Class B	Class C	Class C

Welded splices or mechanical connections used where the area of reinforcement (A_s) provided is less than twice that required by analysis shall meet requirements (b) and (c) above. Where the area of reinforcement provided is at least twice that required by analysis,

(a)   Splices shall be staggered at least 24 in. (600 mm) and in such manner as to develop at every section at least twice the calculated tensile force at that section, but not less than 20,000 psi (140 MPa) for total area of reinforcement provided.

(b)   In computing tensile force developed at each section, spliced reinforcement may be rated at the specified splice strength. Unspliced reinforcement shall be rated at that fraction of f_y defined by the ratio of the shorter actual development length to L_d required to develop the specified yield strength f_y.

Splices in "tension tie members" shall be made with a full-welded splice or full mechanical connection, and splices in adjacent bars shall be staggered at least 30 in. (800 mm).

Deformed Bars in Compression

Minimum length of lap for compression lap splices shall be the development length L_{db} in compression (see formulas on page 131), but at least as large as the greatest of

$0.0005 f_y d_b$ $(0.07 f_y d_b)$

$(0.0009 f_y - 24) d_b$ $((0.13 f_y - 24) d_b)$ if $f_y > 60,000$ psi   (400 MPa)

12 in. (300 mm)

For $f'_c < 3000$ psi (20 MPa), length of lap shall be increased by one-third.

When bars of different size are lap spliced in compression, splice length shall be the larger of the development length of the larger bar and the splice length of the smaller bar. Bar sizes No. 14 (45M) and No. 18 (55M) may be lap spliced to No. 11 (35M) and smaller bars.

In tied reinforced compression members, where ties throughout the lap splice length have an effective area not less than $0.0015 hs$, lap splice length may be multiplied by 0.83, but lap length shall not be less than 12 in. (300 mm). The legs perpendicular to dimension h shall be used in determining the effective area.

In spirally reinforced compression members, lap splice length within a spiral may be multiplied by 0.75, but it shall not be less than 12 in. (300 mm).

END-BEARING SPLICES

In bars required for compression only, compressive stress may be transmitted by bearing of square cut ends held in concentric contact by a suitable device. Bar ends shall terminate in flat surfaces within 1.5° of a right angle to the axis of the bars and shall be fitted within 3° of full bearing after assembly. End-bearing splices shall be used only in members containing closed ties, closed stirrups or spirals.

SPECIAL REQUIREMENTS FOR COLUMNS

(a)   Where factored load stress in longitudinal bars in a column, calculated for various loading combinations, varies from f_y in compression to $(1/2) f_y$ or less in tension, lap splices, butt-welded splices, mechanical connections or end-bearing splices may be used. Total tensile strength provided in each face of the column by splices alone or by splices in combinations with continuing unspliced bars at specified yield strength f_y shall be at least twice the calculated tension in that face, but not less than required by (c) below.

(b)   Where factored load stress in longitudinal bars in a column, calculated for any loading combination, exceeds $(1/2) f_y$ in tension, lap splices designed to develop the specified yield strength f_y in tension, full-welded splices or full mechanical connections shall be used.

(c) At horizontal cross sections of columns where splices are located, a minimum tensile strength in each face of the column equal to one-quarter the area of vertical reinforcement in that face multiplied by f_y shall be provided.

WELDED DEFORMED-WIRE FABRIC IN TENSION

Minimum length of lap for lap splices of welded deformed-wire fabric, measured between the ends of each fabric sheet, shall not be less than $1.7L_d$ or 8 in. (200 mm), and the overlap measured between outermost cross wires of each fabric sheet shall be not less than 2 in. (50 mm). Here, as usual, L_d denotes the development length for the specified yield strength f_y.

Lap splices of welded deformed-wire fabric, with no cross wires within the lap splice length, shall be determined as for deformed wire.

WELDED SMOOTH-WIRE FABRIC IN TENSION

Minimum length of lap for lap splices of welded smooth-wire fabric shall be in accordance with the following:

(a) When the area of reinforcement provided is less than twice that required by analysis at the splice location, the length of overlap measured between the outermost cross wires of each fabric sheet shall be not less than one spacing of cross wires plus 2 in. (50 mm), nor less than $1.5L_d$, nor less than 6 in. (150 mm).

(b) When the area of reinforcement provided is at least twice that required by analysis at the splice location, length of overlap measured between outermost cross wires of each fabric sheet shall not be less than $1.5L_d$ or 2 in. (50 mm).

Solved Problem

7.1. For the part of a continuous frame shown in Fig. 7-3 determine the lengths of the top and bottom bars for the exterior span. The concrete density is 145 lb/ft^3; the factored uniform load is 6.0 kips/ft, including the weight of the beam; the concrete strength is 4000 psi; the steel strength is 60,000 psi; the beam width b is 16 in.; the total depth of the beam h is 22 in.; the concrete cover over the reinforcement is 1.5 in.; the clear span length L_n is 25 ft; No. 4 stirrups are used, with the area of each leg 0.2 in.^{2} and the diameter 0.5 in. Use the 1983 ACI Code approximate coefficients, as given in Chapter 2, for bending moments and shears, and make a soft conversion of all your results into SI units with the aid of Appendix A-1.

At the interior face of the exterior supporting column,

$$M_u = -w_u L_n^2/16 = -(6)(25)^2/16 = -234.4 \text{ kip-ft } (-317.85 \text{ kN} \cdot \text{m}) \tag{1}$$

The positive bending moment near the center of the span is

$$M_u = w_u L_n^2/14 = (6)(25)^2/14 = 267.9 \text{ kip-ft } (363.27 \text{ kN} \cdot \text{m}) \tag{2}$$

At the exterior face of the first interior support, the ultimate (factored) shear force is

$$1.15 w_u L_n/2 = (1.15)(6)(25/2) = 86.3 \text{ kips } (383.86 \text{ kN}) \tag{3}$$

Using the principles of strength design provided in Chapter 5, the reinforcing bar requirements are as follows.

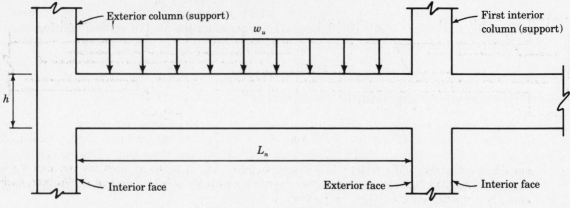

(a)   Elevation view of exterior span beam

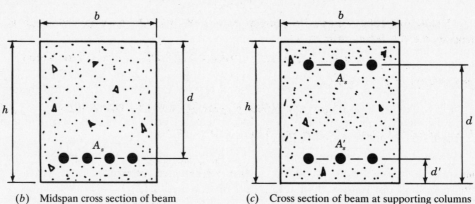

(b)   Midspan cross section of beam   (c)   Cross section of beam at supporting columns

Fig. 7-3

Corresponding to the moment (1), the required area of tension steel in the top of the beam is 2.93 in.^{2} (1890 mm^2). By Table 1.12, 3.16 in.^{2} (2039 mm^2) can be provided by using four No. 8 bars. (Although there is no *exact* metric equivalent to a No. 8 bar, Table 1.13 shows that four 25M bars provide steel area of 2000 mm^2.)

Corresponding to the moment (2), the steel area required is 3.40 in.^{2} (7458 mm^2), and this can be supplied by using two No. 8 bars and two No. 9 bars, which provide a total steel area 3.58 in.^{2} (2310 mm^2). (Five 25M bars provide 2500 m^2.)

For the negative bending moment of 375.0 kip-ft (508.5 kN · m) at the exterior face of the first interior support, the steel area required is 5.01 in.^{2} (3232 mm^2). This can be provided using four No. 10 bars, which corresponds to 5.08 in.^{2} (3277 mm^2). (Five 30M bars provide 3500 mm^2.)

Reinforcement is required for shear and to confine the bars that are in compression. The shear force V_u is calculated at a distance d from the face of the support. The largest shear force occurs at the face of the first interior column; its value is

$$1.15 w_u L_n / 2 = 86.2 \text{ kips (383.9 kN)}$$

At a distance d from the face of the support,

$$V_u = 86.3 - 6(19.4/12) = 76.6 \text{ kips (340.7 kN)}$$

The permissible shear is

$$\phi V_c = 2\phi \sqrt{f_c'} \, b_w d = (2)(0.85)\sqrt{4000}\,(16)(19.4)/1000 = 33.4 \text{ kips (148.6 kN)}$$

For this case, the maximum stirrup spacing to resist shear forces is $d/2 = 9.7$ in. (246.4 mm). Using No. 4 stirrups at 9 in. (229 mm) spacing on centers, the shear force taken by the stirrups is

$$\phi V_s = \phi A_v f_y d/s = (0.85)(0.4)(60)(19.4)/9 = 44.0 \text{ kips (195.7 kN)}$$

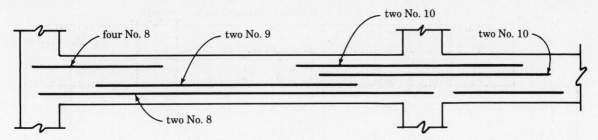

Fig. 7-4

and $\phi V_n = \phi V_c + \phi V_s = 33.4 + 40.0 = 77.4$ kips (344.3 kN), which is greater than the required shear force 76.6 kips (340.74kN) and is satisfactory. Figure 7-4 gives the details. Note that if 10M bars, each with area 100 mm², are used, then, by proportion,

$$\text{spacing} = \frac{(0.2 \text{ in.}^2)(645.2 \text{ mm}^2/\text{in.}^2)}{100 \text{ mm}^2}(229 \text{ mm}) = 295 \text{ mm}$$

Finally, the total length of reinforcing bars is desired. The development length of the bars in tension, the No. 8 bars, is given by

$$L_d = 0.04 A_b f_y / \sqrt{f_c'} = 0.04(0.79)(60,000)/\sqrt{4000} = 29.98 \text{ in. } (761.5 \text{ mm})$$

provided this is not less than

$$0.0004 d_f f_y = 0.0004(1.0)(60,000) = 24.0 \text{ in. } (609.6 \text{ mm})$$

Use 30.0 in. (762 mm). The complete design is shown in Fig. 7-5.

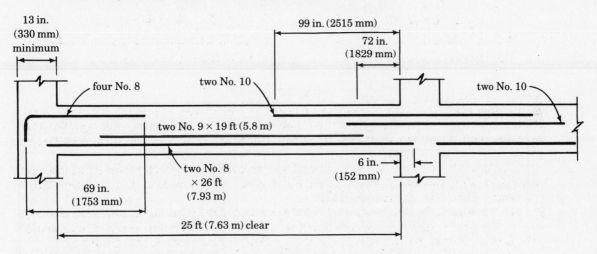

Fig. 7-5

Supplementary Problems

Tables 7.4 through 7.10 list development lengths for bars on splice lengths for Class A splices. The upper portion of each table pertains to U.S. customary units, with f_c' and f_y in kips/in.² and the tabulated lengths in in.; the lower portion pertains to SI units, with f_c' and f_y in MPa and tabulated lengths in mm. The reader is invited to verify selected entries in these tables.

Table 7.4   Non-Top Bars in Tension; Normal-Density Concrete

f'_c / f_y	3.0 / 40.0	4.0 / 40.0	5.0 / 40.0	6.0 / 40.0	3.0 / 50.0	4.0 / 50.0	5.0 / 50.0	6.0 / 50.0	3.0 / 60.0	4.0 / 60.0	5.0 / 60.0	6.0 / 60.0
U.S. Bar No. 3	6.0	6.0	6.0	6.0	7.5	7.5	7.5	7.5	9.0	9.0	9.0	9.0
4	8.0	8.0	8.0	8.0	10.0	10.0	10.0	10.0	12.0	12.0	12.0	12.0
5	10.0	10.0	10.0	10.0	12.5	12.5	12.5	12.5	15.0	15.0	15.0	15.0
6	12.9	12.0	12.0	12.0	16.1	15.0	15.0	15.0	19.3	18.0	18.0	18.0
7	17.5	15.2	14.0	14.0	21.9	19.0	17.5	17.5	26.3	22.8	21.0	21.0
8	23.1	20.0	17.9	16.3	28.8	25.0	22.3	20.4	34.6	30.0	26.8	24.5
9	29.2	25.3	22.6	20.7	36.5	31.6	28.3	25.8	43.8	37.9	33.9	31.0
10	37.1	32.1	28.7	26.2	46.4	40.2	35.9	32.8	55.6	48.2	43.1	39.3
11	45.6	39.5	35.3	32.2	57.0	49.3	44.1	40.3	68.4	59.2	52.9	48.3
14	62.1	53.8	48.1	43.9	77.6	67.2	60.1	54.9	93.1	80.6	72.1	65.8
18	80.3	69.6	62.2	56.8	100.4	87.0	77.8	71.0	120.5	104.4	93.3	85.2

f'_c / f_y	20 / 270	30 / 270	35 / 270	40 / 270	20 / 340	30 / 340	35 / 340	40 / 340	20 / 400	30 / 400	35 / 400	40 / 400
SI Bar No. 10	183	183	183	183	231	231	231	231	271	271	271	271
15	259	259	259	259	326	326	326	326	384	384	384	384
20	362	316	316	316	456	398	398	398	537	468	468	468
25	604	493	456	427	760	621	575	538	894	730	676	632
30	845	690	639	598	1064	869	805	753	1252	1022	947	885
35	1207	986	913	854	1521	1242	1149	1075	1789	1461	1352	1265
45	1509	1232	1141	1067	1901	1552	1437	1344	2236	1826	1690	1581
55	2113	1725	1597	1494	2661	2173	2011	1882	3130	2556	2366	2214

Table 7.5   Non-Top Bars in Tension; All-Low-Density Concrete

f'_c / f_y	3.0 / 40.0	4.0 / 40.0	5.0 / 40.0	6.0 / 40.0	3.0 / 50.0	4.0 / 50.0	5.0 / 50.0	6.0 / 50.0	3.0 / 60.0	4.0 / 60.0	5.0 / 60.0	6.0 / 60.0
U.S. Bar No. 3	8.0	8.0	8.0	8.0	10.0	10.0	10.0	10.0	12.0	12.0	12.0	12.0
4	10.6	10.6	10.6	10.6	13.3	13.3	13.3	13.3	16.0	16.0	16.0	16.0
5	13.3	13.3	13.3	13.3	16.6	16.6	16.6	16.6	19.9	19.9	19.9	19.9
6	17.1	16.0	16.0	16.0	21.4	19.9	19.9	19.9	25.6	23.9	23.9	23.9
7	23.3	20.2	18.6	18.6	29.1	25.2	23.3	23.3	35.0	30.3	27.9	27.9
8	30.7	26.6	23.8	21.7	38.4	33.2	29.7	27.1	46.0	39.9	35.7	32.6
9	38.9	33.6	30.1	27.5	48.6	42.1	37.6	34.3	58.3	50.5	45.1	41.2
10	49.3	42.7	38.2	34.9	61.7	53.4	47.8	43.6	74.0	64.1	57.3	52.3
11	60.6	52.5	46.9	42.9	75.8	65.6	58.7	53.6	90.9	78.7	70.4	64.3
14	82.6	71.5	64.0	58.4	103.2	89.4	79.9	73.0	123.8	107.2	95.9	87.6
18	106.8	92.5	82.8	75.5	133.6	115.7	103.4	94.4	160.3	138.8	124.1	113.3

f'_c / f_y	20 / 270	30 / 270	35 / 270	40 / 270	20 / 340	30 / 340	35 / 340	40 / 340	20 / 400	30 / 400	35 / 400	40 / 400
SI Bar No. 10	243	243	243	243	307	307	307	307	361	361	361	361
15	345	345	345	345	434	434	434	434	511	511	511	511
20	482	420	420	420	607	529	529	529	714	622	622	622
25	803	656	607	568	1011	826	764	715	1190	971	899	841
30	1124	918	850	795	1416	1156	1070	1001	1665	1360	1259	1178
35	1606	1311	1214	1136	2022	1651	1529	1430	2379	1943	1798	1682
45	2007	1639	1517	1419	2528	2064	1911	1787	2974	2428	2248	2103
55	2810	2295	2124	1987	3539	2890	2675	2502	4164	3400	3147	2944

Table 7.6　Non-Top Bars in Tension; Sand-Low-Density Concrete

| f'_c | 3.0 | 4.0 | 5.0 | 6.0 | 3.0 | 4.0 | 5.0 | 6.0 | 3.0 | 4.0 | 5.0 | 6.0 |
f_y	40.0	40.0	40.0	40.0	50.0	50.0	50.0	50.0	60.0	60.0	60.0	60.0
U.S. Bar No.												
3	7.1	7.1	7.1	7.1	8.9	8.9	8.9	8.9	10.6	10.6	10.6	10.6
4	9.4	9.4	9.4	9.4	11.8	11.8	11.8	11.8	14.2	14.2	14.2	14.2
5	11.8	11.8	11.8	11.8	14.7	14.7	14.7	14.7	17.7	17.7	17.7	17.7
6	15.2	14.2	14.2	14.2	19.0	17.7	17.7	17.7	22.8	21.2	21.2	21.2
7	20.7	17.9	16.5	16.5	25.9	22.4	20.6	20.6	31.0	26.9	24.8	24.8
8	27.2	23.6	21.1	19.3	34.0	29.5	26.4	24.1	40.8	35.4	31.6	28.9
9	34.5	29.9	26.7	24.4	43.1	37.3	33.4	30.5	51.7	44.8	40.1	36.6
10	43.8	37.9	33.9	31.0	54.7	47.4	42.4	38.7	65.7	56.9	50.9	46.4
11	53.8	46.6	41.7	38.0	67.2	58.2	52.1	47.5	80.7	69.9	62.5	57.0
14	73.2	63.4	56.7	51.8	91.6	79.3	70.9	64.7	109.9	95.2	85.1	77.7
18	94.8	82.1	73.4	67.0	118.5	102.6	91.8	83.8	142.2	123.1	110.1	100.5

| f'_c | 20 | 30 | 35 | 40 | 20 | 30 | 35 | 40 | 20 | 30 | 35 | 40 |
f_y	270	270	270	270	340	340	340	340	400	400	400	400
SI Bar No.												
10	216	216	216	216	272	272	272	272	320	320	320	320
15	306	306	306	306	385	385	385	385	453	453	453	453
20	427	373	373	373	538	469	469	469	633	552	552	552
25	712	582	539	504	897	732	678	634	1055	862	798	746
30	997	814	754	705	1256	1025	949	888	1478	1206	1117	1045
35	1425	1163	1077	1008	1794	1465	1356	1269	2111	1723	1596	1493
45	1781	1454	1346	1259	2243	1831	1695	1586	2639	2154	1995	1866
55	2493	2036	1885	1763	3140	2564	2374	2220	3694	3016	2792	2612

Table 7.7　Top Bars in Tension; Normal-Density Concrete

| f'_c | 3.0 | 4.0 | 5.0 | 6.0 | 3.0 | 4.0 | 5.0 | 6.0 | 3.0 | 4.0 | 5.0 | 6.0 |
f_y	40.0	40.0	40.0	40.0	50.0	50.0	50.0	50.0	60.0	60.0	60.0	60.0
U.S. Bar No.												
3	8.4	8.4	8.4	8.4	10.5	10.5	10.5	10.5	12.6	12.6	12.6	12.6
4	11.2	11.2	11.2	11.2	14.0	14.0	14.0	14.0	16.8	16.8	16.8	16.8
5	14.0	14.0	14.0	14.0	17.5	17.5	17.5	17.5	21.0	21.0	21.0	21.0
6	18.0	16.8	16.8	16.8	22.5	21.0	21.0	21.0	27.0	25.2	25.2	25.2
7	24.5	21.3	19.6	19.6	30.7	26.6	24.5	24.5	36.8	31.9	29.4	29.4
8	32.3	28.0	25.0	22.8	40.4	35.0	31.3	28.6	48.5	42.0	37.5	34.3
9	40.9	35.4	31.7	28.9	51.1	44.3	39.6	36.1	61.3	53.1	47.5	43.4
10	51.9	45.0	40.2	36.7	64.9	56.2	50.3	45.9	77.9	67.5	60.3	55.1
11	63.8	55.3	49.4	45.1	79.7	69.1	61.8	56.4	95.7	82.9	74.1	67.7
14	86.9	75.3	67.3	61.5	108.6	94.1	84.1	76.8	130.4	112.9	101.0	92.2
18	112.5	97.4	87.1	79.5	140.6	121.7	108.9	99.4	168.7	146.1	130.7	119.3

| f'_c | 20 | 30 | 35 | 40 | 20 | 30 | 35 | 40 | 20 | 30 | 35 | 40 |
f_y	270	270	270	270	340	340	340	340	400	400	400	400
SI Bar No.												
10	256	256	256	256	323	323	323	323	380	380	380	380
15	363	363	363	363	457	457	457	457	538	538	538	538
20	507	442	442	442	639	557	557	557	751	655	655	655
25	845	690	639	598	1064	869	805	753	1252	1022	947	885
30	1183	966	895	837	1490	1217	1126	1054	1753	1431	1325	1240
35	1690	1380	1278	1195	2129	1738	1609	1505	2504	2045	1893	1771
45	2113	1725	1597	1494	2661	2173	2011	1882	3130	2556	2366	2214
55	2958	2415	2236	2092	3725	3042	2816	2634	4383	3578	3313	3099

Table 7.8   Top Bars in Tension; All-Low-Density Concrete

| f'_c | 3.0 | 4.0 | 5.0 | 6.0 | 3.0 | 4.0 | 5.0 | 6.0 | 3.0 | 4.0 | 5.0 | 6.0 |
f_y	40.0	40.0	40.0	40.0	50.0	50.0	50.0	50.0	60.0	60.0	60.0	60.0
3	11.2	11.2	11.2	11.2	14.0	14.0	14.0	14.0	16.8	16.8	16.8	16.8
4	14.9	14.9	14.9	14.9	18.6	18.6	18.6	18.6	22.3	22.3	22.3	22.3
5	18.6	18.6	18.6	18.6	23.3	23.3	23.3	23.3	27.9	27.9	27.9	27.9
6	23.9	22.3	22.3	22.3	29.9	27.9	27.9	27.9	35.9	33.5	33.5	33.5
7	32.6	28.3	26.1	26.1	40.8	35.3	32.6	32.6	49.0	42.4	39.1	39.1
8	43.0	37.2	33.3	30.4	53.7	46.5	41.6	38.0	64.5	55.8	49.9	45.6
9	54.4	47.1	42.1	38.5	68.0	58.9	52.7	48.1	81.6	70.7	63.2	57.7
10	69.1	59.8	53.5	48.8	86.3	74.8	66.9	61.1	103.6	89.7	80.3	73.3
11	84.9	73.5	65.7	60.0	106.1	91.9	82.2	75.0	127.3	110.2	98.6	90.0
14	115.6	100.1	89.5	81.7	144.5	125.1	111.9	102.2	173.4	150.1	134.3	122.6
18	149.6	129.5	115.9	105.8	187.0	161.9	144.8	132.2	224.4	194.3	173.8	158.7

| f'_c | 20 | 30 | 35 | 40 | 20 | 30 | 35 | 40 | 20 | 30 | 35 | 40 |
f_y	270	270	270	270	340	340	340	340	400	400	400	400
10	341	341	341	341	429	429	429	429	505	505	505	505
15	483	483	483	483	608	608	608	608	715	715	715	715
20	674	588	588	588	849	741	741	741	999	871	871	871
25	1124	918	850	795	1416	1156	1070	1001	1665	1360	1259	1178
30	1574	1285	1190	1113	1982	1618	1498	1401	2332	1904	1763	1649
35	2248	1836	1700	1590	2831	2312	2140	2002	3331	2720	2518	2355
45	2810	2295	2124	1987	3539	2890	2675	2502	4164	3400	3147	2944
55	3935	3213	2974	2782	4955	4045	3745	3503	5829	4759	4406	4122

(U.S. Bar No. for first part; SI Bar No. for second part)

Table 7.9   Top Bars in Tension; Sand-Low-Density Concrete

| f'_c | 3.0 | 4.0 | 5.0 | 6.0 | 3.0 | 4.0 | 5.0 | 6.0 | 3.0 | 4.0 | 5.0 | 6.0 |
f_y	40.0	40.0	40.0	40.0	50.0	50.0	50.0	50.0	60.0	60.0	60.0	60.0
3	9.9	9.9	9.9	9.9	12.4	12.4	12.4	12.4	14.9	14.9	14.9	14.9
4	13.2	13.2	13.2	13.2	16.5	16.5	16.5	16.5	19.8	19.8	19.8	19.8
5	16.5	16.5	16.5	16.5	20.6	20.6	20.6	20.6	24.8	24.8	24.8	24.8
6	21.2	19.8	19.8	19.8	26.5	24.8	24.8	24.8	31.9	29.7	29.7	29.7
7	29.0	25.1	23.1	23.1	36.2	31.3	28.9	28.9	43.4	37.6	34.7	34.7
8	38.1	33.0	29.5	27.0	47.7	41.3	36.9	33.7	57.2	49.5	44.3	40.4
9	48.3	41.8	37.4	34.1	60.3	52.2	46.7	42.7	72.4	62.7	56.1	51.2
10	61.3	53.1	47.5	43.3	76.6	66.3	59.3	54.2	91.9	79.6	71.2	65.0
11	75.3	65.2	58.3	53.2	94.1	81.5	72.9	66.5	112.9	97.8	87.5	79.8
14	102.5	88.8	79.4	72.5	128.2	111.0	99.3	90.6	153.8	133.2	119.2	108.8
18	132.7	114.9	102.8	93.8	165.9	143.7	128.5	117.3	199.1	172.4	154.2	140.8

| f'_c | 20 | 30 | 35 | 40 | 20 | 30 | 35 | 40 | 20 | 30 | 35 | 40 |
f_y	270	270	270	270	340	340	340	340	400	400	400	400
10	302	302	302	302	381	381	381	381	448	448	448	448
15	428	428	428	428	539	539	539	539	634	634	634	634
20	598	522	522	522	754	657	657	657	887	773	773	773
25	997	814	754	705	1256	1025	949	888	1478	1206	1117	1045
30	1396	1140	1056	987	1758	1436	1329	1243	2069	1689	1564	1463
35	1995	1629	1508	1411	2512	2051	1899	1776	2955	2413	2234	2090
45	2493	2036	1885	1763	3140	2564	2374	2220	3694	3016	2792	2612
55	3491	2850	2639	2468	4396	3589	3323	3108	5172	4223	3909	3657

(U.S. Bar No. for first part; SI Bar No. for second part)

Table 7.10　Bars in Compression

| | f'_c | 3.0 | 4.0 | 5.0 | 6.0 | 3.0 | 4.0 | 5.0 | 6.0 | 3.0 | 4.0 | 5.0 | 6.0 |
	f_y	40.0	40.0	40.0	40.0	50.0	50.0	50.0	50.0	60.0	60.0	60.0	60.0
U.S. Bar No.	3	5.5	4.7	4.5	4.5	6.8	5.9	5.6	5.6	8.2	7.1	6.7	6.7
	4	7.3	6.3	6.0	6.0	9.1	7.9	7.5	7.5	11.0	9.5	9.0	9.0
	5	9.1	7.9	7.5	7.5	11.4	9.9	9.4	9.4	13.7	11.9	11.2	11.2
	6	11.0	9.5	9.0	9.0	13.7	11.9	11.2	11.2	16.4	14.2	13.5	13.5
	7	12.8	11.1	10.5	10.5	16.0	13.8	13.1	13.1	19.2	16.6	15.7	15.7
	8	14.6	12.6	12.0	12.0	18.3	15.8	15.0	15.0	21.9	19.0	18.0	18.0
	9	16.5	14.3	13.5	13.5	20.6	17.8	16.9	16.9	24.7	21.4	20.3	20.3
	10	18.5	16.1	15.2	15.2	23.2	20.1	19.1	19.1	27.8	24.1	22.9	22.9
	11	20.6	17.8	16.9	16.9	25.7	22.3	21.1	21.1	30.9	26.8	25.4	25.4
	14	24.7	21.4	20.3	20.3	30.9	26.8	25.4	25.4	37.1	32.1	30.5	30.5
	18	33.0	28.5	27.1	27.1	41.2	35.7	33.9	33.9	49.4	42.8	40.6	40.6

| | f'_c | 20 | 30 | 35 | 40 | 20 | 30 | 35 | 40 | 20 | 30 | 35 | 40 |
	f_y	270	270	270	270	340	340	340	340	400	400	400	400
SI Bar No.	10	171	139	129	122	215	175	162	154	253	206	191	181
	15	241	197	183	173	304	248	230	218	358	292	270	256
	20	294	240	222	211	371	303	280	265	436	356	330	312
	25	380	311	288	272	479	391	362	343	563	460	426	403
	30	451	368	341	323	568	464	430	407	669	546	505	478
	35	539	440	407	386	679	554	513	486	798	652	603	571
	45	660	539	499	472	831	678	628	594	977	798	739	699
	55	851	695	644	609	1072	875	810	767	1261	1030	953	902

Chapter 8

Serviceability and Deflections

NOTATION

A_g = gross area of section, in^2 (mm^2)

A_s = area of nonprestressed tension reinforcement, in^2 (mm^2)

A_s' = area of compression reinforcement, in^2 (mm^2)

d = effective depth from compression face to centroid of tension steel, in (mm)

d' = distance from extreme compression fiber to centroid of compression reinforcement, in (mm)

d_s = distance from extreme tension fiber to centroid of tension reinforcement, in (mm)

E_c = modulus of elasticity of concrete, psi (MPa)

f_c' = specified compressive strength of concrete, psi (MPa)

f_r = modulus of rupture of concrete, psi (MPa)

f_y = specified yield strength of nonprestressed reinforcement, psi (MPa)

h = overall thickness of member, in (mm)

I_{cr} = moment of inertia of cracked section transformed to concrete, in^4 (mm^4)

I_e = effective moment of inertia for computation of deflection, in^4 (mm^4)

I_g = moment of inertia of gross concrete section about centroidal axis, neglecting reinforcement, in^4 (mm^4)

L = span length of beam or one-way slab

L_n = length of clear span in long direction of two-way construction, measured face-to-face of supports in slabs without beams and face-to-face of beams or other supports in other cases

M_a = maximum moment in member at stage deflection is computed

M_{cr} = cracking moment

w_c = density of concrete, lb/ft^3 (kg/m^3)

y_t = distance from centroidal axis of gross section, neglecting reinforcement, to extreme fiber in tension, in (mm)

α = ratio of flexural stiffness of beam section to flexural stiffness of a width of slab bounded laterally by centerline of adjacent panel (if any) on each side of beam

α_m = average value of α for all beams on edges of a panel

β = ratio of clear spans in long to short direction of two-way slabs

β_s = ratio of length of continuous edges to total perimeter of a slab panel

λ = multiplier for additional long-time deflection

ξ = time-dependent factor for sustained load

ρ = reinforcement ratio for tension reinforcement = A_s/bd

ρ' = reinforcement ratio for nonprestressed compression reinforcement = A_s'/bd

ϕ = strength reduction factor

GENERAL

Serviceability deals primarily with aesthetics and human comfort, and is only indirectly related to the strength of a structure. While a structure might be perfectly safe from the point of view of strength, it might be unsatisfactory insofar as serviceability is concerned. Two extreme examples may be cited. Visible cracks in members, or vibration-producing deflections, in a storage warehouse might be perfectly acceptable, as long as the strength of the structure is not impaired. But neither cracks nor vibrations would be acceptable in an art museum.

In Chapter 11 *building drift* (horizontal deflections of a structure due to lateral loads) is discussed in detail. Back-and-forth drift of a building from wind gusts could cause motion sickness to the inhabitants. Thus, it is important to be able to establish the desired degree of serviceability, and to be able to control cracking, vertical deflection, vibration and horizontal drift by appropriate means.

CRACK CONTROL

The 1983 ACI Code contains empirical equations for establishing control of cracks in the concrete. After the bending moment has exceeded the *cracking moment*, the concrete will crack due to tension stresses. Instead of visible cracks, it is desirable to have numerous *hairline cracks*, or *microcracks*, so called because they are visible only under a high-powered microscope.

For normal-density concrete, the cracking moment is given by

$$M_{cr} = f_r I_g / y_t$$

where $f_r = 7.5\sqrt{f'_c}$ $\quad (0.7\sqrt{f'_c})$. For other types of concrete, f_r is modified per "Other Provisions for Shear," Chapter 6. The 1983 ACI Code gives the following empirical formula for crack control, assuming that f_y for the steel exceeds 40,000 psi (300 MPa):

$$Z = f_s \sqrt[3]{d_c A}$$

in which

Z = index of crack width, kips/in (kN/m)

f_s = stress in tension steel at service loads, as calculated according to the alternate design method (Chapter 4) or else as $0.6f_y$, kips/in^2 (MPa)

d_c = the distance of the centers of the *lower layer* of tension reinforcement (when one or more layers are used) to the tension face of the member, in (mm)

A = the effective area *per bar*, calculated relative to Fig. 8-1, in^2 (mm^2)

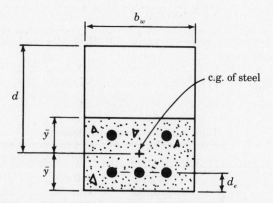

Fig. 8-1   Crack Control (Distribution of Reinforcement)

The value of Z is limited to 175 kips/in (30 000 kN/m) for *interior exposure* (member not exposed to weather) and 145 kips/in (20 000 kN/m) for *exterior exposure* (member exposed to weather). If the member is exposed to chemicals that could corrode the reinforcing steel, more restrictive limits on Z are necessary.

In Fig. 8-1, the distance $\bar{y}$ is that from the tension face of the member to the centroid of steel. The total affected area, A_t, equals $b_w(2\bar{y})$. The area A per bar is equal to A_t/N, where N is the total number of *equivalent* larger bars. For example, if the three bars in the lower layer each have area A_1 and the two bars in the upper layer each have area A_2, then $N = (3A_1 + 2A_2)/A_1$.

When the flanges of T-beams are in tension, part of the flexural reinforcement for tension shall be distributed over an effective flange width equal to 1/10 of the span length, b_w, plus eight times the slab thickness or one-half the distance to the next web, whichever is the lesser distance. For beams with a flange on one side only (*spandrel beams*), the distance is 1/12 of the span length, b_w, plus six times the flange thickness or one-half the distance to the next web, whichever is the lesser distance.

In general, small bars produce better crack control than larger bars. However, per unit weight or mass, small bars are more expensive than larger bars (because of the higher labor costs involved in the placement of the larger number of small bars necessary to provide a given area). Thus, economy of the structure becomes a factor for consideration.

MINIMUM THICKNESSES, MAXIMUM DEFLECTIONS

The 1983 ACI Code provides tables and equations for determining *minimum depths* of beams and slabs. If those minimum depths are used, the Code does not require that deflections be calculated. Table 8.1 provides such minimum thicknesses for beams and one-way slabs. The values are very conservative, and, for buildings of many stories, such thicknesses will be extremely uneconomical. It would therefore be appropriate to use lesser depths that will satisfy strength requirements, and also to compare calculated deflections with the permissible maximum deflections provided in Table 8.2.

Table 8.1 applies directly to members with normal-density concrete—$w_c = 145$ lb/ft^3 (2300 kg/m^3)—and Grade 60 (Grade 400M) reinforcement—$f_y = 60,000$ psi (400 MPa). For low-density concrete—$90 < w_c < 120$ lb/ft^3 ($1500 < w_c < 2000$ kg/m^3)—tabular values are to be multiplied by the greater of

$$1.65 - 0.005 w_c \quad \text{and} \quad 1.09$$

For other values of f_y, tabular values are to be multiplied by

$$0.4 + \frac{f_y}{100,000} \quad \left(0.4 + \frac{f_y}{700}\right)$$

The following material relating to two-way (nonprestressed) slab construction is adapted from the 1983 ACI Code, by permission.

Table 8.1 Minimum Thicknesses

Member	Minimum thickness, h			
	Simply supported	One end continuous	Both ends continuous	Cantilever
	Members not supporting or attached to partitions or other construction likely to be damaged by large deflections			
Solid one-way slabs	$L/20$	$L/24$	$L/28$	$L/10$
Beams or ribbed one-way slabs	$L/16$	$L/18.5$	$L/21$	$L/8$

Table 8.2　Maximum Permissible Computed Deflections

Type of member	Deflection to be considered	Deflection limitation
Flat roofs not supporting or attached to nonstructural elements likely to be damaged by large deflections	Immediate deflection due to live load	$L/180$*
Floors not supporting or attached to nonstructural elements likely to be damaged by large deflections	Immediate deflection due to live load	$L/360$
Roof or floor construction supporting or attached to nonstructural elements likely to be damaged by large deflections	That part of the total deflection occurring after attachment of nonstructural elements (sum of the long-time deflections due to all sustained loads and the immediate deflection due to any additional live load)‡	$L/480$†
Roof or floor construction supporting or attached to nonstructural elements not likely to be damaged by large deflections		$L/240$§

* Limit not intended to safeguard against ponding. Ponding should be checked by suitable calculations of deflection, including added deflections due to ponded water, and considering long-time effects of all sustained loads, camber, construction tolerances, and reliability of provisions for drainage.

† Limit may be exceeded if adequate measures are taken to prevent damage to supported or attached elements.

‡ Long-time deflection shall be determined in accordance with Section 9.5.2.5 or 9.5.4.2 of the 1983 ACI Code but may be reduced by amount of deflection calculated to occur before attachment of nonstructural elements. This amount shall be determined on basis of accepted engineering data relating to time-deflection characteristics of members similar to those being considered.

§ But not greater than tolerance provided for nonstructural elements. Limit may be exceeded if camber is provided so that total deflection minus camber does not exceed limit.

The minimum thickness, in in. (mm), of slabs designed in accordance with Chapter 13 (of the 1983 ACI Code) and having aspect ratio $\beta \leqq 2$ shall be the greater of

$$\frac{L_n(800 + 0.005f_y)}{36{,}000 + 5000\beta[\alpha_m - 0.5(1 - \beta_s)(1 + 1/\beta)]}$$
$$\left(\frac{L_n(800 + f_y/1.5)}{36\,000 + 5000\beta[\alpha_m - 0.51(1 - \beta_s)(1 + 1/\beta)]}\right) \qquad (8.1)$$

and

$$\frac{L_n(800 + 0.005f_y)}{36{,}000 + 5000\beta(1 + \beta_s)} \qquad \left(\frac{L_n(800 + f_y/1.5)}{36\,000 + 5000\beta(1 + \beta_s)}\right) \qquad (8.2)$$

If (8.1) is the greater and exceeds

$$\frac{L_n(800 + 0.005f_y)}{36{,}000} \qquad \left(\frac{L_n(800 + f_y/1.5)}{36\,000}\right) \qquad (8.3)$$

then (8.3) may be used instead to define the thickness. In no case shall the thickness be less than: 5 in (120 mm), for slabs without beams or drop panels; 4 in (100 mm), for slabs without beams but with drop panels conforming to Section 9.5.3.2; 35 in (90 mm), for slabs with beams on all four edges with $\alpha_m \geqq 2.0$.

COMPUTATION OF DEFLECTIONS

Hand computation of deflections for two-way systems is close to impossible. Only with the aid of computer programs using the *finite element method* can these deflections be calculated and compared with the permissible deflections listed in Table 8.2 for the appropriate conditions. However, deflections for one-way slabs and beams are not difficult to calculate, even when the slab or beam is continuous over many supports (Fig. 8-2).

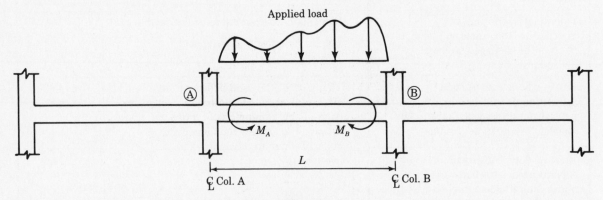

Fig. 8-2

The final bending moment in the member A–B in Fig. 8-2 can be obtained using superposition of the cases due to the applied load and the resulting final bending moments M_A and M_B at the ends of the member. The span length L is taken as the center-to-center distance between columns A and B. Figure 8-3 illustrates the procedure for superimposing the deflections due to the various conditions. The common practice is to determine the deflection at the center of a span as $y_c = y_1 + y_2 + y_3$, but in some cases it may be necessary to calculate deflections at some intermediate point. Figure 8-4 illustrates common loading conditions and expresses the deflections caused thereby. Consistent sets of units for the deflection formulas are as follows: D in in (mm); w in lb/in (kN/m); M in in-lb (N·mm); x and L in in (mm); E in psi (MPa); I in in^4 (mm^4). If other units are chosen, the formulas must be modified by appropriate conversion factors. As a particular application of Fig. 8-4, we have for the total deflection at the centerline of a continuous beam subjected to uniform loading

$$D_{c.l.} = D_c + D'_c + D''_c \qquad (8.4)$$

In the formulas of Fig. 8-4, we set $E = E_c$, the modulus of elasticity for the concrete. E_c is calculated from (1.3) of Chapter 1; for normal-density concrete, a close approximation of (1.3) is

$$E_c = 57{,}000\sqrt{f'_c} \qquad (4700\sqrt{f'_c}) \qquad (8.5)$$

As for the moment of inertia I, the 1983 ACI Code requires that $I = I_e$, where the *effective moment of inertia* of the member is given by

$$I_e = (M_{cr}/M_a)^3 I_g + [1 - (M_{cr}/M_a)^3]I_{cr} \qquad (8.6)$$

In equation (8.6),

M_{cr} = cracking moment for the section

M_a = the maximum (positive) bending moment on the member, normally taken at the centerline of the span

I_g = the gross moment of inertia of the concrete cross section, disregarding the steel reinforcement

I_{cr} = the cracked-section (transformed-section) moment of inertia of the member, including the modular ratio $n = E_s/E_c$

E_s = 29,000 psi (200 000 MPa)

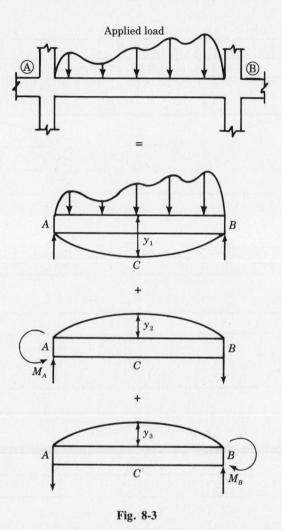

Fig. 8-3

The 1983 ACI Code states that: "For continuous members, I_e *may* be taken as the average of the values obtained for the critical positive and negative moment sections. For prismatic members, I_e *may* be taken at midspan for simple spans and continuous spans, and at the support for cantilevers." The term *may* here is important, because it implies that this is *not mandatory*. In other words, the designer *may* use judgment in these cases. (When the 1983 ACI Code *mandates* a requirement, the term *shall* is used.)

The cracked-section moment of inertia, I_{cr}, needed in equation (8.6) is easily calculated once the neutral axis has been located in the transformed section. This last was accomplished for several types of members in Chapter 4, via the parameter k. Thus, for a **rectangular beam reinforced for tension only** (see Fig. 4-2),

$$I_{cr} = \frac{b(kd)^3}{3} + nA_s(d - kd)^2 \qquad (8.7)$$

in which k is given by (4.3) with p replaced by $\rho = A_s/bd$.

For a **beam with both tension and compression reinforcement** (see Fig. 4-8),

$$I_{cr} = \frac{b(kd)^3}{3} + (2n - 1)A_s'(kd - d')^2 + nA_s(d - kd)^2 \qquad (8.8)$$

in which k is given by the upper equation (4.21) with p and p' replaced by $\rho = A_e/bd$ and $\rho' = A_s'/bd$.

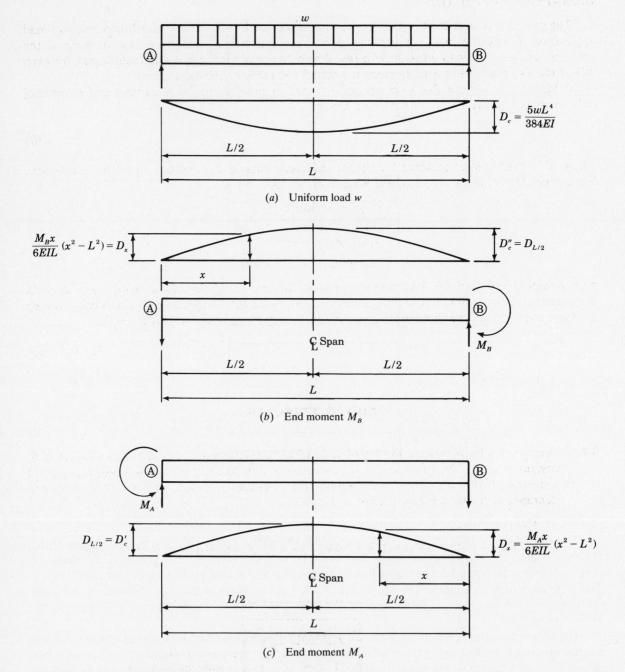

$$D_c = \frac{5wL^4}{384EI}$$

(a) Uniform load w

$$\frac{M_B x}{6EIL}(x^2 - L^2) = D_x$$

$$D_c'' = D_{L/2}$$

(b) End moment M_B

$$D_{L/2} = D_c'$$

$$D_x = \frac{M_A x}{6EIL}(x^2 - L^2)$$

(c) End moment M_A

Fig. 8-4

For a **T-beam** (see Fig. 4-6) such that $kd > h_f$,

$$I_{cr} = \frac{b_w(kd)^3}{3} + \frac{(b - b_w)h_f^3}{12} + (b - b_w)h_f\left(kd - \frac{h_f}{2}\right)^2 + nA_s(d - kd)^2 \qquad (8.9)$$

in which k is given by (4.12) with p replaced by $\rho = A_s/bd$. If $kd \leq h_f$, use (8.7) instead of (8.9), for the reasons given in Chapter 4.

LONG-TERM DEFLECTIONS

The procedures described above concern *immediate* deflections due to dead load and live load application. Long-term loads, or *sustained loads*, cause additional deflections due to *creep* of the concrete. A sustained load normally consists of only the dead load, but it may include part (or even all) of the live load if live load remains in place for extended periods of time.

Additional deflections due to creep are obtained by multiplying the instantaneous (immediate) deflections due to *dead load* by the *creep factor*

$$\lambda = \frac{\xi}{1 + 50\rho'} \qquad\qquad (8.10)$$

where ρ' is evaluated at midspan for simple and continuous spans, and at support for cantilevers. Time-dependent factor ξ for sustained loads may be taken equal to

5 years or more	2.0
12 months	1.4
6 months	1.2
3 months	1.0

with interpolation used for intermediate periods. Since ρ' is proportional to A_s', the area of compression steel, (8.10) implies that an effective way to reduce long-term deflections due to creep is to use compression reinforcement, even though it may not be required for strength.

Solved Problems

8.1. Use Fig. 8-5 to investigate the distribution of reinforcement (crack control), employing U.S. customary units. The lower layer (A_{s1}) consists of four No. 9 bars and the upper layer (A_{s2}) consists of two No. 8 bars; $f_y = 60,000$ psi. No. 4 stirrups are used, with $b = 12$ in. One-inch clearance is required between bars.

From the figure,

$$x = 1.0 + (1.128/2) + (1.0/2) = 2.064 \text{ in} \qquad \text{and} \qquad d_c = 1.5 + 0.5 + (1.128/2) = 2.564 \text{ in}$$

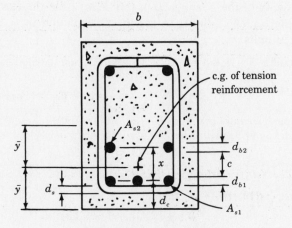

Fig. 8-5 Beam Cross Section

Total $A_s = 4.0 + 1.58 = 5.58$ in^2. Equivalent number of No. 9 bars is $N = 5.58/1.0 = 5.58$; thus,

$$\bar{y} = \frac{(4)(1.0)(2.564) + (2)(0.79)(2064 + 2.564)}{(4)(1.0) + (2)(0.79)} = (10.256 + 7.3/2)/5.58 = 3.1484 \text{ in}$$

$$A = 2(3.1484)(12)/5.58 = 13.695 \text{ in}^2/\text{bar}$$

Take $f_s = 0.6 f_y = 0.6(60) = 36.0$ kips/in^2, as allowed by the 1983 ACI Code. Then,

$$Z = f_s \sqrt[3]{d_c A} = 36 \sqrt[3]{(2.564)(13.695)} = 117.9 \text{ kips/in}$$

This is satisfactory for interior exposure ($Z \leq 175$ kips/in) and for exterior exposure ($Z \leq 145$ kips/in).

8.2. Use Fig. 8-5 to investigate the distribution of reinforcement (crack control) using SI units. Both layers of steel consist of four No. 25M bars. Stirrups are No. 10M bars, $f_y = 400$ MPa, and $b = 250$ mm. Use 40.0 mm clear cover over the stirrups and 25 mm clear distance between the bars.

Proceed as in Problem 8.1 to obtain:

$$x = 25 + 2(25.2/2) = 50.2 \text{ mm}$$
$$d_c = 40.0 + 11.3 + (25.2/2) = 63.9 \text{ mm}$$

$$\bar{y} = \frac{(4)(500)(63.9) + (4)(500)(63.9 + 50.2)}{4(500) + 4(500)} = (127.800 + 228.200)/4000 = 89 \text{ mm}$$

$$A = (2)(89)(250)/8 = 5562.5 \text{ mm}^2/\text{bar}$$
$$f_s = 0.6(400) = 240 \text{ MPa}$$

$$Z = f_s \sqrt[3]{d_c A} = 240 \sqrt[3]{(63.9)(5562.5)} = (240)(70.806) = 16\,993.4 \text{ kN/m}$$

This is satisfactory for interior exposure ($Z \leq 30\,000$ kN/m) and for exterior exposure ($Z \leq 20\,000$ kN/m).

8.3. The deflection of a beam in a continuous frame can be calculated by considering the beam to be simply supported, with applied loads and the end bending moments from the frame analysis superimposed thereon. Let a beam have span $L = 20$ ft (6.1 m), width $b = 12$ in (305 mm), and total depth $h = 22$ in (560 mm). Properties of the concrete are: strength $f'_c = 4000$ psi (27.58 MPa), density $w_c = 145$ lb/ft^3 (2323 kg/m^3). Area of reinforcing steel in the bottom of the beam is 3 in^2 (1935 mm^2); for the steel, $E_s = 29 \times 10^6$ psi (2×10^5 MPa). The uniformly distributed load is 6 kips/ft (87.56 kN/m), and the frame analysis shows that the negative moments on the ends of the beam are of magnitude 100 ft-kips (135.6 kN·m). The effective depth of the beam is 20 in (508 mm) from the top of the beam. Determine the deflection at the center of the span (*a*) by a calculation in U.S. customary units, (*b*) by a calculation in SI units.

(*a*) The steel ratio is $\rho = A_s/bd = 3/(12)(20) = 0.0125$, and the modulus of elasticity of the concrete is, by (*1.3*),

$$E_c = 33 w_c^{1.5} \sqrt{f'_c} = 33(145)^{1.5} \sqrt{4000} = 3.644 \times 10^6 \text{ psi}$$

Thus, the modular ratio is $n = E_s/E_c = 29/3.644 = 7.958$.

By equation (*4.3*) (with ρ instead of p), the neutral axis location below the compression face is kd, where

$$k = \sqrt{2\rho n + (\rho n)^2} - \rho n = \sqrt{2(0.0125)(7.958) + [(0.0125)(7.958)]^2} - (0.0125)(7.958) = 0.3575$$

so $kd = (0.3575)(20) = 7.15$ in. Then,

$$I_{cr} = b(kd)^3/3 + nA_s(d - kd)^2$$
$$= (12)(7.15)^3/3 + (7.958)(3.0)(20.0 - 7.15)^2$$
$$= 1462 + 3942 = 5406 \text{ in}^4$$

The gross cross-sectional moment of inertia is

$$I_g = bh^3/12 = 12(22)^3/12 = 10{,}648 \text{ in}^4$$

The maximum bending moment at the center of the span equals the simple beam bending moment minus the average of the absolute values of the end moments:

$$M_a = \frac{(6)(20)^2}{8} - \frac{1}{2}(100 + 100) = 200 \text{ ft-kips}$$

The bending moment that will theoretically cause the concrete section to crack is $M_{cr} = f_r I_g(h/2)$, where

$$f_r = 7.5\sqrt{f_c'} = 7.5\sqrt{4000} = 474.34 \text{ psi}$$

is the tensile strength of the concrete, and

$$I_g = bh^2/12 = (12)(22)^3/12 = 10{,}648 \text{ in}^4$$

is the gross sectional moment of inertia. Then,

$$M_{cr} = 474.34(10{,}648)/11 = 459{,}161 \text{ in-lb} = 38.26 \text{ ft-kips}$$

The "effective moment of inertia" is

$$\begin{aligned}I_e &= (M_{cr}/M_a)^3 I_g + [1 - (M_{cr}/M_a)^3]I_{cr}\\ &= (0.007)(10{,}648) + (0.993)(5406) = 5443 \text{ in}^4\end{aligned}$$

which is smaller than I_g; therefore, $E_c I_e = (3644)(5443) = 19.83 \times 10^6 \text{ kips-in}^2$.

The downward deflection at the center of the span due to the uniformly distributed load is

$$D_w = \frac{5wL^4}{384(E_c I_e)} = \frac{5(6/12)(20 \times 12)^4}{(384)(19.83 \times 10^6)} = 1.09 \text{ in}$$

and the upward deflection due to equal negative end bending moments is

$$D_M = -\frac{ML^2}{8(E_c I_e)} = -\frac{(100 \times 12)(20 \times 12)^2}{8(19.83 \times 10^6)} = -0.436 \text{ in}$$

The net deflection at the center of the span is therefore

$$D_c = D_w + D_M = 1.09 - 0.436 = 0.654 \text{ in downward}$$

It should be noted that the 1983 ACI Code permits use of the average of the effective moments of inertia at the center of a span and the ends of the span, if the cross sections differ.

(b) Repeating the calculations of (a) in SI units:

$$\rho = A_s/bd = 1935/(305)(508) = 0.0125$$
$$E_c = 0.043w_c^{1.5}\sqrt{f_c'} = 0.043(2314)^{1.5}\sqrt{27.58} = 25\,137 \text{ MPa}$$
$$n = E_s/E_c = 200\,000/25\,137 = 7.956$$
$$\begin{aligned}k &= \sqrt{2\rho n + (\rho n)^2} - \rho n\\ &= \sqrt{2(0.0125)(7.956) + [(0.0125)(7.956)]^2} - (0.0125)(7.956) = 0.3575\end{aligned}$$
$$kd = (0.3575)(508) = 181.6 \text{ mm}$$
$$\begin{aligned}I_{cr} &= b(kd)^3/3 + nA_s(d - kd)^2\\ &= (305)(181.6)^3/3 + (7.956)(1935)(508 - 181.6)^2 = 22.49 \times 10^8 \text{ mm}^4\end{aligned}$$
$$I_g = bh^3/12 = (305)(560)^3/12 = 44.64 \times 10^8 \text{ mm}^4 = 44.64 \times 10^{-4} \text{ m}^4$$
$$M_a = \frac{(87.56)(6.1)^2}{8} - \frac{1}{2}(135.6 + 135.6) = 271.8 \text{ kN} \cdot \text{m}$$
$$f_r = 0.7\sqrt{f_c'} = 0.7\sqrt{27.58} = 3.676 \text{ MPa}$$
$$\begin{aligned}M_{cr} &= f_r I_g/(h/2) = (3.676 \times 10^3)(44.64 \times 10^{-4})/(280 \times 10^{-3})\\ &= 58.61 \text{ kN} \cdot \text{m}\end{aligned}$$
$$\begin{aligned}I_e &= (M_{cr}/M_a)^3 I_g + [1 - (M_{cr}/M_a)^3]I_{cr}\\ &= (0.0094)(44.64 \times 10^8) + (0.9906)(22.49 \times 10^8)\\ &= 22.70 \times 10^8 \text{ mm}^4 < 44.64 \times 10^8 \text{ mm}^4\end{aligned}$$

$$E_c I_e = (25.137 \text{ kN/mm}^2)(22.70 \times 10^8 \text{ mm}^4) = 570.61 \times 10^8 \text{ kN} \cdot \text{mm}^2$$

$$D_w = \frac{5wL^4}{384(E_c I_e)} = \frac{5(87.56 \times 10^{-3})(6.1 \times 10^3)^4}{384(570.61 \times 10^8)} = 27.67 \text{ mm}$$

$$D_M = -\frac{ML^2}{8(E_c I_e)} = -\frac{(135.6 \times 10^3)(6.1 \times 10^3)^2}{8(570.61 \times 10^8)} = -11.05 \text{ mm}$$

$$D_c = D_w + D_M = 27.67 - 11.05 = 16.62 \text{ mm downward}$$

As determined in (a), $D_c = (0.654)(25.4) = 16.61$ mm.

Supplementary Problems

8.4. Verify the solutions to subproblems (a)–(o) as given in Table 8.3. Bars 1 are above and bars 2 are below (see Fig. 8-5). The two layers of bars are separated by 1 in, clear cover is 1.5 in, and No. 4 stirrups are used.

Table 8.3

	GIVEN			SOLUTION
	Bars 1	Bars 2	b, in	Crack Index Z, kips/in
(a)	4 No. 9	2 No. 8	12	117.4
(b)	3 No. 8	2 No. 7	10	117.6
(c)	4 No. 8	None	12	120.4
(d)	4 No. 10	2 No. 8	12	120.9
(e)	4 No. 7	4 No. 7	12	104.8
(f)	4 No. 8	2 No. 6	12	116.7
(g)	4 No. 10	4 No. 7	14	124.2
(h)	4 No. 9	2 No. 8	12	117.4
(i)	5 No. 8	None	15	120.4
(j)	4 No. 9	2 No. 8	12	117.4
(k)	3 No. 9	3 No. 7	10	118.2
(l)	4 No. 11	4 No. 7	14	128.1
(m)	3 No. 11	3 No. 8	12	132.1
(n)	4 No. 9	2 No. 9	12	116.1
(o)	4 No. 10	2 No. 8	14	127.2

8.5. Verify the solutions to subproblems (a)–(n) as given in Table 8.4. Bars 1 are above and bars 2 are below (see Fig. 8-5). The two layers of bars are separated by 25 mm, clear cover is 40 mm, and No. 10M stirrups are used.

8.6. Verify the solutions to subproblems (a)–(m) as given in Table 8.5. The notation is as in Fig. 8-4 and equation (8.4). Assume that $f'_c = 4000$ psi, concrete density $w_c = 145$ lb/ft^3, $E_s = 29,000,000$ psi.

Table 8.4

	GIVEN			SOLUTION
	Bars 1	Bars 2	b, mm	Crack Index Z, kN/m
(a)	4 No. 25M	4 No. 25M	250	17 600
(b)	4 No. 35M	2 No. 25M	350	22 800
(c)	4 No. 30M	2 No. 30M	300	20 500
(d)	4 No. 35M	None	350	23 400
(e)	3 No. 30M	3 No. 30M	250	19 900
(f)	4 No. 30M	4 No. 25M	300	19 900
(g)	4 No. 35M	None	350	23 400
(h)	4 No. 25M	4 No. 25M	250	17 600
(i)	4 No. 30M	2 No. 25M	275	20 200
(j)	4 No. 35M	2 No. 35M	300	21 200
(k)	3 No. 25M	None	250	21 800
(l)	3 No. 25M	3 No. 25M	275	21 300
(m)	4 No. 25M	2 No. 25M	300	19 900
(n)	4 No. 35M	2 No. 35M	350	22 300

Table 8.5

	GIVEN								SOLUTIONS			
	b, in	h, in	d, in	A_s, in^2	w, kips/ft	L, ft	M_A, ft-kips	M_B, ft-kips	D_c, in	D_c', in	D_c'', in	$D_{c.l.}$, in
(a)	10	18	16	3 No. 11 4.68	8.0	14	75.0	90.0	0.4638	−0.1065	−0.1278	0.2295
(b)	12	24	22	4 No. 8 3.16	5.5	15	80.0	60.0	0.2140	−0.0664	−0.0498	0.0978
(c)	13	24	22	4 No. 10 5.08	7.0	15	50.0	80.0	0.1740	−0.0265	−0.0424	0.1051
(d)	14	26	24	4 No. 11 6.24	4.5	20	60.0	70.0	0.3012	−0.0482	−0.0562	0.1968
(e)	12	20	18	4 No. 8 3.16	3.8	18	80.0	90.0	0.5176	−0.1614	−0.1816	0.1745
(f)	10	18	16	3 No. 10 3.81	4.0	20	100.0	120.0	1.0951	−0.3285	−0.3942	0.3723
(g)	14	22	20	4 No. 11 6.24	6.5	24	150.0	200.0	1.4137	−0.2719	−0.3625	0.7793
(h)	16	24	22	6 No. 10 7.62	7.0	22	250.0	300.0	0.7060	−0.2500	−0.3001	0.1559
(i)	10	16	14	3 No. 10 3.81	6.0	16	120.0	140.0	0.9217	−0.3456	−0.4032	0.1728
(j)	12	18	16	3 No. 11 4.59	6.5	20	300.0	320.0	1.1010	−0.6098	−0.6505	−0.1592
(k)	13	20	18	5 No. 9 5.0	7.5	16	110.0	140.0	0.4847	−0.1333	−0.1697	0.1818
(l)	11	22	20	3 No. 11 4.59	6.5	18	115.0	160.0	0.5903	−0.1547	−0.2153	0.2203
(m)	12	24	22	5 No. 9 5.0	8.0	19	200.0	220.0	0.7044	−0.2204	−0.2424	0.2416

8.7.   Verify the solutions to subproblems (a)–(k) as given in Table 8.6. The notation is as in Fig. 8-4 and equation (8.4). Assume that $f'_c = 30$ MPa, concrete density $w_c = 2320$ kg/m^3, $E_s = 200\,000$ MPa.

Table 8.6

	b, mm	h, mm	d, mm	A_s, mm^2	w, kN/m	L, m	M_A, kN·m	M_B, kN·m	D_c, mm	D'_c, mm	D''_c, mm	$D_{c.l.}$, mm
					GIVEN						SOLUTIONS	
(a)	250	460	400	3000	120	4.3	100	120	12.85	−2.78	−3.34	6.73
(b)	300	610	560	2000	80	4.6	110	80	5.16	−1.61	−1.17	2.38
(c)	355	660	610	4000	66	6.0	80	95	6.97	−1.13	−1.34	4.50
(d)	300	510	460	2000	55	5.5	110	120	12.40	−3.94	−4.29	4.17
(e)	250	460	400	2500	58	6.0	140	160	26.14	−8.41	−9.61	8.11
(f)	355	560	510	4000	95	7.3	200	270	34.91	−6.62	−8.94	19.35
(g)	250	400	360	2500	88	5.0	160	190	24.75	−8.64	−10.26	5.85
(h)	300	460	410	3000	95	6.0	410	430	25.03	−14.40	−15.11	−4.48
(i)	330	510	460	3200	110	5.0	150	190	13.29	−3.48	−4.41	5.41
(j)	280	560	510	3000	95	5.5	160	220	14.56	−3.89	−5.35	5.32
(k)	300	610	560	3200	120	5.8	270	300	17.11	−5.49	−6.10	5.51

Chapter 9

General Provisions for Columns

NOTATION AND DEFINITIONS

a = depth of equivalent rectangular stress block = $\beta_1 c$, in. (mm)

a_b = depth of equivalent rectangular stress block for balanced conditions = $\beta_1 c_b$, in. (mm)

A_c = area of core of spirally reinforced column, measured to the outside diameter of the spiral, in.^{2} (mm^2)

A_g = gross area of section, in.^{2} (mm^2)

A_s = area of tension reinforcement, in.^{2} (mm^2)

A_s' = area of compression reinforcement, in.^{2} (mm^2)

A_{st} = total area of longitudinal reinforcement, in.^{2} (mm^2)

b = width of compression face of flexural member, in. (mm)

$\beta_1 = a/c$

c = distance from extreme compression fiber to neutral axis, in. (mm)

c_b = distance from extreme compression fiber to neutral axis for balanced conditions, in. (mm)

d = distance from extreme compression fiber to centroid of tension reinforcement, in. (mm)

d_b = diameter of a reinforcing bar, in. (mm)

d_c = diameter of core or outside diameter of spiral hoop, in. (mm)

d' = distance from extreme compression fiber to centroid of compression reinforcement, in. (mm)

d'' = distance from tension face to centroid of tension reinforcement, in. (mm)

D = overall diameter of circular section, in. (mm)

D_s = diameter of the circle through centers of reinforcement arranged in a circular pattern (also symbolized γD or γh), in. (mm)

e = eccentricity of axial load at end of member measured from plastic centroid of the section, calculated by conventional methods of frame analysis, ($e = M/P$), in. (mm)

e' = eccentricity of axial load at end of member measured from the centroid of the tension reinforcement, calculated by conventional methods of frame analysis, in. (mm)

e_b = eccentricity of load P_b measured from the centroid of a section, in. (mm)

E_c = modulus of elasticity of concrete per (1.3)

E_s = modulus of elasticity of steel = 29×10^6 psi (200 000 MPa)

f_c' = compressive strength of concrete, psi (MPa)

f_s = stress in tension reinforcement, psi (MPa)

f_s' = stress in compression steel, psi (MPa)

f_y = yield stress for reinforcing steel, psi (MPa)

γ = the ratio, D_s/h, D_s/D or $(d - d'')/h$

h = larger dimension of column cross section, in. (mm)

h' = effective length of column, in. (mm)

157

I = moment of inertia of beam or column cross section, in.^{4} (mm^4)

j = ratio of distance between centroid of compression force and centroid of tension force to the depth, d

$k' = P_n/f'_c A_g$

K = stiffness factor = EI/L

l_u = actual unsupported length of column, in. (mm)

L = span length of slab or beam, ft (m)

L' = clear span for positive moment and shear and the average of the two adjacent clear spans for negative moment, ft (m)

$\mu = f_y/0.85f'_c$

$\mu' = \mu - 1$

M = bending moment, ft-kips (kN·m)

M_b = ultimate moment capacity at simultaneous crushing of concrete and yielding of tension steel (balanced conditions, $M_b = P_b e_b$), ft-kips (kN·m)

M_n = nominal bending moment, M_u/ϕ, ft-kips (kN·m)

M_o = bending moment when pure bending exists, ft-kips (kN·m)

M_u = ultimate moment capacity under combined axial load and bending, ft-kips (kN·m)

n = ratio of modulus of elasticity of steel to that of concrete, or E_s/E_c

ρ = ratio of area of tension reinforcement to effective area of concrete

ρ' = ratio of area of compression reinforcement to effective area of concrete

ρ_g = ratio of area of vertical reinforcement to the gross area, A_g

ρ_s = ratio of volume of spiral reinforcement (per turn) to volume of core (per turn) of a spirally reinforced concrete or composite column

$\rho_t = A_{st}/A_g$

P_b = axial load capacity at simultaneous crushing of concrete and yielding of tension steel (balanced conditions), kips (kN)

P_n = nominal axial load, P_u/ϕ, kips (kN)

P_o = ultimate axial load capacity of actual member when concentrically loaded, kips (kN)

P_u = ultimate axial load capacity under combined axial load and bending, kips (kN)

ψ = the ratio of ΣK of columns to ΣK of floor members in a plane at one end of a column (stiffness ratio)

r = radius of gyration of gross concrete area of a column, in. (mm)

R = a reduction factor for long columns

s = pitch or center-to-center distance between successive turns of a spiral bar, in. (mm)

ϕ = capacity reduction factor or resistance factor

GENERAL PROVISIONS FOR CAST-IN-PLACE COLUMNS

A number of general requirements are stipulated in the 1983 ACI Code relative to columns. Some of the more important factors which should be remembered are listed in the paragraphs which follow.

Principal columns which support a floor or roof are subject to the following limitations:

(a) For columns having *single ties* (rectangular or round), the minimum number of bars is 4. For columns in which *continuous spirals* are used, at least 6 vertical bars must be used. The area of the vertical bars must be at least 1 percent and not more than 8 percent of the gross area of the column; that is, $(0.01)A_g \leq \rho_t \leq (0.08)A_g$.

(b) When *spirals* are used, the minimum *ratio of spiral steel* (ρ_s) shall be given by the equation:

$$\rho_s = 0.45[(A_g/A_c) - 1](f_c'/f_y) \qquad (9.1)$$

in which $f_y \leq 60{,}000$ psi (400 MPa).

Spirals may not be made of bars or wire less than $\frac{3}{8}$ in. (10 mm) in diameter for cast-in-place construction.

Spirals shall consist of evenly spaced continuous bar or wire of such size and so assembled to permit handling and placing without distortion from designed dimensions.

Spirals shall be held firmly in place and true to line by vertical spacers. For spiral bar or wire smaller than $\frac{5}{8}$ in. (16 mm) diameter, a minimum of two spacers shall be used for spirals less than 20 in. (500 mm) in diameter, three spacers for spirals 20 to 30 in. (500 to 800 mm) in diameter, and four spacers for spirals greater than 30 in. (800 mm) in diameter.

For spiral bar or wire $\frac{5}{8}$ (16 mm) or larger in diameter, a minimum of three spacers shall be used for spirals 24 in. (600 mm) or less in diameter, and four spacers for spirals greater than 24 in. (600 mm) in diameter.

Clear spacing between turns of the spiral shall not exceed 3 in. (80 mm) or be less than 1 in. (25 mm). Clear spacing shall not be less than $1\frac{1}{3}$ times the maximum aggregate size.

Clear cover for spirals for cast-in-place concrete shall be at least $1\frac{1}{2}$ in. (40 mm) or $1\frac{1}{3}$ times the maximum aggregate size when not exposed to weather or earth, and at least 2 in. (50 mm) when so exposed.

Anchorage of spirals shall be at least $1\frac{1}{2}$ extra turns of spiral bar or wire at each end of a spiral unit. Splices shall be lap splices of 48 bar or wire diameters (d_b of spiral) but not less than 12 in. (300 mm), or welded.

For tied columns, all nonprestressed U.S. customary bars shall be enclosed by lateral ties, at least No. 3 in size for longitudinal bars No. 10 or smaller and at least No. 4 in size for No. 11, No. 14 or No. 18 and bundled longitudinal bars. For metric bars, all longitudinal bars and bundled nonprestressed bars shall be enclosed by No. 10M metric bars. Deformed-wire or welded-wire fabric of equivalent areas may be used for all spirals.

Vertical spacing of ties shall not exceed 16 longitudinal bar diameters, 48 tie bar diameters, or the least dimension of the column.

Ties shall be so arranged such that every corner and alternate longitudinal bar shall have lateral support provided by the corner of a tie with an included angle of not more than 135 degrees ($3\pi/4$ rad) and no bar shall be farther than 6 in. (150 mm) clear on each side along the tie from such a laterally supported bar.

For tied columns in cast-in-place concrete the clear cover shall be at least $1\frac{1}{2}$ in. (40 mm) or $1\frac{1}{3}$ times the maximum aggregate size when not exposed to earth or weather. The clear cover is at least 2 in. (50 mm) when exposed to earth or weather.

Note. Circular ties may be used when the vertical bars are arranged in a circle, but shall conform to the conditions stated above for ties. (It is usually assumed that tied columns refer to rectangular ties surrounding the vertical bars. However, round ties are sometimes used.)

A column may be designed as a circular section of diameter D and built as a section of *any* shape having a least dimension equal to D. The steel ratio shall be based on the assumed circular section diameter.

In columns, the clear distance between vertical bars shall be the *greater* of (a) $1\frac{1}{2}$ times the bar diameter, (b) $1\frac{1}{3}$ times the maximum size of coarse aggregate or (c) $1\frac{1}{2}$ in. (40 mm). (This also applies to the distance between contact splices and adjacent bars or splices.)

For columns formed *below grade* and in contact with earth after removal of the forms or for columns *in a corrosive atmosphere*, the cover shall be *at least* 2 in. (50 mm).

For precast columns manufactured under plant control conditions, the clear cover for concrete exposed to earth or weather shall be 2 in. (50 mm) for No. 14 (45M) and No. 18 (55M) bars, $1\frac{1}{2}$ in. (40 mm) for No. 6 (20M) through No. 11 (35M) bars and $1\frac{1}{4}$ in. (30 mm) for No. 5 (15M) bars, W31 wire or D31 wire, and smaller. For concrete not exposed to earth or weather, cover shall be one bar diameter (d_b), but not less than $\frac{5}{8}$ in. (15 mm) and need not be more than $1\frac{1}{2}$ in. (40 mm). For ties and spirals the minimum cover is $\frac{3}{8}$ in. (10 mm).

In Problem 9.8 the upper limit

$$s \leqq A_b(d_c - d_b)(\pi f_y)/[0.45 f_c'(A_g - A_c)] \tag{9.2}$$

is derived for the vertical spacing (s) of spiral bars.

HEIGHT OF COLUMNS

As the height of a column increases, the allowable load decreases because of the tendency of long columns to *buckle*. The height of a column is therefore an important factor in the design. Figure 9-1 indicates the height that must be used in the design for different situations.

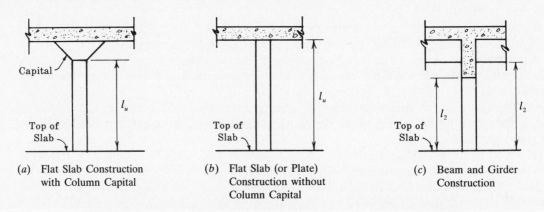

| (a) | Flat Slab Construction with Column Capital | (b) | Flat Slab (or Plate) Construction without Column Capital | (c) | Beam and Girder Construction |

Note: In beam and girder construction, use the height in the plane of bending. When the design is made for *minimum eccentricity*, use the particular value of l_u which will provide the *largest l_u/r ratio*. (Minimum eccentricity and l_u/r are discussed in subsequent paragraphs.)

Fig. 9-1

TRANSMISSION OF COLUMN LOADS

When the specified compressive strength of concrete in a column is greater than 1.4 times that specified for a floor system, transmission of load through the floor system shall be provided by one of the following.

Concrete of strength specified for the column shall be placed in the floor about the column for an area 4 times the column area. Column concrete shall be well integrated into floor concrete, and shall be placed in accordance with Sections 6.4.5 and 6.4.6 of the 1983 ACI Code.

Strength of a column through a floor system shall be based on the lower value of concrete strength with vertical dowels and spirals as required.

For columns laterally supported on four sides by beams of approximately equal depth or by slabs, strength of the column may be based on an assumed concrete strength in the column joint equal to 75 percent of column concrete strength plus 35 percent of floor concrete strength.

DESIGN OF COLUMNS

Modern, computer-developed, design charts, based on the strength method as required in the 1983 ACI Code, will be presented in Chapter 10.

AXIAL LOADS AND MOMENTS

The 1983 ACI Code requires that the axial loads and moments should be obtained using *acceptable elastic design* methods. The axial loads and moments obtained from elastic analysis must be increased by appropriate load factors in order to obtain the ultimate values for use in ultimate strength design. The understrength (resistance) factors ϕ must also be applied.

Solved Problems

9.1. A round spirally reinforced concrete column has an overall diameter of 24″ and a core diameter of 20″. Determine the required pitch s for a No. 5 spiral bar. Maximum aggregate size is 1″, $f'_c = 3000$ psi and $f_y = 40,000$ psi.

The spiral steel ratio ρ_s is defined as

$$\rho_s = \frac{\text{volume of spiral bar}}{\text{volume of concrete core per turn of spiral}}$$

and may not be less than

$$\rho_s \geqq 0.45[(A_g/A_c) - 1](f'_c/f_y)$$

Here $A_g = \pi(24)^2/4 = 452.0$ in.^{2}, $A_c = \pi(20)^2/4 = 314.2$ in.^{2} and $f'_c/f_y = 3/40 = 0.075$. Hence

$$\rho_s \geqq 0.45[(452.0/314.2) - 1](0.075) \geqq 0.0149$$

The bar volume per turn is equal to $(\pi d_s/4)A_b$, so

$$V_s = \pi(20.0 - 0.625)(0.31) = 18.9 \text{ in.}^3/\text{turn}$$

The concrete volume per turn of spiral will be $V_c = A_c(\text{pitch}) = 314.2s$. Thus $V_s/V_c = 18.9/(314.2) \geqq 0.0149$, from which $s \leqq 4.03″$.

Other requirements to be satisfied are:

(a) $s \leqq 3.0″ + d_b = 3.0 + 0.625 = 3.625″$

(b) $s \geqq 1.0″ + d_b = 1.0 + 0.625 = 1.625″$

(c) $s \geqq (1\frac{1}{3} \times \text{max. aggregate size}) + d_b = (1.33)(1) + 0.625 = 1.955″$

All requirements are satisfied if $s = 3.0″$.

9.2. A rectangular tied column measures $10″ \times 20″$. The ties will be No. 4 bars and the main vertical bars will be No. 9 size. Determine the required tie spacing.

The tie spacing may not exceed the least of (a) 16 vertical bar diameters $= (16)(1.128) = 18.05″$, (b) 48 tie bar diameters $= (48)(\frac{1}{2}) = 24.0″$, (c) the least dimension of the column, $10.0″$. Thus the tie spacing may not exceed $10.0″$. Use $s = 10.0″$.

9.3. A column is subjected to direct force and flexure. The analysis indicates that a steel ratio $\rho_t = 0.036$ will satisfy load requirements. Determine whether or not this design will satisfy the 1983 ACI Code.

$\rho_{\min} = 0.01$, $\rho_{\max} = 0.08$
$0.01 < 0.036 < 0.08$. Code is satisfied.

9.4. A 20″ square column is to be constructed above the ground. The vertical steel consists of No. 9 bars and the maximum aggregate size is 1″. Determine the cover required over the tie bars.

The minimum clear cover over the ties must be the largest of: (a) $1\frac{1}{2}$″; (b) the diameter of the vertical bars, 1.128″; (c) $1\frac{1}{3} \times$ the maximum aggregate size $= (4/3)(1) = 1.33$″. Thus a minimum clear cover of $1\frac{1}{2}$″ must be used.

9.5. A 23″ square column contains 12 No. 9 bars arranged in a circle within a No. 5 spiral bar. The clear cover over the spiral is $1\frac{1}{2}$″. Check the clearance between the vertical bars for the 1983 ACI Code requirements. Maximum size of aggregate $= 1.0$″.

The diameter of the circle through the centers of the vertical bars will be

$$D_s = \gamma D = 23.0 - (2)(1.5 + 0.625) - 1.0 = 17.75''$$

The circumference of that circle will be $\pi\gamma D = \pi(17.75) = 55.76''$.
For the 12 bars the center-to-center spacing will be $55.76/12 = 4.646''$. The clear spacing will then be $4.646 - 1.128 = 3.518''$.
The spacing is governed by the largest of (a) $1.5 \times$ vertical bar diameter $= 1.692''$, (b) $1.33 \times$ maximum aggregate size $= 1.33''$, (c) $1\frac{1}{2}$″.
Therefore the 3.518″ clear spacing provided will be satisfactory.

9.6. Check the clear cover provided for the spiral in Problem 9.5 for compliance with Code requirements.

The minimum permissible cover will be the greater of (a) $1\frac{1}{2}$″, (b) $1\frac{1}{3} \times$ maximum aggregate size $= 1.33''$.
The $1\frac{1}{2}$″ clear cover provided is therefore satisfactory.

9.7. Determine the spiral pitch s required for the column described in Problem 9.5. Use a No. 5 spiral bar.

The governing relation is

$$\rho_s \geq 0.45[(A_g/A_c) - 1](f_c'/f_y) = V_s/V_c$$

where $A_g = (23)(23) = 529$ in.^{2}, $A_c = (\pi/4)(20)^2 = 314.2$ in.^{2} and $f_c'/f_y = 5/60 = 0.0833$. By direct substituting of these values, obtain $\rho_s \geq 0.0257$.

The length of the spiral bar per turn will be $L_s = \pi d_s = \pi(20.0 - 0.625) = 60.87''$.
The volume of the spiral bar per turn will be $V_s = A_b L_s = (0.31)(60.87) = 18.87$ in.^{3}
The volume of the concrete core within the outer limits of the spiral will be $V_c = A_c \times$ pitch $= 314.2s$. Thus $s \leq 2.36''$.

The pitch must also satisfy requirements (a), (b), and (c) of Problem 9.1. Using $s = 2\frac{1}{4}$″ will be satisfactory.

9.8. Derive a general expression for determining the spiral pitch s in terms of the appropriate variables to satisfy the condition

$$\rho_s \geqq 0.45[(A_g/A_c) - 1](f_c'/f_y)$$

Let L_s = the length of the spiral bar per turn
 A_b = area of the spiral bar
 V_s = volume of the spiral bar per turn
 V_c = volume of the concrete core per turn of the spiral bar
 d_c = core diameter
 d_s = mean diameter of the spiral hoop
  s = vertical pitch of the spiral bar
 d_b = diameter of the spiral bar

By definition, and to satisfy the stated condition,

$$\rho_s = V_s/V_c = \pi d_s A_b/(A_c s) \geqq 0.45[(A_g/A_c) - 1](f_c'/f_y)$$

from which it follows that

$$A_b/s \geqq 0.45[(A_g/A_c) - 1](f_c'/f_y)[A_c/(\pi d_s)]$$

Now, since $d_s = d_c - d_b$, the final relation is

$$s \leqq A_b(d_c - d_b)(\pi f_y)/[0.45f_c'(A_g - A_c)] \qquad (9.2)$$

The resulting spiral pitch must also satisfy the clearance requirements and the maximum spacing previously discussed.

9.9. Use equation (9.2) to solve for the theoretical minimum spiral pitch s required in Problem 9.7.

Referring to Problem 9.7 and substituting into equation (9.2), obtain

$$s \leqq (0.31)(20.0 - 0.625)(60\pi)/[(0.45)(5)(529.0 - 314.2)]$$

from which $s \leqq 2.36''$, as previously determined.

9.10. Calculate the tie spacing for a rectangular section $18'' \times 24''$ reinforced with No. 9 vertical bars and No. 4 *circular hoops*.

The circular hoops must satisfy the same requirements as those for rectangular ties. Thus the tie spacing must be the lesser dimension of (a) 16 diameters of the vertical bars = 16", (b) 48 diameters of the tie bars = 24", (c) the least dimensions of the column = 18". Therefore the spacing of the circular hoops may not exceed 16".

9.11. A round, spirally reinforced, concrete column has an overall diameter of 600 mm and a core diameter of 500 mm. Determine the required pitch s for a No. 10M spiral bar. Maximum size of aggregate is 25 mm. Use $f_c' = 20$ MPa and $f_y = 270$ MPa.

The spiral steel ratio ρ_s may not be less than $0.45[(A_g/A_c) - 1](f_c'/f_y)$. Here,

$$A_g = \pi(600)^2/4 = 282\,743 \text{ mm}^2$$
$$A_c = \pi(500)^2/4 = 196\,350 \text{ mm}^2$$
$$f_c'/f_y = 20/270 = 0.074$$

Hence, $\rho_s \geqq 0.45[(282\,743/196\,350) - 1](0.074) = 0.01465$.

The diameter of a No. 10M spiral bar is 11.3 mm and its cross-sectional area is 100 mm², so the volume of spiral bar per turn is

$$V_s = \pi(d_c - d_b)A_b = \pi(500 - 11.3)(100) = 153\,529 \text{ mm}^3$$

The volume of concrete per turn is $V_c = A_c \times \text{pitch} = (196\,350)s$ mm³. Thus,

$$\frac{V_s}{V_c} = \frac{153\,529}{196\,350s} \geqq 0.01465 \qquad \text{or} \qquad s \leqq \frac{0.7819}{0.01465} = 53.4 \text{ mm}$$

(Direct use of (9.2) would give the same limit for s.) The other requirements are:

  (a) $s \leqq (80 \text{ mm}) + d_b = 80 + 11.3 = 91.3$ mm

  (b) $s \geqq (25 \text{ mm}) + d_b = 25 + 11.3 = 36.3$ mm

  (c) $s \geqq 1\frac{1}{3} \times \text{maximum aggregate size} + d_b = (1.33)(25) + 11.3 = 44.55$ mm

Using $s = 50$ mm will be satisfactory.

9.12. A rectangular tied column measures 250 mm wide and 500 mm deep. The vertical bars are No. 35M and the ties are No. 10M. Determine the spacing required for the ties.

The 1983 ACI Code requires the least of:

  (a) 16 longitudinal bar diameters $= (16)(35.7) = 571$ mm

  (b) 48 tie bar diameters $= (48)(11.3) = 542$ mm

  (c) minimum cross-sectional dimension $= 250$ mm

The center-to-center spacing of tie bars is therefore 250 mm.

9.13. A 600-mm-diameter column contains fifteen No. 35M longitudinal bars arranged in a circle. Determine whether or not the clear spacing between bars satisfies the 1983 ACI Code, if the No. 10M spiral bar has a clear cover of 40 mm. The maximum aggregate size is 25 mm.

The cover to the center of the No. 35M longitudinal bars is equal to

(40 mm) + the diameter of the spiral bar + one-half the diameter of the No. 45M bar

or $40 + 11.3 + 35.7/2 = 69.15$ mm. The center-to-center distance of bars along a diameter within the spiral is then $600 - (2)(69.15) = 46.17$ mm. The perimeter of a circle through the bars is 461.7π or 1450.5 mm. There are 15 spaces between bars, so the center-to-center spacing of No. 35M longitudinal bars is $1450.5/15 = 96.7$ mm. The clear spacing is therefore $96.7 - 35.7 = 61.0$ mm.

The clear spacing must satisfy the criteria:

  (a) $x \geqq 1.5d_b = (1.5)(35.7) = 53.55$ mm

  (b) $x \geqq 40$ mm

  (c) $x \geqq 1\frac{1}{3} \times \text{maximum aggregate size} = (1.33)(25) = 33.25$ mm

The spacing provided is therefore satisfactory.

9.14. Investigate whether the steel ratio implied in Problem 9.13 satisfies the 1983 ACI Code.

The gross area of the column is $A_g = (\pi/4)(600)^2 = 282\,744$ mm² and the total cross-sectional area of fifteen No. 35M bars is

$$A_{st} = (15)(1000) = 15\,000 \text{ mm}^2$$

The steel ratio is therefore

$$\rho_t = A_{st}/A_g = 15\,000/282\,744 = 0.053$$

Since $0.01 \leqq 0.053 \leqq 0.08$, the 1983 ACI Code is satisfied.

Supplementary Problems

In the following problems, consider the maximum aggregate size to be $1''$ (25 mm).

9.15. Solve Problem 9.1 using a $26''$ diameter column having a $20''$ diameter core, all other data remaining the same. Check all clearances and minimum requirements.   *Ans.*   Pitch $= 3.0''$

9.16. Using No. 4 ties, determine the tie spacing for a $20''$ square column reinforced with No. 11 vertical bars. *Ans.* $s = 22.5''$

9.17. Solve Problem 9.4 considering No. 11 vertical bars.   *Ans.*   Cover $= 1.5''$

9.18. Solve Problem 9.7 using $f'_c = 3000$ psi and $f_y = 50{,}000$ psi.   *Ans.* $s = 3.0''$

9.19. Solve Problem 9.10 using No. 11 vertical bars.   *Ans.* $s = 18.0''$

9.20. Solve Problem 9.11 for an overall column diameter of 500 mm and a core diameter of 400 mm. *Ans.* $s = 85$ mm

9.21. Solve Problem 9.20 using equation (9.2).   *Ans.* $s = 85$ mm

Chapter 10

Short Columns: Strength Design

NOTATION

The notation and definitions relating to this chapter are identical to those listed for Chapter 9. Additional definitions are provided in connection with the pertinent material.

SPECIAL CONDITIONS

The following paragraphs list the special conditions imposed by the 1983 ACI Code on the strength design of (short) columns. Some of these conditions have already been presented under "Load Factors and Understrength Factors," Chapter 5.

(1) The forces and moments due to dead load D, live load L, and wind load W, as calculated using elastic procedures for analysis, shall be increased to an ultimate value U using the most severe of the following conditions:

$$(a) \quad U = 1.4D + 1.7L \tag{10.1}$$
$$(b) \quad U = 0.75(1.4D + 1.7L + 1.7W) \tag{10.2}$$
$$(c) \quad U = 0.9D + 1.3W \tag{10.3}$$

Earthquake (E) or other phenomena which produce conditions which can be converted to an *equivalent wind force* shall be included in the same manner as wind forces. Creep, elastic deformation, shrinkage and temperature shall be considered on the same basis as dead load. Horizontal forces due to soil pressures or liquid pressures (as in storage tanks) are normally included in live load. Vertical pressures due to soil, liquids, grain, etc., are live or dead load depending on whether they increase or decrease the effect of given live loads. *Impact loads* due to vibrating machinery, elevators or moving vehicles are always considered to be live loads.

(2) Designs may not be based on steel yield strength in excess of 80,000 psi (550 MPa).

(3) For flexural members and for members subject to combined flexure and compressive axial load when the design axial load ϕP_n is less than the smaller of $0.10f'_c A_g$ and ϕP_b, the ratio of reinforcement ρ provided shall not exceed 0.75 of the ratio ρ_b that would produce balanced strain conditions for the section under flexure without axial load. For members with compression reinforcement, the portion of ρ_b equalized by compression reinforcement is not to be reduced by the 0.75 factor.

(4) For spirally reinforced columns, the design maximum axial compression load shall be

$$\phi P_{n(\text{max})} = 0.85\phi[0.85f'_c(A_g - A_{st}) + f_y A_{st}] \tag{10.4}$$

(5) For columns with ties, the design maximum axial compression load shall be

$$\phi P_{n(\text{max})} = 0.80\phi[0.85f'_c(A_g - A_{st}) + f_y A_{st}] \tag{10.5}$$

(6) The total longitudinal steel ratio $\rho_t = A_{st}/A_g$ shall not be less than 0.01 nor more than 0.08.

In equations (10.4) and (10.5) above, ϕ is taken as 0.7 for tied columns and 0.75 for spirally reinforced columns. Recent research (by Nielson at Cornell University) indicates that calculated values of $P_{n(\text{max})}$ should be multiplied by 0.85 when low-density aggregate is used in the design.

ASSUMPTIONS FOR STRENGTH DESIGN

For convenience, the following assumptions are reproduced from earlier chapters.

(1) Strength design of members for bending and axial load shall be based on the assumptions given herein, and on satisfaction of the applicable conditions of equilibrium and compatibility of strains.

(2) Strain in the concrete shall be assumed directly proportional to the distance from the neutral axis. Except in anchorage regions, strain in reinforcing bars shall be assumed equal to the strain in the concrete at the same position.

(3) The maximum strain at the extreme compression fiber at ultimate strength shall be assumed equal to 0.003.

(4) Stress in reinforcing bars below the yield strength f_y for the grade of steel used shall be taken as 29,000,000 psi (200 000 MPa) times the steel strain. For strain greater than that corresponding to the design yield strength f_y, the reinforcement stress shall be considered independent of strain and equal to the design yield strength f_y.

(5) Tensile strength of the concrete shall be neglected in flexural calculations.

(6) At ultimate strength, concrete stress is not proportional to strain. The diagram of compressive concrete stress distribution may be assumed to be a rectangle, trapezoid, parabola, sine wave or any other shape which results in predictions of ultimate strength in reasonable agreement with the results of comprehensive tests.

(7) The requirements of (6) may be considered satisfied by the equivalent rectangular concrete stress distribution which is defined as follows: At ultimate strength, a concrete stress intensity of $0.85f'_c$ shall be assumed uniformly distributed over an equivalent compression zone bounded by the edges of the cross section and a straight line located parallel to the neutral axis at a distance $a = \beta_1 c$ from the fiber of maximum compressive strain. The distance c from the fiber of maximum strain to the neutral axis is measured in the direction perpendicular to that axis. The fraction β_1 shall be taken as 0.85 for strengths f'_c up to 4000 psi (30 MPa) and shall be reduced continuously at a rate of 0.05 for each 1000 psi (0.008 for each 1.0 MPa) of strength in excess of 4000 psi (30 MPa); but β_1 shall not be taken less than 0.65.

SAFETY PROVISIONS

As stated above, the *resistance factor* ϕ is to equal 0.7 for tied columns, and 0.75 for spiral columns. This rule has the following exception: For members subjected to axial load and bending in which f_y does not exceed 60,000 psi (400 MPa) with symmetric reinforcement, and with $(h - d' - d_s)/h$ (also called γ) not less than 0.7, ϕ may be increased linearly to 0.9 as ϕP_n decreases from $0.10f'_c A_g$ to zero. For other reinforced members, ϕ may be increased linearly to 0.9 as ϕP_n decreases from $0.10f'_c A_g$ or ϕP_b, whichever is smaller, to zero.

Here, P_b is the axial load that corresponds to *balanced conditions*, which exist when the concrete strain at the compression face is 0.003 and the strain in the reinforcing bars farthest from the neutral axis is exactly f_y/E_s. The corresponding bending moment is M_b, and the related eccentricity of the axial load (e_b) is equal to M_b/P_b.

The increase in the ϕ-factor is permitted because, as the compressive axial load approaches zero, the member behaves less as a column and more as a beam, for which $\phi = 0.9$ for pure bending moment.

LIMITATIONS

(1) All members subjected to a compression load shall be designed for the eccentricity e corresponding to the maximum moment which can accompany this loading condition, but not less than $e_{min} = 0.6 + 0.03h$ $(15 + 0.03h)$, about each principal axis.

(2) The maximum load capacities for members subject to axial load as determined by the requirements of this chapter apply only to *short members* and shall be reduced for the effects of length according to the requirements of Chapter 11, when appropriate.

(3) Members subjected to small compressive loads may be designed for the maximum moment $P_u e$, in accordance with the provisions of Chapter 5 and disregarding the axial load, but the resulting section shall have a capacity P_n greater than the applied compressive load. (The *column interaction diagrams*, Figs. 10-2 through 10-13, will show that disregard of small axial load is on the conservative side.)

GENERAL CASE

For any shape section, reinforced in any manner, the ultimate load shall be computed using the equations of equilibrium and strain compatibility with the general assumptions concerning stress distribution over the section.

BIAXIAL BENDING

The 1983 ACI Code does not contain specific procedures for design of sections subject to bending about two axes. However, experimental results indicate that a reciprocal type of interaction equation, such as

$$(1/P_u) = (1/P'_x) + (1/P'_y) - (1/P_o) \qquad (10.6)$$

provides conservative results. In (10.6),

P_u = ultimate load capacity of the section with eccentricities e_x and e_y (see Fig. 10-1)

P'_x = ultimate load capacity of the section with e_x only $(e_y = 0)$

P'_y = ultimate load capacity of the section with e_y only $(e_x = 0)$

P_o = axial load capacity of the section $(e_x = e_y = 0)$

Figures 10-6 through 10-13 can be used to evaluate the dimensionless ratio $K = P/f'_c bh$, and so P'_x, P'_y, and P_o in (10.6) can be computed as $P = K(f'_c bh)$. (Compare Problem 10.7, which involves a circular section.)

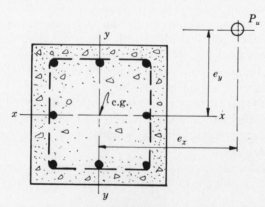

Fig. 10-1

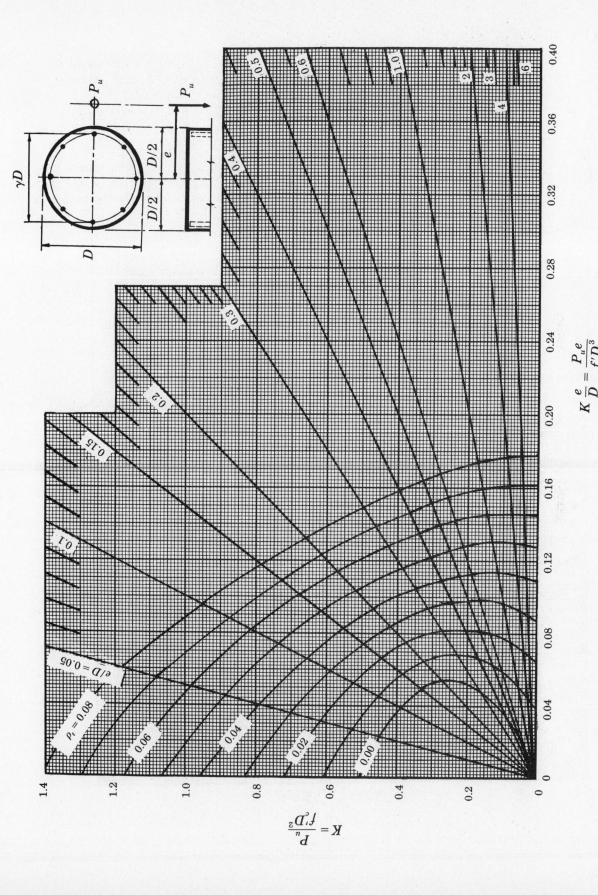

Fig. 10-2. $\gamma = 0.6;\ \phi = 0.75$

$$K\frac{e}{D} = \frac{P_u e}{f_c' D^3}$$

$$K = \frac{P_u}{f_c' D^2}$$

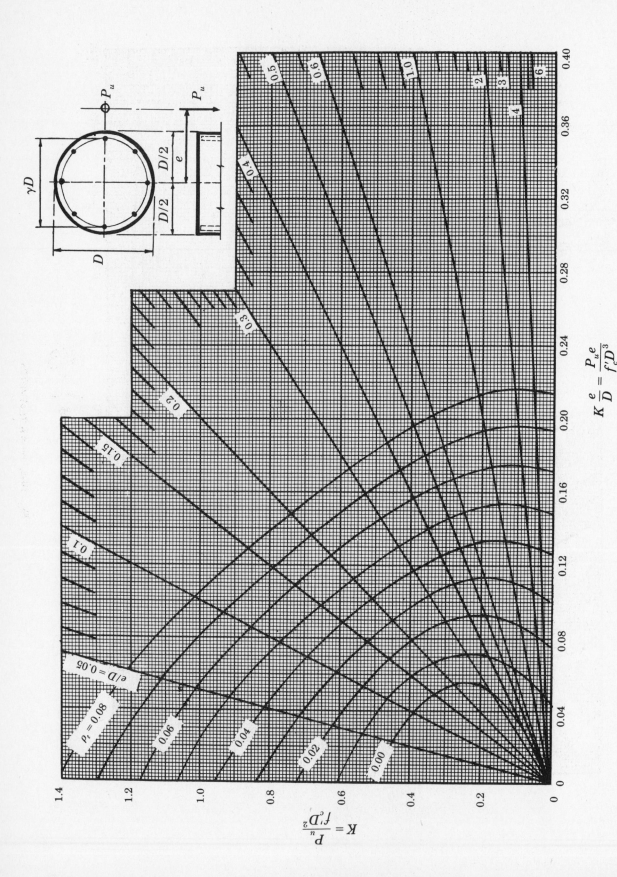

Fig. 10-3. $\gamma = 0.7$; $\phi = 0.75$

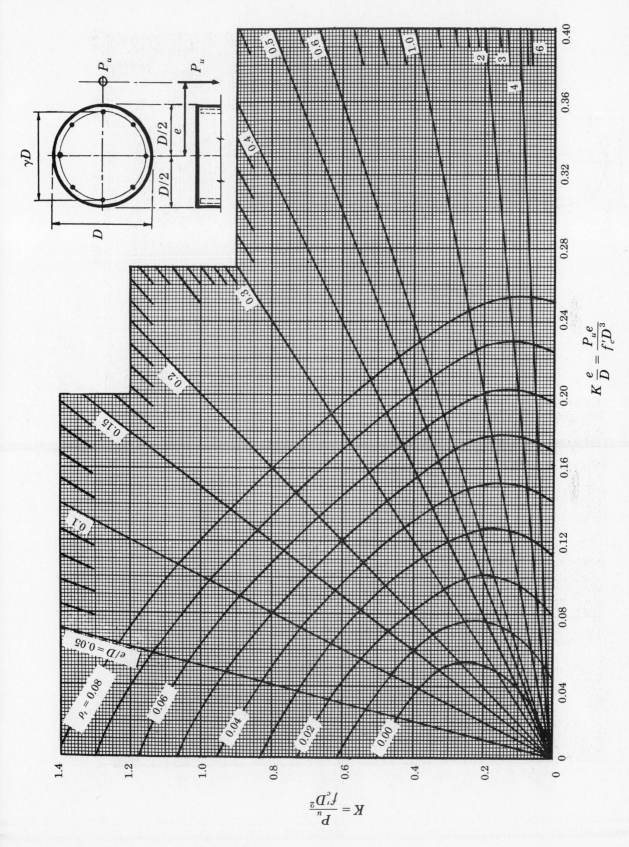

$$K\frac{e}{D} = \frac{P_u e}{f'_c D^3}$$

Fig. 10-4. $\gamma = 0.8$; $\phi = 0.75$

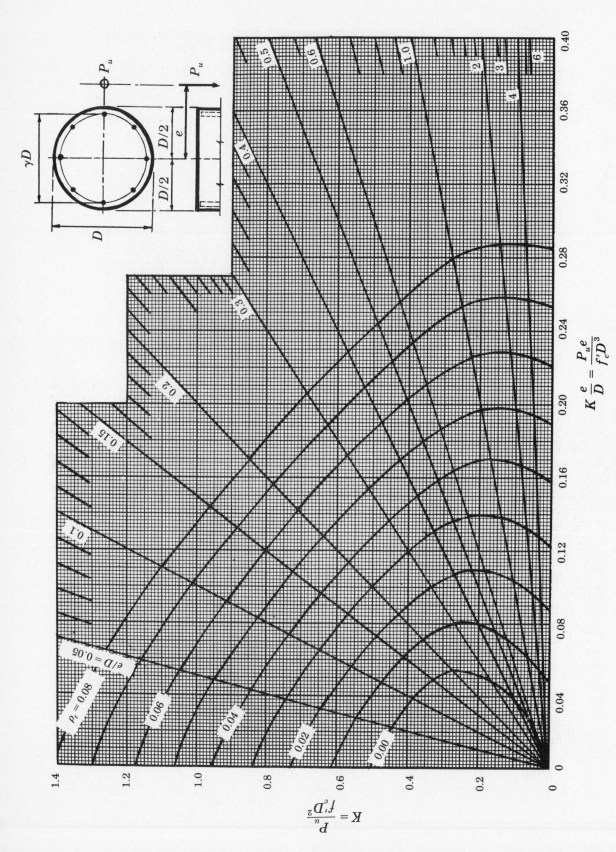

Fig. 10-5. $\gamma = 0.9$; $\phi = 0.75$

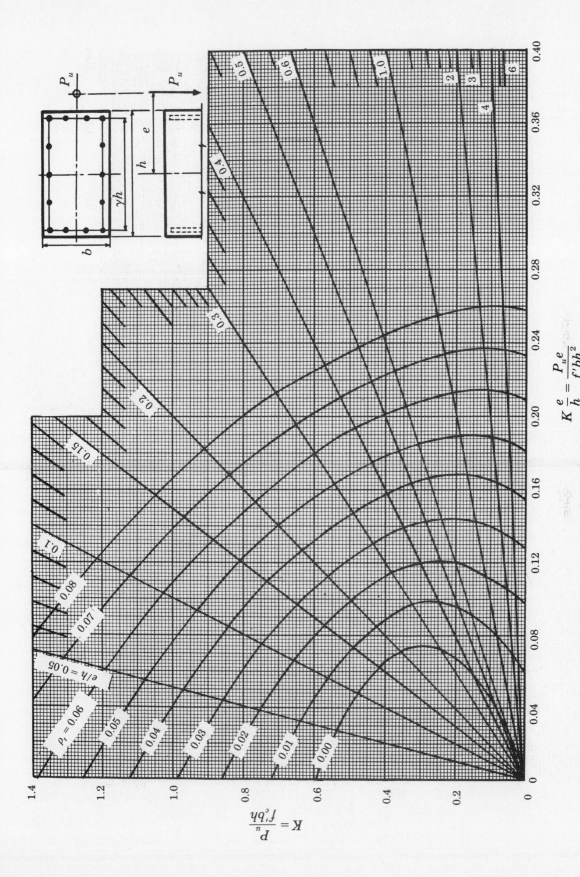

$$K \frac{e}{h} = \frac{P_u e}{f'_c b h^2}$$

Fig. 10-6. $\gamma = 0.6; \ \phi = 0.7$

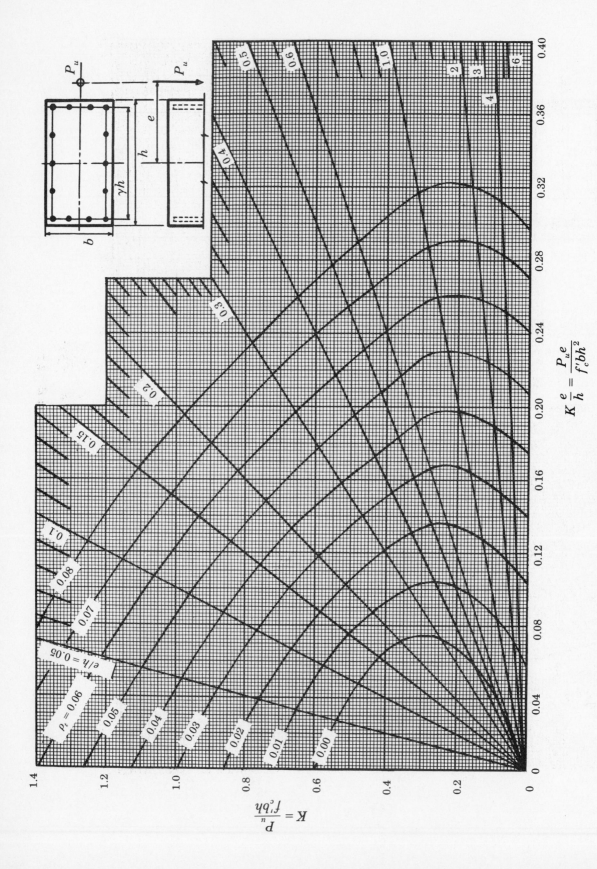

$$K \frac{e}{h} = \frac{P_u e}{f'_c b h^2}$$

Fig. 10-7. $\gamma = 0.7$; $\phi = 0.7$

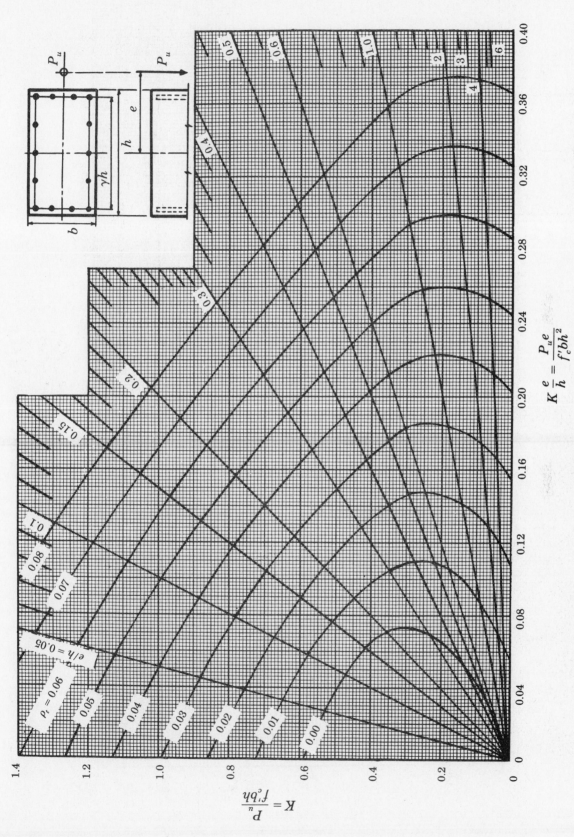

$$K\frac{e}{h} = \frac{P_u e}{f'_c bh^2}$$

Fig. 10-8. $\gamma = 0.8$; $\phi = 0.7$

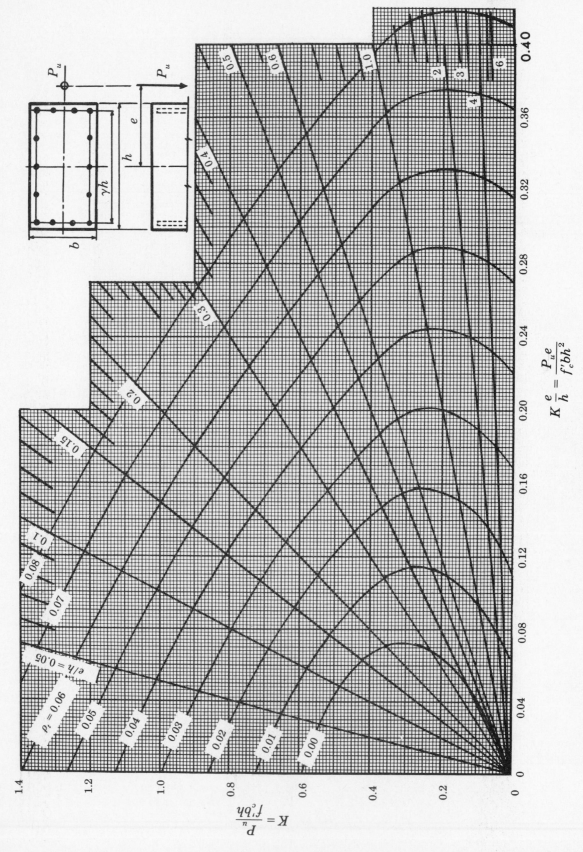

$$K\frac{e}{h} = \frac{P_u e}{f'_c b h^2}$$

Fig. 10-9. $\gamma = 0.9;\ \phi = 0.7$

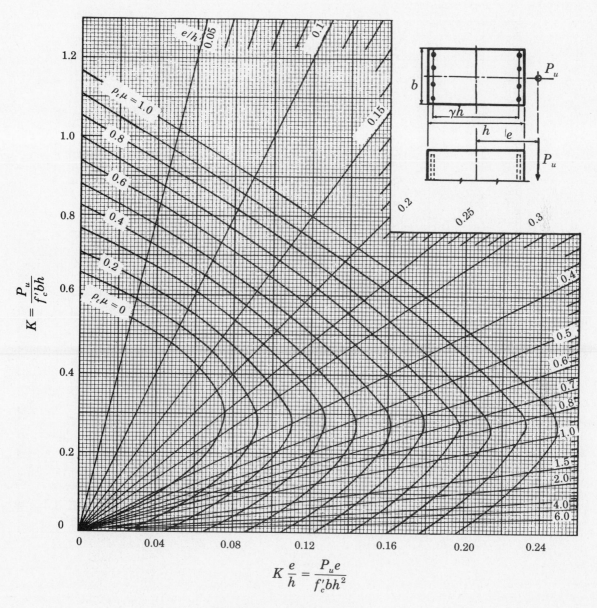

$$K \frac{e}{h} = \frac{P_u e}{f'_c b h^2}$$

Fig. 10-10. $\gamma = 0.6$; $\phi = 0.7$

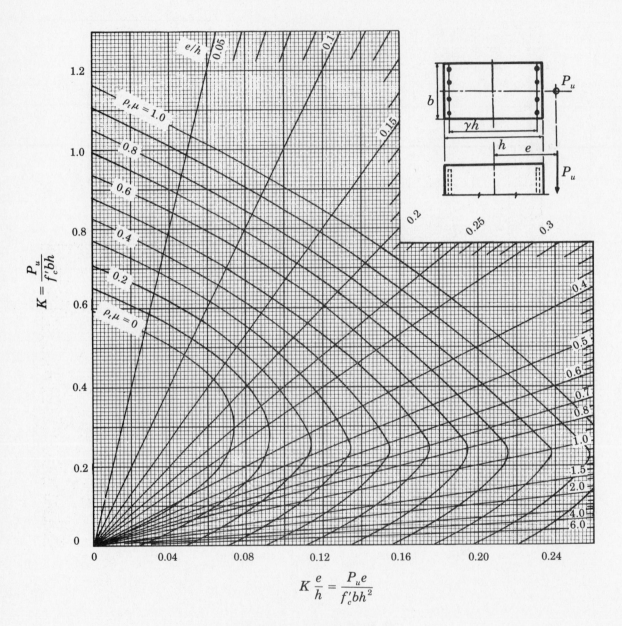

Fig. 10-11. $\gamma = 0.7$; $\phi = 0.7$

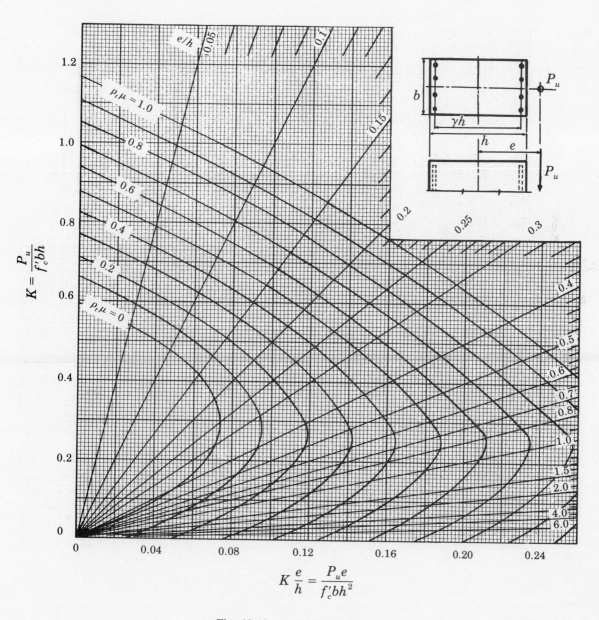

$$K \frac{e}{h} = \frac{P_u e}{f_c' b h^2}$$

Fig. 10-12. $\gamma = 0.8$; $\phi = 0.7$

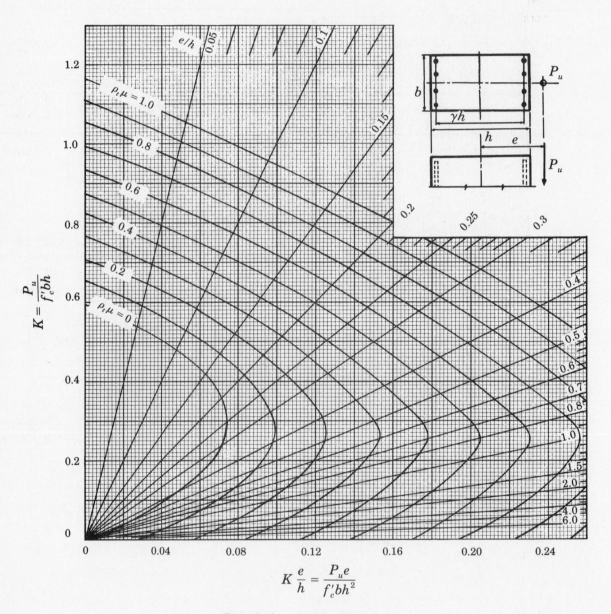

Fig. 10-13. $\gamma = 0.9$; $\phi = 0.7$

COLUMN DESIGN CHARTS (INTERACTION DIAGRAMS)

The basic problem of column design—to determine ρ_t and A_{st} corresponding to given loading conditions and satisfying all provisions of the 1983 ACI Code—may be solved with a high degree of accuracy by use of computer-generated interaction diagrams developed at the University of Texas at Arlington. Twelve of these diagrams are here reproduced as Figs. 10-2 through 10-13; their use is illustrated in Problem 10.6.

Several observations concerning the diagrams are appropriate here:

(1) All twelve interaction diagrams apply to the case $f'_c \le 4000$ psi (27.58 MPa), $f_y = 60,000$ psi (413.7 MPa) (soft conversion to SI units).

(2) The value of the *understrength factor* ϕ given below an interaction diagram is that which is included in K via $P_u = \phi P_n$. Revised *resistance factors* (call them λ for the moment) are now being defined and will eventually be introduced into all Building Codes. The interaction diagram will still be valid provided P_u is replaced by $(\lambda/\phi)P_u$.

(3) Figures 10-2 through 10-9 give ρ_t directly; but Figs. 10-10 through 10-13 give the combination

$$\rho_t \mu \equiv \rho_t \frac{f_y}{0.85 f'_c}$$

The reasons for this minor inconvenience are historical.

(4) All variables in the interaction diagrams are dimensionless, so that the column design can be carried out in any consistent set of units. It is recommended that a diagram be entered with values of K and Ke/D (or Ke/h), rather than with values of K and e/D (or e/h). The latter procedure—unfortunately used in many textbooks—will normally require an angular interpolation that becomes very inaccurate as the e/D (e/h) rays converge.

Solved Problems

10.1. An analysis was made using elastic procedures. The axial loads and moments obtained therefrom are listed below. Determine the combinations of P_u and M_u required for use in strength design.

Dead Load:	$P = 150$ kips	$M = 75$ ft-kips
Live Load:	$P = 200$ kips	$M = 85$ ft-kips
Wind Load:	$P = 80$ kips	$M = 130$ ft-kips

The load factors must be applied to the forces and moments calculated using the elastic theory. Thus:

(a) For $U = 1.4D + 1.7L$,

$$\begin{cases} P_u = (1.4)(150) + (1.7)(200) = 550.0 \text{ kips} \\ M_u = (1.4)(75) + (1.7)(85) = 249.5 \text{ ft-kips} \end{cases}$$

(b) For $U = 0.75(1.4D + 1.7L + 1.7W)$

$$\begin{cases} P_u = 0.75[1.4(150) + 1.7(200) + 1.7(80)] = 514.5 \text{ kips} \\ M_u = 0.75[1.4(75) + 1.7(85) + 1.7(130)] = 352.9 \text{ ft-kips} \end{cases}$$

(c) For $U = 0.9D + 1.3W$,

$$\begin{cases} P_u = (0.9)(150) + (1.3)(80) = 239.0 \text{ kips} \\ M_u = (0.9)(75) + (1.3)(130) = 236.5 \text{ ft-kips} \end{cases}$$

All three of the combinations must be investigated in order to provide a satisfactory design.

10.2. The column cross section shown in Fig. 10-14 contains steel having a yield stress of 50,000 psi. Determine the actual stresses in the reinforcement, considering $0.85f'_c$ to be deducted from the steel stress in the compression zone. Use $f'_c = 3000$ psi.

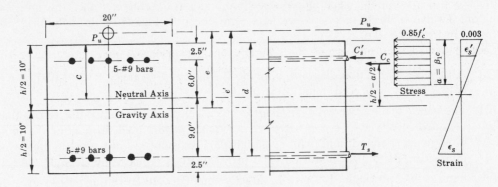

Fig. 10-14

Using similar triangles compute the unit stress in the compression steel based on strain compatibility and the assumption that the maximum strain in the concrete is 0.003 in/in. Thus

$$\epsilon'_s = \epsilon_u(c - d')/c = (0.003)(6.0)/8.5 = 0.00212 \text{ in/in}$$

and

$$f'_s = \epsilon'_s E_s = (0.00212)(29 \times 10^6) = 61,400 \text{ psi}$$

Note that the calculated stress exceeds f_y, so limit the stress to f_y. The net effective compressive stress in the steel will be

$$(f'_s - 0.85f'_c) = 50,000 - (0.85)(3000) = 47,450 \text{ psi}$$

Similarly, the strain in the steel on the tension side will be

$$\epsilon_s = \epsilon_u(d - c)/c = (0.003)(9.0)/8.5 = 0.00318 \text{ in/in}$$

Therefore $f_s = \epsilon_s E_s = (0.00318)(29 \times 10^6) = 92,000$ psi.

Note that the calculated stress exceeds f_y, so limit f_s to f_y. Thus, $f_s = 50,000$ psi.

10.3. Determine the ultimate load P_u that will satisfy statics in Fig. 10-14. Use the results of Problem 10.2. ($P_u = \phi P_n$ where $\phi = 0.7$ for tied columns.)

Since statics requires that $\Sigma F_x = 0$, then $P'_u = C_c + C'_s - T_s$ where T_s is in the negative direction. The forces are calculated as the stresses (in kips/in^2) times the areas; thus

$$C_c = 0.85f'_c \beta_1 cb = 0.85(3)(0.85)(8.5)(20) = 368.0 \text{ kips}$$
$$C'_s = A'_s(f'_s - 0.85f'_c) = (5.0)(47.45) = 237.3 \text{ kips}$$
$$T_s = A_s f_s = (5.0)(50.0) = 250.0 \text{ kips}$$

Hence

$$P_n = 368.0 + 237.3 - 250.0 = 355.3 \text{ kips} \quad \text{and} \quad P_u = \phi P_n = 0.7(355.3) = 248.7 \text{ kips}$$

10.4. Determine the eccentricity of P_n required to establish the conditions of equilibrium shown in Fig. 10-14.

The moment of the forces may be summed about the gravity axis or the neutral axis or any axis parallel thereto. Using the neutral axis, and transferring to the centroid,

$$M_c = C_c(h/2 - a/2) = 368(8.5 - 3.61) = 1800 \text{ in-kips}$$
$$M'_s = C'_s(c - d') = (237.3)(6.0) = 1424.0 \text{ in-kips}$$
$$M_s = T_s(d - d' - c) = (250.0)(9.0) = 2250.0 \text{ in-kips}$$

The moment about the centroid is

$$\sum M_{\text{centroid}} = 6007.0 \text{ in-kips}$$

Then $\phi M_n = M_u = (6007)(0.7) = 4204.9$ in-kips.
About the centroid, $e = M_u/P_u = 16.9$ in

10.5. Use statics and strain compatibility to determine the ultimate load P_u for the spiral column shown in Fig. 10-15. Pertinent data are as follows: $f'_c = 3000$ psi, $f_y = 60,000$ psi, $b = h = 20''$. Total steel consists of 12 No. 9 bars, or $A_{st} = 12.0$ in^2; thus $\rho_t = 0.03$. Also, $\gamma = 0.8$, $a = \beta_1 c = 8.925''$ and the neutral axis lies $5.537''$ below the top of the section.

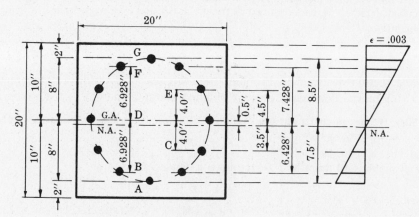

Fig. 10-15

Using strain compatibility and statics, the steel forces may be computed as in Table 10.1.

Table 10.1

Row	A_s, in^2	$\epsilon_s \times 10^2$, in/in	f_s, kips/in^2	$f_s - 0.85f'_c$, kips/in^2	Force, kips	Moment arm, in	Moment, in-kips
G	1.0	0.243	60.00	57.45	57.45	8.000	460.0
F	2.0	0.212	60.0	57.45	114.90	6.925	796.0
E	2.0	0.129	37.40	34.85	69.70	4.000	279.0
D	2.0	0.014	4.15	—	8.30	0.000	0.0
C	2.0	-0.100	-29.00	—	-58.00	-4.000	232.0
B	2.0	-0.183	-53.00	—	-106.00	-6.925	734.0
A	1.0	-0.214	-60.00	—	-60.00	-8.000	480.0

$$\sum F_s = 26.35 \text{ kips} \qquad \sum M_s = 2981.0 \text{ in-kips}$$

The compressive force in the concrete and its moment about the gravity axis will be

$$C_c = 0.85 \times 3.0 \times 8.925 \times 20.0 = 455 \text{ kips} \qquad \text{and} \qquad M_c = 455 \times 5.537 = 2520 \text{ in-kips}$$

Hence
$$P_u = \phi P_n = \phi(C_c + F_s) = 0.75(455.00 + 26.35) = 360 \text{ kips}$$
$$M_u = \phi M_n = \phi(M_c + M_s) = 0.75(2520 + 2981) = 4125 \text{ in-kips}$$

Thus
$$e = M_u/P_u = 11.55 \text{ in} \qquad \text{and} \qquad e/h = 11.55/20 = 0.5775$$

10.6. Figure 10-4 is a typical interaction design diagram for use in strength design of columns. Use the chart to design a column to withstand an ultimate moment of 2880 in-kips and an axial load of 720 kips. Take $f'_c = 3000$ psi and $f_y = 60,000$ psi.

Assume that a 20″ diameter column will be used. Calculate

$$e = M_u/P_u = 2880/720 = 4.0'', \quad e/D = 4.0/20.0 = 0.2, \quad K = P_u/f'_c D^2 = 720/[(3.0)(20)^2] = 0.6$$

Enter the chart with $K = 0.6$ and $Ke/D = (0.6)(0.2) = 0.12$, and at the point $(0.12, 0.6)$ read $\rho_t = 0.05$. Then

$$A_{st} = \rho_t A_g = 0.05(\pi)(20)^2/4 = 15.71 \text{ in}^2$$

Any size bars may be used to provide the required steel area, but the clearances between bars must be maintained in accord with the detailed requirements of the 1983 ACI Code. If the required amount of steel cannot be provided because of clearance requirements, it will be necessary to increase the size of the column and redesign the steel bar sizes.

It should be noted that the minimum number of vertical bars shall be 4 for bars confined by circular or rectangular ties, 3 for bars confined by triangular ties and 6 for bars confined by spirals or circular ties in columns.

10.7. Use the section designed in Problem 10.6 to check the validity of the reciprocal equation (10.6) for biaxial bending. Consider bending about a line $45°$ to the x axis, with $e = 5.656''$.

Since the design charts already contain ϕ, the primes are dropped and the equation becomes

$$1/P_u = 1/P_x + 1/P_y - 1/P_o$$

We may multiply all terms by $f'_c D^2$ to obtain a dimensionless form

$$1/K = 1/K_x + 1/K_y - 1/K_o$$

Since $e = 5.656''$ along a line $45°$ to the x axis, $e_x = e_y = 5.656/1.414 = 4.0''$ as in Problem 10.6. The section is circular, so $K_x = K_y = 0.6$ as determined in Problem 10.6.

The term K_o refers to pure axial load and may be obtained as 1.065 at the intersection of the vertical ($Ke/D = 0$) with $\rho_t = 0.05$. Thus

$$1/K = 1/0.6 + 1/0.6 - 1/1.065 \quad \text{or} \quad K = 0.418$$

Now, using $e/D = 5.656/20 = 0.2828$ and $\rho_t = 0.05$, return to the chart to find $K = 0.49$. This value is the correct solution, and indicates that the reciprocal formula provides a solution that is incorrect by 14.3% on the conservative side.

10.8. For a circular column with closely spaced spiral confinement (lateral reinforcement) determine the maximum permissible value of $P_u = \phi P_n$, if $f'_c = 4000$ psi, $f_y = 60,000$ psi and the vertical steel ratio is $\rho_t = 0.02$.

Equation (10.4) applies, with $\phi = 0.75$:

$$\phi P_{n(max)} = 0.85\phi[0.85f'_c(A_g - A_{st}) + f_y A_{st}]$$

But $A_g = \pi D^2/4 = (0.7854)(20)^2 = 314 \text{ in}^2$ and $A_{st} = \rho_t A_g = 0.02(314) = 6.28 \text{ in}^2$; so

$$\phi P_{n(max)} = 0.85(0.75)[0.85(4.0)(314.0 - 6.28) + (60.0)(6.28)] = 907.2 \text{ kips}$$

10.9. For a 20″ square column, calculate the maximum axial load permitted by the 1983 ACI Code. Lateral confinement reinforcement is provided by ties; $A_{st} = 10.0 \text{ in}^2$, $f'_c = 3000$ psi, $f_y = 50,000$ psi, and $\phi = 0.7$.

$A_g = (20)(20) = 400.0 \text{ in}^2$ and equation (10.5) applies. Therefore,

$$\phi P_{n(max)} = (0.8)(0.7)[0.85(3.0)(400.0 - 10.0) + (50.0)(10.0)] = 836.9 \text{ kips}$$

10.10. Refer to Fig. 10-4. For a 20″ diameter column with service-level axial compressive forces $P_D = 200$ kips and $P_L = 100$ kips, accompanied by service-level bending moments $M_D = 75$ ft-kips and $M_L = 100$ ft-kips, determine the required area of vertical steel reinforcement.

The factored design values are:

$$P_u = \phi P_n = 1.4(200) + 1.7(100) = 450 \text{ kips}$$
$$M_u = \phi M_n = 1.4(75) + 1.7(100) = 275 \text{ ft-kips} = P_u e$$

The values required to enter the chart are:

$$K = \frac{P_u}{f'_c D^2} = \frac{450}{(4.0)(400)} = 0.281$$

$$Ke/D = \frac{P_u e}{f'_c D^3} = \frac{275}{(4.0)(8000)} = 0.103$$

By intersection of these values on the chart, and by interpolation between curves for ρ_t, find $\rho_t = 0.022$. Thus,

$$A_{st} = \rho_t A_g = 0.022(0.7854)(400) = 6.91 \text{ in}^2$$

10.11. Determine the reinforcing steel area required for a rectangular tied column with $b = 12$ in, $h = 20$ in., $f'_c = 4000$ psi, $f_y = 60{,}000$ psi and $\gamma = 0.8$. $P_u = 200$ kips and $P_u e = M_u = 300$ ft-kips; steel is to be placed along end faces only.

$$K = \frac{P_u}{f'_c bh} = \frac{200}{(4.0)(12)(20)} = 0.2083$$

$$\frac{K_e}{h} = \frac{P_u e}{f'_c bh^2} = \frac{300 \times 12}{(4.0)(12)(20)^2} = 0.1875$$

Figure 10-12 applies: at $(0.1875, 0.2083)$ find $\rho_t \mu = 0.52$. But

$$\mu = \frac{f_y}{0.85 f'_c} = \frac{60.0}{(0.85)(4.0)} = 17.647$$

so $\rho_t = 0.52/17.647 = 0.0295$ and $A_{st} = \rho_t A_g = (0.0295)(12)(20) = 7.08 \text{ in}^2$.

10.12. Determine the steel area required for a 510 mm × 510 mm square tied column having the steel equally distributed along all four faces. Use $f'_c = 27.58$ MPa and $f_y = 413.7$ MPa. $P_u = 1245$ kN and $M_u = 434$ kN·m $= P_u e$.

Figure 10-7 applies. To calculate the entering values, choose the compatible units newtons and meters:

$$K = \frac{P_u}{f'_c bh} = \frac{1245 \times 10^3}{(27.58 \times 10^6)(510 \times 10^{-3})^2} = 0.174$$

$$\frac{Ke}{h} = \frac{M_u}{f'_c bh^2} = \frac{434 \times 10^3}{(27.58 \times 10^6)(510 \times 10^{-3})^3} = 0.119$$

From the chart, by interpolation, $\rho_t = 0.017$ and

$$A_{st} = \rho_t A_g = (0.017)(510)^2 = 4422 \text{ mm}^2$$

10.13. A short rectangular tied column, 12 in wide and 20 in deep, is to withstand service-level axial loads $P_D = 90$ kips, $P_L = 130$ kips and service-level bending moments $M_D = 70$ ft-kips, $M_L = 70$ ft-kips. The reinforcement is distributed equally along the two 12 in end faces, so that the center-to-center distance between the two layers of bars is 16 in. If $f'_c = 3000$ psi and $f_y = 60{,}000$ psi, find how many No. 8 bars are required.

Since $\gamma h = 16$ in, $\gamma = 16/20 = 0.8$ and Fig. 10-12 is applicable. We have:

$$P_u = (1.4)(90) + (1.7)(130) = 347 \text{ kips}$$
$$M_u = P_u e = (1.4)(70) + (1.7)(70) = 217 \text{ ft-kips}$$

$$K = \frac{P_u}{f'_c bh} = \frac{347}{(3.0)(12)(20)} = 0.482$$

$$\frac{Ke}{h} = \frac{M_u}{f'_c bh^2} = \frac{217 \times 12}{(3.0)(12)(20)^2} = 0.181$$

and Fig. 10-12 gives $\rho_t \mu = 0.69$. Since

$$\mu = \frac{f_y}{0.85 f'_c} = \frac{60.0}{(0.85)(3.0)} = 23.53$$

$\rho_t = 0.69/23.53 = 0.029$ and $A_{st} = \rho_t bh = (0.029)(12)(20) = 6.96 \text{ in}^2$. From Chapter 1 or Appendix A, eight No. 8 bars—four along each end face—will provide 7.07 in^2.

Supplementary Problems

10.14. Use the general method of statics and strain compatibility to determine P_u and M_u for the section shown in Fig. 10-16. The total steel area is 8.5 in^2 and the distance of the neutral axis from the compression face is $c = 14.0''$.   *Ans.* $P_u = 537$ kips, $M_u = 3085$ in-kips

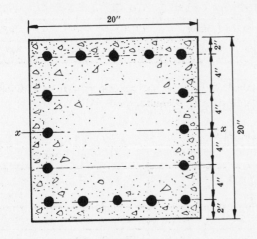

Fig. 10-16

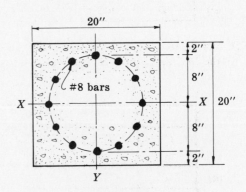

Fig. 10-17

10.15. Solve Problem 10.14 for $c = 8.0''$. Use $A_{st} = 6.8 \text{ in}^2$.   *Ans.* $P_u = 202$ kips, $M_u = 3189$ in-kips

10.16. Solve Problem 10.14 for $c = 8.0''$ and $A_{st} = 8.5 \text{ in}^2$.   *Ans.* $P_u = 191$ kips, $M_u = 3583$ in-kips

10.17. Solve Problem 10.14 for $A_{st} = 5.1 \text{ in}^2$ and $c = 16.0''$.   *Ans.* $P_u = 578$ kips, $M_u = 2274$ in-kips

10.18. Using Fig. 10-17 with No. 9 bars, $c = 16.0''$, $f'_c = 3000$ psi and $f_y = 60,000$ psi, calculate the ultimate load and ultimate moment using statics and strain compatibility.
Ans. $P_u = 771.0$ kips, $M_u = 2940.0$ in-kips

10.19. For the spirally reinforced columns described in Table 10.2, calculate the maximum axial loads allowed by the 1983 ACI Code. Be sure to use a consistent set of units in equation (10.4).

Table 10.2

	GIVEN				SOLUTION		
	f'_c, MPa	f_y, MPa	Diameter D, mm	ρ_t	A_g, mm^2	A_{st}, mm^2	$\phi P_{n((max))}$, kN
(a)	20	400	500	0.025	196 350	4909	3326
(b)	30	300	320	0.035	80 425	2815	1800
(c)	20	400	400	0.030	125 664	3770	2282
(d)	20	300	350	0.015	96 211	1443	1303
(e)	30	400	380	0.020	113 412	2268	3551

10.20. For the rectangular tied columns described in Table 10.3, calculate the maximum axial loads allowed by the 1983 ACI Code. Be sure to use a consistent set of units in equation (*10.5*).

Table 10.3

	GIVEN					SOLUTION		
	f'_c, MPa	f_y, MPa	Width b, mm	Depth h, mm	ρ_t	A_g, mm^2	A_{st}, mm^2	$\phi P_{n(max)}$, kN
(a)	20	400	300	500	0.025	150 000	3650	2232
(b)	30	300	320	320	0.035	102 400	3584	2013
(c)	20	400	300	400	0.030	120 000	3600	1915
(d)	20	300	350	350	0.015	122 500	1838	1457
(e)	30	400	380	380	0.020	144 400	2888	2668

10.21. For the rectangular tied columns described in Table 10.4, calculate the maximum axial loads allowed by the 1983 ACI Code. Be sure to use a consistent set of units in equation (*10.5*).

Table 10.4

	GIVEN					SOLUTION		
	f'_c, psi	f_y, psi	Width b, in.	Depth h, in.	ρ_t	A_g, in^2	A_{st}, in^2	$\phi P_{n(max)}$, kips
(a)	3000	60,000	12	20	0.025	240	6.00	536
(b)	4000	50,000	16	16	0.018	256	4.61	608
(c)	5000	60,000	18	18	0.025	324	8.10	1024
(d)	4000	60,000	15	15	0.030	225	6.75	642
(e)	3000	50,000	16	20	0.022	320	7.04	644

Chapter 11

Long Columns

NOTATION

Notation pertinent to this chapter is listed in Chapter 9. Additional definitions are provided immediately below or as required with the related material.

C_m = a factor relating actual moment diagram to uniform moment diagram

D = dead-load symbol

E = earthquake force or moment symbol

I_g = gross moment of inertia of compression member cross section about the centroidal axis

I_{se} = moment of inertia of reinforcing bars about the centroidal axis

k_b = effective length factor of a compression member under gravity loads

k_s = effective length factor of a compression member under lateral loads

L = live-load symbol

L_u = unsupported length of a compression member

M_c = design bending moment

M_{1b} = smallest factored bending moment at end of a compression member under gravity loads

M_{2b} = largest factored bending moment at end of a compression member under gravity loads

M_{1s} = smallest factored bending moment at end of a compression member under lateral loads

M_{2s} = largest factored bending moment at end of a compression member under lateral loads

M_u = ultimate bending moment

P_c = critical buckling axial load on a compression member

P_u = ultimate axial load on an axially loaded member

U = ultimate axial load or ultimate bending moment

W = wind symbol

β_d = creep factor = (sustained factor load moment)/(total factored load moment)

δ_b = moment magnifier for gravity loads

δ_s = moment magnifier for lateral loads

ϕ = understrength factor (resistance factor)

FRAME ANALYSIS

Prior to the invention of the electronic computer, frames were commonly analyzed by the *cantilever method* or by the *portal method*. Both methods assumed a point of contraflexure *at the center of* each member; both were dangerously inaccurate (cf. Chapter 3), and were "saved" only by the circumstance that the old straight-line theory prescribed very large safety factors to be used along with them.

The vast majority of structural reinforced concrete frames in modern practice are analyzed using electronic computers. In North America, such computer programs as SAP (Structural Analysis Program) and STRUDL (Structural Design Language) are used to analyze structures. While the computer programs are rather exotic, they do not take account of the secondary moments produced by the axial load P acting at an eccentricity Δ (due to relative displacement between the upper and lower ends of the columns). This so-called *P-Delta effect* is illustrated in Fig. 11-1. These secondary moments are added to the primary moments that are determined from the first-order computer analysis.

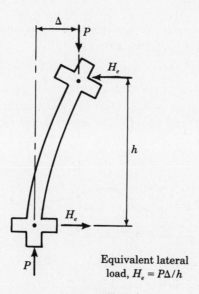

Equivalent lateral
load, $H_e = P\Delta/h$

Fig. 11-1

The 1983 ACI Code requires that a P-Delta analysis must take into account the influence of axial loads and variable moment of inertia on member stiffnesses and fixed-end moments. Compressive axial loads *decrease* stiffness and *increase* fixed-end moments; tension axial loads *increase* stiffness and *decrease* fixed-end moments.

Structural engineers currently use two methods for including the secondary effects in a structural analysis.

Iterative Method

A first run of the computer program is executed to obtain the axial loads on the columns and the horizontal deflections of the joints. Then the column axial loads are multiplied by the relative displacements between the tops and bottoms of the columns to obtain the P-Δ moments. Those moments are divided by the column heights to obtain *equivalent lateral loads*. These fictitious lateral (horizontal) loads are then applied to the structure along with the actual loads, and the computer program is executed again. This process of *iteration* is continued until the difference in horizontal displacements of the joints for computer run i do not differ appreciably from those obtained in run $i - 1$. The secondary moments from run i are incorporated in the final solution.

Drift Analysis

The initial assumption is made that the horizontal deflection at the roof level, or *drift*, lies between $h/500$ and $h/400$, where h is the total building height. This, plus the assumption that the building is cantilevered from the foundation, allows the horizontal deflection at any given floor level to be inferred from a parabolic law. The relative deflections (Δ) are then obtained by subtraction, and the axial loads (P) on the columns are hand- or computer-calculated. The corresponding equivalent lateral loads, computed as in the iterative method, are now applied to the structure along with the actual loads, and a solution is obtained using the computer. The resulting horizontal deflections of the joints are then compared to the assumed deflections, and if they do not exceed the assumed deflections, the solution is considered to be satisfactory, conservative, and inclusive of the secondary moments.

MOMENT-MAGNIFIER METHOD

When the secondary P-Δ moments are *not* included in the analysis, a conservative approximate method is permitted by the 1983 ACI Code, provided

$$\frac{kL_u}{r} \leq 100 \qquad (11.1)$$

Here, the dimension L_u is defined as in Fig. 11-2; $r = \sqrt{I/A}$ is the *radius of gyration* of the column cross section (see Appendix A-7); and the *effective length factor k* depends on the bracing conditions (to be treated below) of the column joints. When $kL_u/r > 100$, a second-order analysis is required.

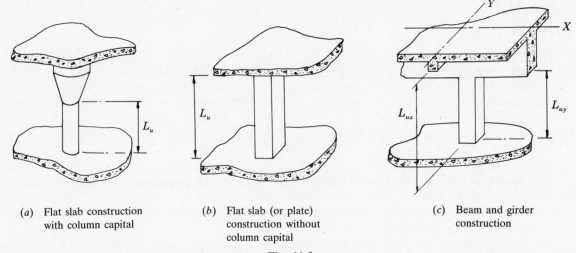

(a) Flat slab construction   (b) Flat slab (or plate)   (c) Beam and girder
     with column capital   construction without   construction
                                          column capital

Fig. 11-2

Before giving the details of this *moment-magnifier method*, we point out that it is ultraconservative, and therefore extremely uneconomical, when applied to very-high-rise structures. Such buildings require the secondary-moment approach.

Column Stiffness

The *stiffness EI* of a column, in kips-in^2 (N·mm^2), is computed as

$$EI = \frac{E_c I_g / 2.5}{1 + \beta_d} \qquad (11.2)$$

or more accurately as

$$EI = \frac{E_c I_g / 5 + E_s I_{se}}{1 + \beta_d} \qquad (11.3)$$

In (11.2) and (11.3), E_c and E_s have the usual values, in kips/in^2 (MPa);

I_g = gross moment of inertia of the concrete cross section (reinforcement excluded) about the centroidal axis perpendicular to the direction of bending, in^4 (mm^4);

I_{se} = moment of inertia of the area of the reinforcing bars about the centroidal axis of the column perpendicular to the direction of bending, in^4 (mm^4);

β_d = *creep factor* = (sustained factored load moment)/(total factored load moment)

The usual formula for β_d is:

$$\beta_d = \frac{1.4M_D}{1.4M_D + 1.7M_L} \qquad (11.4)$$

The remarks under "Long-Term Deflections," Chapter 8, apply to the numerator of (11.4).

Critical Buckling Load

The stiffness EI (see above) and the *effective length* kL_u (see below) of a column codetermine, via

$$P_c = \pi^2 \frac{EI}{(kL_u)^2} \qquad (11.5)$$

the *initial load*, in kips (N), at which a *pinned-end column* equivalent to the given column may be expected to *buckle*. Failure by buckling would precede failure due to stresses in the concrete and steel set up by the factored ultimate load, $P_u = 1.4P_D + 1.7P_L$, and the factored ultimate bending moment, $M_u = 1.4M_D + 1.7M_L$.

Effective Length Factor

Figure 11-3 indicates the relationship between the unsupported length L_u and the effective length kL_u of a pinned-end column (which can undergo different horizontal displacements at its upper and lower ends). Due to the lateral loads or unsymmetrical gravity loads or lack of symmetry of the structure, joint $(i-1)$ deflects horizontally to point $(i-1)'$ and joint (i) deflects horizontally to point $(i)'$. If a curve is drawn through the deflected points and is continued with the same curvature, it will intersect a vertical line drawn through point $(i-1)'$ (at point "A" in Fig. 11-3), thereby defining the *equivalent pinned-end column*. This results in an effective length kL_u. The magnitude of the k-factor depends on whether the column is part of a *braced frame* or an *unbraced frame*. In general, columns may be considered to be *braced* (*against sidesway*) if *shear walls* are provided such that, at any given level,

$$(E_c I_g / L_u)_{\text{walls}} \geqq 6 \sum_{\text{columns}} (E_c I_g / L_u)$$

In braced frames, the shear walls may be assumed to take *all of the lateral load*, and the columns themselves *none*. In all other cases, with one exception, columns are considered to be *braced* under *gravity loads* and *unbraced* under *lateral loads*. The exceptional case occurs when the computer solution for the primary moments gives a value of Δ (see Fig. 11-3) in excess of $L_u/1500$; the columns

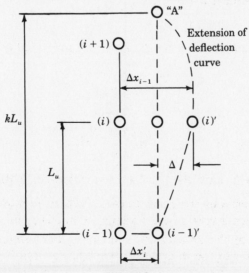

Fig. 11-3

are then considered to be subjected to *appreciable sidesway* and are treated as *unbraced* under both gravity and lateral loads.

The most accurate method for computing k is by computer solution of the trigonometric equation

$$\frac{\psi_A \psi_B}{4} (\pi/k_b)^2 + \frac{\psi_A + \psi_B}{2} \left[1 - \frac{\pi/k_b}{\tan(\pi/k_b)}\right] + \frac{2\tan(\pi/2k_b)}{\pi/k_b} = 1 \qquad (11.6)$$

for braced frames, or

$$\frac{\psi_A \psi_B (\pi/k_s)^2 - 36}{6(\psi_A + \psi_B)} = \frac{\pi/k_s}{\tan(\pi/k_s)} \qquad (11.7)$$

for unbraced frames. In these equations the parameter ψ is defined by

$$\psi = \frac{\text{sum of } EI/L \text{ for columns (at a joint)}}{\text{sum of } EI/L \text{ for beams (at a joint)}} \qquad (11.8)$$

and ψ_A and ψ_B are, respectively, the values of ψ at the upper and lower ends of the column. Figures 11-4 and 11-5 are nomographic solutions of (11.6) and (11.7): a straight line connecting the values of ψ_A and ψ_B intersects the middle axis in the value of k.

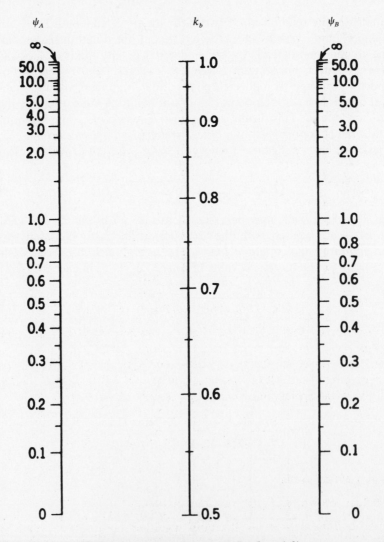

Fig. 11-4.   Braced Frames $(0.5 < k_b < 1.0)$

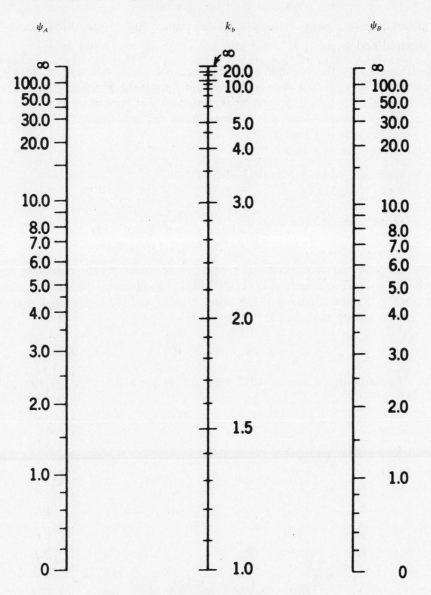

Fig. 11-5. Unbraced Frames ($k_s > 1.0$)

It is the intent of the 1983 ACI Code that the column stiffnesses EI required in (11.8) be computed (in rough approximation) from (11.2) or (more refined) from (11.3), with β_d set equal to zero in either equation. Moreover, in the calculation of the beam stiffnesses, the transformed section (Chapter 4) is to be employed. All this may be simplified if $kL_u/r \leqq 60$, *a condition satisfied by the great majority of practical frames.* For then (11.8) may be replaced by

$$\psi = \frac{\text{sum of } EI_g/L \text{ for columns (at a joint)}}{(\frac{1}{2}) \text{ sum of } EI_g/L \text{ for beams (at a joint)}} \qquad (11.9)$$

Long versus Short Column Design

With the k-values known, and with the four quantities

M_{1b} = factored end moment of smaller absolute value due to gravity loads (dead + live)

M_{2b} = factored end moment of larger absolute value due to gravity loads

M_{1s} = factored end moment of smaller absolute value due to lateral loads

M_{2s} = factored end moment of larger absolute value due to lateral loads

given for each column by the preliminary computer analysis, it can now be decided whether the columns are *short* or *long*. If they turn out to be *short*, moment magnification will be unnecessary, and the columns may be designed directly by the methods of Chapter 10. If they are *long*, moments must be magnified; however, once this is done, they become "short," and the design may be completed by the methods of Chapter 10.

The length criteria are as follows:

(1)   In an *unbraced frame*, a column is *long* if

$$k_s L_u / r > 22$$

(2)   In a *braced frame*, a column is *long* if

$$k_b L_u / r > 34 - 12(M_{1b}/M_{2b})$$

where M_{1b}/M_{2b} is taken positive or negative according as the curvature of the column is single (Fig. 11-6) or double (Fig. 11-7). If $|M_{2b}|$ as given by the computer is very small, set $M_{1b}/M_{2b} = +1$ in the criterion. Likewise, if $|M_{1b}|$ and $|M_{2b}|$ are small and if the axial load P_{ub} due to gravity is such that

$$\frac{M_u}{P_{ub}} = e < e_{\min} = 0.6 + 0.03h \qquad (15 + 0.03h)$$

(see "Limitations," Chapter 10), then set $M_{1b} = M_{2b} = P_{ub}e_{\min}$; so that, once more, $M_{1b}/M_{2b} = +1$.

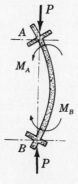

Fig. 11-6

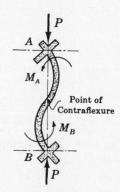

Fig. 11-7

Moment Magnification

For long columns, the largest moments (M_2) at a joint must be magnified to obtain the design moment, M_c:

$$M_c = \delta_b M_{2b} + \delta_s M_{2s} \qquad\qquad (11.10)$$

Here, δ_b is the magnification factor for gravity loads under a braced condition at the given story level and δ_s is the magnification factor for lateral loads under an unbraced condition. The magnification factors are computed as follows:

$$\delta_b = \frac{C_{mb}}{1 - P_{ub}/(\phi P_{cb})} \qquad\qquad (11.11)$$

where

$C_{mb} = 0.6 + 0.4(M_{1b}/M_{2b}) \geqq 0.4$

P_{ub} = ultimate axial load on the individual column under gravity loads = $1.4P_{Db} + 1.7P_{Lb}$

P_{cb} = critical buckling load for the individual column under gravity loads, as given by (11.5) with $k = k_b$

ϕ = understrength factor, considered as 0.7 for tied columns and 0.75 for axially loaded columns, when P_{ub} is greater than $0.1f'_c A_g$

A_g = gross cross-sectional area of the individual column

and
$$\delta_s = \frac{1}{1 - \left(\sum P_{us}\right) \Big/ \left(\sum \phi P_{cs}\right)} \tag{11.12}$$

The summands in (11.12) are:

P_{us} = ultimate axial load on a column due to lateral loads

ϕP_{cs} = product of understrength factor for column and critical buckling load for column under lateral loads, as given by (11.5) with $k = k_s$

and the summations extend over all columns at the considered story level.

The understrength factors may be modified in the case of small axial load ($P_{ub} \leqq 0.1f'_c A_g$) to

$$\phi = 0.9 - 2.0P_{ub}/f'_c A_g \tag{11.13}$$

for tied columns, and

$$\phi = 0.9 - 0.15P_{ub}/f'_c A_g \tag{11.14}$$

for spirally reinforced columns.

WALLS

Walls may be designed as columns, according to the equation

$$\phi P_{nw} = 0.55\phi f'_c A_g \left[1.0 - \left(\frac{kL_c}{32h}\right)^2\right] \tag{11.15}$$

where $\phi = 0.7$ and k is (a) 0.8, when the wall is restrained against rotation at top and/or bottom; (b) 1.0, when the wall is unrestrained against rotation at both ends; (c) 2.0, when the wall is not braced against lateral translation. In (11.15), both the gross concrete area (A_g) and the axial load capacity (ϕP_{nw}) are on a per unit width basis.

Unless justified by a detailed analysis, the horizontal length of wall to be considered as effective for each concentrated load shall not exceed the center-to-center distance between loads or the width of bearing plus four times the wall thickness. Walls shall be anchored to intersecting elements such as floors, roofs, or to columns, pilasters, buttresses and intersecting walls, and footings.

The minimum vertical reinforcement area (per unit width) shall be 0.0012 times the gross concrete area for deformed bars not larger than No. 5 (No. 15M) with a specified yield strength not less than 60,000 psi (400 MPa); it shall be 0.0015 times the gross concrete area for other deformed bars, or 0.0012 times the gross concrete area for smooth or deformed welded wire fabric not larger than W31 or D31.

The minimum ratio of horizontal reinforcement area (per unit height) to the gross area of concrete shall be 0.0020 for deformed bars not larger than No. 5 (No. 15M) with a specified yield strength not less than 60,000 psi (400 MPa); it shall be 0.0025 for other deformed bars, or 0.0020 for smooth or deformed welded wire fabric not larger than W31 or D31.

However, the quantities of reinforcement and limits of thickness may be waived where the structural analysis has shown adequate strength and stability.

Walls more than 10 in (250 mm) thick, except for basement walls, shall have the reinforcement in both directions placed in two layers parallel with the faces of the wall in accordance with the following:

(1) One layer consisting of not less than one-half nor more than two-thirds of the total reinforcement required in each direction shall be placed not less than 2 in (50 mm) nor more than one-third the wall thickness from the exterior surface.

(2) The other layer, consisting of the remainder of the reinforcement in that direction, shall be placed not less than 3/4 in (20 mm) nor more than one-third the wall thickness from the interior surface.

Vertical and horizontal reinforcement on any face shall not be spaced farther apart than three times the wall thickness or 18 in (500 mm).

Vertical reinforcement need not be enclosed by lateral ties if the area of the vertical reinforcement is not greater than 0.01 times the gross area of the concrete cross section, or where the vertical reinforcement is not required as compression reinforcement.

In addition to the minimum reinforcement specified above, no fewer than two No. 5 (No. 15M) bars shall be provided around all window and door openings. Such bars shall be extended to develop the full strength of the bars beyond the corners of the openings, but not less than 24 in (600 mm).

The minimum thickness of bearing walls shall not be less than 1/25 the supported height or length, whichever is smaller, nor less than 4 in (100 mm). Thickness of exterior basement walls and foundation walls shall not be less than $7\frac{1}{2}$ in (190 mm). The thickness of nonbearing walls shall not be less than 4 in (100 mm) or less than 1/30 the distance between members that provide lateral support. Portions of grade beam walls exposed above grade shall also meet the requirement for minimum reinforcement.

If structural analysis indicates appropriate strength and stability, the minimum reinforcement and thickness requirements may be waived.

Compression members built integrally with walls shall be designed as columns. Transfer of forces at the base of walls shall conform with that required for columns. Shear forces must be resisted by the walls.

Solved Problems

All problems refer to Figs. 11-8, 11-9 and 11-10.

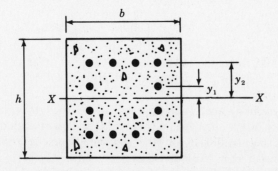

Fig. 11-8. Column Cross Section

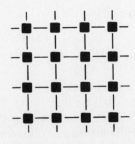

Fig. 11-9. Column Plan

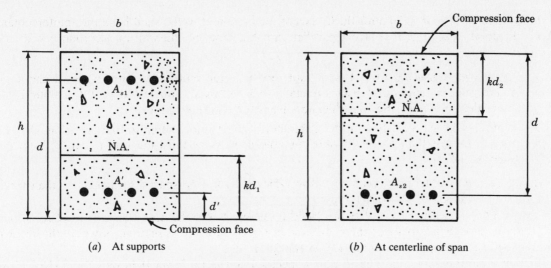

(a) At supports   (b) At centerline of span

Fig. 11-10. Beam Cross Sections

11.1. Find the equivalent length factors k_b and k_s, given $f'_c = 3000$ psi, beam width $b = 12$ in, $d' = 2$ in, $A_{s1} = 4.0$ in^2, $A_{s2} = 4.0$ in^2, $A'_s = 2.0$ in^2, $w_c = 145$ lb/ft^3, column dimensions $b = h = 20$ in, $y_1 = 4$ in, and $y_2 = 8$ in. Use the transformed section to compute beam stiffness. Consider an interior column at an interior floor level. The twelve column bars are No. 9. Column heights are $L_u = 12$ ft and beam spans are 20 ft.

The modulus of elasticity of the steel is $29,000$ kips/in^2 and the modulus of elasticity of the concrete is

$$E_c = (33.0)w_c^{3/2}\sqrt{f'_c} = (33.0)(145)^{3/2}\sqrt{3000}/1000 = 3156 \text{ kips/in}^2$$

giving the modular ratio as $n = E_s/E_c = 29,000/3156 = 9.189$. The 1983 ACI Code recommends using the nearest whole number for n; so $n = 9$.

For twelve No. 9 bars, with A_s for each bar 1.0 in^2, the moment of inertia of the steel is

$$I_{se} = (8)(1.0)(8)^2 + (4)(1.0)(4)^2 = 512 + 64 = 576 \text{ in}^4$$

and (11.3) gives, with $\beta_d = 0$, the column stiffness

$$EI = \tfrac{1}{5}E_cI_g + E_sI_{se}$$

$$= \tfrac{1}{5}(3156)\frac{(20)(20)^3}{12} + (29,000)(576) = 25.12 \times 10^6 \text{ kips-in}^2$$

and stiffness ratio

$$\frac{EI}{L} = \frac{25.12 \times 10^6}{12 \times 12} = 174,440 \text{ in-kips} \tag{1}$$

For the beams, the transformed section at the centerline of the span is shown in Fig. 11-11. Summing moments of area about the neutral axis,

$$(12kd)(kd/2) = (36)(20 - kd) \quad \text{or} \quad (kd)^2 + 6(kd) - 120 = 0$$

from which

$$kd = \frac{-6 \pm \sqrt{36 + (4)(120)}}{2} = -3 + 11.36 = 8.36 \text{ in}$$

Hence, $I_t = (12)(8.36)^3/3 + (9)(4)(20 - 8.36)^2 = 2337 + 4878 = 7215$ in^4, and the stiffness ratio at the centerline is

$$\frac{E_cI_t}{L} = \frac{(3156)(7215)}{20 \times 12} = 94,880 \text{ in-kips}$$

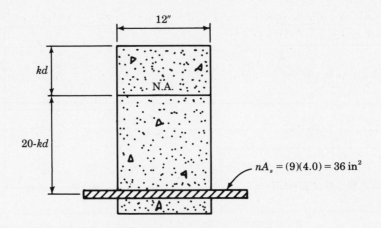

Fig. 11-11

Figure 11-12 shows the transformed section for the beams at the support. The 1983 ACI Code requires the use of $(2n-1)A'_s$ to transform compression steel "for stress computations"; but no provisions are given by the Code for stiffness calculations. Here, $(2n-1)A'_s$ will be used. Summing moments of area about the neutral axis,

$$(12)(kd)(kd/2) + (34)(kd-2) = 36(20-kd)$$
$$6(kd)^2 + 70(kd) - 788 = 0$$
$$(kd)^2 + 11.67(kd) - 131.33 = 0$$

$$kd = \frac{-11.67 \pm \sqrt{(11.67)^2 + (525.32)}}{2} = -5.835 + 12.86 = 7.025 \text{ in}$$

Hence,

$$I_t = (12)(7.025)^3/3 + (34)(5.025)^2 + (36)(12.975)^2 = 1386.8 + 858.5 + 6060.6 = 8305.9 \text{ in}^4$$

and

$$\frac{E_c I_t}{L} = \frac{(3156)(8305.9)}{20 \times 20} = 109{,}200 \text{ in-kips}$$

The average stiffness ratio at the center of the span and at the face of the support is

$$EI/L = \frac{94{,}880 + 109{,}200}{2} = 102{,}040 \text{ in-kips}$$

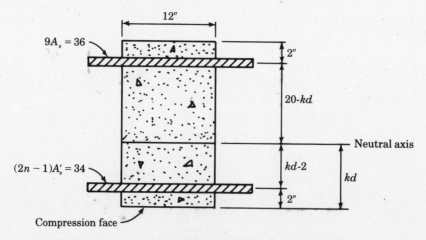

Fig. 11-12

For two beams (uniaxial bending) at each joint, the sum of EI/L is $2(102,040) = 204,080$ in-kips. From (1), the sum of EI/L for two columns is

$$2(174,440) = 348,880 \text{ in-kips}$$

Equation (11.8) now gives $\psi = 348,880/204,080 = 1.71$, and Figs. 11-4 and 11-5 give, for $\psi_A = \psi_B = 1.71$, $k_b = 0.84$, $k_s = 1.52$.

11.2. For Problem 11.1, determine k_b and k_s using gross moments of inertia.

For the columns, $I_g = (20)(20)^3/12 = 13,333 \text{ in}^4$.
For the beams, $I_g = (12)(22)^3/12 = 10,648 \text{ in}^4$.
By (11.9), and the fact that ψ is a pure number,

$$\psi_A = \psi_B = \frac{\displaystyle\sum_{\text{cols}} (E_c I_g/L)}{\frac{1}{2}\displaystyle\sum_{\text{beams}} (E_c I_g/L)} = \frac{2E_c(13,333)/12}{\frac{1}{2}2E_c(10,648)/20} = 4.17$$

Figures 11-4 and 11-5 give $k_b = 0.92$ and $k_s = 2.1$, which are conservative estimates of the values found in Problem 11.1.

11.3. For the interior columns of Fig. 11-9, let the ultimate column moments be $M_{1b} = 485$ ft-kips, $M_{2b} = 500$ ft-kips; the columns are bent in single curvature. The unsupported column length is 12 ft and A_s consists of twelve No. 9 bars. Interior column axial loads are $P_{ub} = 400$ kips and $P_{us} = 400$ kips. In Fig. 11-8, take $b = h = 20$ in and $y_2 = 2y_1 = 7.5$ in. Assume that side columns take one-half of the axial loads in the interior columns and that corner columns take one-fourth of the axial loads in the interior columns. (This has been assumed to simplify the data, and would not normally occur in practice.) If the creep factor is $\beta_d = 0.4$, find the magnified design bending moment. For simplicity, bending moments in all columns are the same, and use $k_b = 0.92$ and $k_s = 2.1$ (Problem 11.2).

The equivalent number of interior columns is $4 + 8/2 + 4/4 = 9$. For the gravity load condition on an interior column,

$$P_{ub} = 0.75(1.4P_{Db} + 1.7P_{Lb}) = 400 \text{ kips}$$

and the largest bending moments on the columns are $M_{2b} = 500$ ft-kips and $M_{2s} = 580$ ft-kips. Those values must be compared with the moment corresponding to *minimum eccentricity*,

$$M_{\min} = P_{ub}(0.6 + 0.03h)$$

where h, the cross-sectional dimension of the column in the direction of bending, is in this case 20 in. Thus,

$$M_{\min} = 400[0.6 + (0.03)(20)] = 480 \text{ in-kips} = 40 \text{ ft-kips}$$

Since all bending moments in the columns exceed this minimum value, the actual moments are to be used in the design. Then,

$$C_{mb} = 0.6 + (0.4)(M_{1b}/M_{2b}) = 0.6 + (0.4)(485/500) = 0.988$$

The columns may be designed as *short columns* if

$$\frac{k_b L_u}{r} = \frac{k_b L_u}{0.3h} = \frac{(0.92)(12 \times 12)}{0.3(20)} = 22.08$$

does not exceed $34 - 12(M_{1b}/M_{2b}) = 22.36$, *and* if

$$\frac{k_s L_u}{r} = \frac{(2.1)(12 \times 12)}{0.3(20)} = 50.4$$

does not exceed 22. In this case, the unbraced lateral load condition requires design as *long columns*.

Since neither $k_b L_u/r$ nor $k_s L_u/r$ exceeds 100, the moment magnifier method may be used.
As in Problem 11.1, $E_c = 3156$ kips/in^2 and $I_g = 13,333$ in^4; but now

$$I_{se} = 8(1.0)(7.5)^2 + 4(1.0)(3.75)^2 = 506.25 \text{ in}^4$$

Hence, for lateral loads ($\beta_d = 0$), (11.3) gives

$$(EI)_s = E_c I_g/5 + E_s I_{se}$$

$$= \frac{(3156)(13,333)}{5} + (29,000)(506.25) = 23.096 \times 10^6 \text{ kips-in}^2$$

while, for gravity loads ($\beta_d = 0.4$),

$$(EI)_b = \frac{23.096 \times 10^6}{1.4} = 16.497 \times 10^6 \text{ kips-in}^2$$

The corresponding critical buckling loads are, from (11.5),

$$P_{cs} = \pi^2 \frac{(EI)_s}{(k_s L_u)^2} = (9.87) \frac{23.096 \times 10^6}{[(2.1)(12 \times 12)]^2} = 2493 \text{ kips}$$

$$P_{cb} = \pi^2 \frac{(EI)_b}{(k_f L_u)^2} = (9.87) \frac{16.497 \times 10^6}{[(0.92)(12 \times 12)]^2} = 9277 \text{ kips}$$

Then, for lateral loads, (11.12) gives

$$\delta_s = \frac{1}{1 - \left(\sum P_{us} \Big/ \sum \phi P_{cs} \right)} = \frac{1}{1 - \dfrac{9 \times 400}{9 \times 0.7 \times 2493}} = 1.2974$$

and, for gravity loads, (11.11) gives

$$\delta_b = \frac{C_{mb}}{1 - P_{ub}/(\phi P_{cb})} = \frac{0.988}{1 - \dfrac{400}{(0.7)(9277)}} = 1.0529$$

Finally, by (11.10), the design column moment is

$$M_c = (1.0529)(500) + (1.2974)(580) = 1279 \text{ ft-kips}$$

11.4. Rework Problem 11.3 (in hard translation), given $b = 510$ mm, $h = 510$ mm, $L_u = 4$ m, $f'_c = 20$ MPa, $y_2 = 2y_1 = 200$ mm, $A_s = 700$ mm^2 per bar, $M_{1b} = 660$ kN·m, $M_{2b} = 680$ kN·m, $M_{1s} = M_{2s} = 800$ kN·m, $w_c = 2320$ kg/m^3, $P_{ub} = P_{us} = 1800$ kN for interior columns.

Proceeding as in Problem 11.3, calculate

$$M_{\min} = P_{ub}(15 + 0.03h) = 1800[15 + 0.03(510)]/10^3 = 54.54 \text{ kN·m}$$

Actual moment values are to be used. Then,

$$C_{mb} = 0.6 + (0.4)(M_{1b}/M_{2b}) = 0.6 + (0.4)(660/680) = 0.988$$

Long-column design is required and permitted because

$$22 < \frac{k_s L_u}{r} = \frac{(2.1)(4 \times 10^3)}{(0.3)(510)} = 54.9 < 100$$

Calculate stiffnesses as follows:

$$E_c = 0.043 w_c^{1.5} \sqrt{f'_c} = 0.043(2320)^{1.5} \sqrt{20} = 21\,488 \text{ MPa}$$

$$I_g = \frac{(510)(510)^3}{12} = 5636 \times 10^6 \text{ mm}^4$$

$$I_{se} = 8(700)(200)^2 + 4(700)(100)^2 = 252 \times 10^6 \text{ mm}^4$$

$$(EI)_s = \frac{(21\,488)(5636 \times 10^6)}{5} + (200\,000)(252 \times 10^6) = 74.63 \times 10^{12} \text{ N} \cdot \text{mm}^2$$

$$(EI)_b = \frac{74.63 \times 10^{12}}{1.4} = 53.307 \times 10^{12} \text{ N} \cdot \text{mm}^2$$

From (11.5),

$$P_{cs} = \pi^2 \frac{(EI)_s}{(k_s L_u)^2} = 9.87 \frac{74.63 \times 10^{12}}{[(2.1)(4 \times 10^3)]^2} (10^{-3}) = 10\,439 \text{ kN}$$

$$P_{cb} = \pi^2 \frac{(EI)_b}{(k_b L_u)^2} = 9.87 \frac{53.307 \times 10^{12}}{[(0.92)(4 \times 10^3)]^2} (10^{-3}) = 38\,852 \text{ kN}$$

and from (11.12) and (11.11),

$$\delta_s = \frac{1}{1 - \left(\sum P_{us} / \sum \phi P_{cs} \right)} = \frac{1}{1 - \dfrac{9 \times 1800}{9 \times 0.7 \times 10\,439}} = 1.327$$

$$\delta_b = \frac{C_{mb}}{1 - P_{ub}/(\phi P_{cb})} = \frac{0.988}{1 - \dfrac{1800}{(0.7)(38\,852)}} = 1.058$$

The design moment is therefore

$$M_c = (1.058)(680) + (1.327)(800) = 1781 \text{ kN} \cdot \text{m}$$

11.5. A concentrically loaded reinforced concrete wall has height $L_c = 12$ ft and thickness $h = 6$ in. It is restrained at both ends. If $f_c' = 4000$ psi and $f_y = 60,000$ psi, determine the axial load per foot of width it can support and calculate the reinforcement required.

Substitute $A_g = (6)(12) = 72$ in^2/foot width, $k = 0.8$, and $\phi = 0.7$ in (11.15) to obtain

$$\phi P_{nw} = (0.55)(0.7)(4.0)(72)\left\{ 1 - \left[\frac{(0.8)(12)(12)}{(32)(6)} \right]^2 \right\} = 70.96 \text{ kips/ft}$$

For vertical steel with No. 5 or smaller deformed bars having f_y at least equal to 60,000 psi, the vertical bar area is

$$A_{sv} = 0.0012 A_g = (0.0012)(72) = 0.0864 \text{ in}^2/\text{ft}$$

Bar spacing may not exceed $3h$ or 18 in (here, $3h = 18$ in). For bars at 18 in (1.5 ft) on centers, $A_{sv} = 1.5(0.0864) = 0.1296$ in^2 required; No. 4 bars at 18 in on centers provide 0.13 in^2. For horizontal reinforcement with No. 5 or smaller deformed bars having f_y at least equal to 60,000 psi, $A_{sh} = 0.0020 A_g = (0.0020)(72) = 0.144$ in^2/ft; No. 4 bars at 16 in on centers provide $A_{sh} = 0.15$ in^2/ft. Both horizontal and vertical bars are placed at the center of the wall, and are fastened together with bailing wire at every bar intersection.

Supplementary Problems

All problems refer to Figs. 11-8 and 11-9.

11.6. Table 11.1 presents data for subproblems (a) through (k). Verify the listed solutions, assuming that all columns are square, so $h = b$ and $y_2 = 2y_1$; $w_c = 145$ lb/ft^3; all columns are identical in cross section; side columns take one-half as much axial load as interior columns; corner columns take one-fourth as much axial load as interior columns. A_s is the area of one reinforcing bar.

Table 11.1

					GIVEN										SOLUTIONS		
	b, in	L_u, ft	f'_c, psi	β_d	M_{1b}, ft-kips	M_{2b}, ft-kips	M_{1s}, ft-kips	M_{2s}, ft-kips	P_{ub}, kips	P_{us}, kips	k_b	k_s	A_s, in^2	y_1, in	δ_b	δ_s	M_c, ft-kips
(a)	16	12	3000	0.490	485	500	510	580	400	400	0.92	2.10	1.27	3.00	1.096	1.525	1432.4
(b)	22	16	5000	0.355	220	400	185	220	250	250	0.87	1.40	1.56	4.00	1.000	1.066	634.6
(c)	20	14	3000	0.480	485	525	510	600	485	485	0.98	2.40	1.27	3.75	1.082	1.729	1605.7
(d)	22	20	4000	0.367	180	210	280	350	340	340	1.00	1.20	1.56	4.00	1.043	1.113	608.5
(e)	19	17	3000	0.300	210	280	260	320	450	450	0.95	1.40	1.0	3.00	1.119	1.486	788.8
(f)	21	18	4000	0.485	500	530	510	630	400	400	0.84	2.40	1.0	4.00	1.085	2.201	1961.6
(g)	20	16	5000	0.474	270	300	450	580	100	100	0.95	1.30	1.0	3.75	1.000	1.037	901.2
(h)	22	18	4000	0.474	360	400	200	360	100	100	0.93	2.40	1.56	4.50	1.000	1.090	792.3
(i)	24	22	3000	0.490	485	500	540	580	400	400	0.92	1.80	1.00	5.00	1.119	1.429	1388.2
(j)	20	22	5000	0.461	325	380	400	425	325	325	1.00	2.40	1.56	3.75	1.098	2.269	1381.5
(k)	20	15	5000	0.400	200	300	320	350	300	300	0.85	1.50	1.56	3.75	1.000	1.103	686.2

Table 11.2

					GIVEN										SOLUTIONS		
	b, mm	L_u, m	f'_c, MPa	β_d	M_{1b}, kN·m	M_{2b}, kN·m	M_{1s}, kN·m	M_{2s}, kN·m	P_{ub}, kN	P_{us}, kN	k_b	k_s	A_s, mm^2	y_1, mm	δ_b	δ_s	M_c, kN·m
(a)	530	4.00	20	0.490	660	680	700	800	1800	1800	0.87	2.40	800	110	1.038	1.325	1765.7
(b)	540	5.50	35	0.523	620	740	615	620	445	445	1.00	2.20	700	100	1.000	1.117	1432.4
(c)	600	3.66	20	0.490	725	760	680	805	1790	1790	0.92	2.30	700	150	1.010	1.132	1678.3
(d)	560	4.00	20	0.463	660	710	650	800	1800	1800	0.88	2.00	800	115	1.015	1.176	1661.4
(e)	610	5.50	35	0.400	600	920	620	800	445	445	1.00	1.80	900	150	1.000	1.031	1744.7
(f)	540	3.66	20	0.470	630	715	715	820	1850	1850	0.92	2.30	600	110	1.006	1.290	1777.2
(g)	525	4.00	20	0.400	660	680	700	800	1800	1800	0.83	1.90	800	105	1.034	1.199	1662.3
(h)	540	5.50	35	0.450	420	510	600	700	445	445	1.00	1.20	900	110	1.000	1.024	1227.0
(i)	615	4.00	20	0.510	665	700	720	815	1850	1850	0.86	2.50	700	130	1.052	1.238	1682.5
(j)	580	4.00	20	0.460	580	680	700	800	1800	1800	0.92	2.10	800	120	1.000	1.175	1619.9

11.7. Verify the solutions to subproblems (*a*) through (*j*) of Table 11.2, making the same assumptions as in Problem 11.6 and taking $w_c = 2400$ kg/m^3.

11.8. Verify the concrete wall solutions of subproblems (*a*) through (*l*) as given in Table 11.3. Use $f_y = 60,000$ psi.

Table 11.3

	GIVEN				SOLUTIONS		
	f'_c, psi	k	L_c, ft	h, in	ϕP_{nw}, kips/ft	A_{sv}, in^2/ft	A_{sh}, in^2/ft
(*a*)	4000	0.8	12	6	71.0	0.086	0.144
(*b*)	3000	0.8	14	6	42.4	0.086	0.144
(*c*)	4000	1.0	13	7	66.6	0.101	0.168
(*d*)	5000	1.0	16	8	80.9	0.115	0.192
(*e*)	3000	0.8	15	7	56.9	0.101	0.168
(*f*)	4000	1.0	11	6	58.5	0.086	0.144
(*g*)	3000	0.8	17	6	23.1	0.086	0.144
(*h*)	4000	1.0	18	7	9.1	0.101	0.168
(*i*)	5000	1.0	12	8	126.3	0.115	0.192
(*j*)	3000	0.8	16	7	51.4	0.101	0.168
(*k*)	4000	0.8	17	8	87.8	0.115	0.192
(*l*)	3000	1.0	20	8	13.4	0.115	0.192

11.9. Verify the concrete wall solutions to subproblems (*a*) through (*j*) as given in Table 11.4. Use $f_y = 400$ MPa.

Table 11.4

	GIVEN				SOLUTIONS		
	f'_c, MPa	k	L_c, m	h, mm	ϕP_{nw}, kN/m	A_{sv}, mm^2/m	A_{sh}, mm^2/m
(*a*)	30	1.0	4.0	175	990.0	210	350
(*b*)	35	1.0	5.0	200	1050.1	240	400
(*c*)	30	0.8	4.5	175	1185.9	210	350
(*d*)	35	1.0	3.5	150	946.6	180	300
(*e*)	30	0.8	5.0	150	529.4	180	300
(*f*)	35	1.0	5.5	175	83.5	210	350
(*g*)	35	1.0	3.8	200	1744.9	240	400
(*h*)	20	0.8	4.8	175	713.9	210	350
(*i*)	30	0.8	5.0	200	1407.7	240	400
(*j*)	20	1.0	6.0	200	186.5	240	400

Chapter 12

Footings

NOTATION

a = column width, in. (mm)

ϕa_u = a coefficient for strength design of steel as given in Table 5.2 (Table 5.3)

A_F = area of footing, ft^2 (m^2)

A_s = steel area, in.^{2} (mm^2)

A_{sB} = steel area in band width B, in.^{2} (mm^2)

A_{st} = total steel area, in.^{2} (mm^2)

A_p = pile area, ft^2 (m^2)

b = width of a beam or smaller dimension of a column cross section, in. (mm)

b_o = perimeter of a shear section, in. (mm)

B = footing width, ft (m)

c = distance between centers of columns, ft (m)

c = distance from neutral axis to a stressed point, in. (mm)

d = effective depth of a footing, in. (mm)

d = distance between pile centers, ft (m)

d_p = diameter of a pile, in. (mm)

D = diameter of a reinforcing bar, in. (mm)

e = eccentricity of load, ft (m)

f_c' = ultimate strength of concrete, psi (MPa)

f_s = steel stress, psi (MPa)

f_y = yield stress for steel, psi (MPa)

F = any force, kips (kN)

h = distance from footing base to horizontal force, ft (m)

h = greater dimension of a column cross section, in. (mm)

H = horizontal force, kips (kN)

I = moment of inertia of footing base, ft^4 (m^4)

L = length of footing base, ft (m)

M = bending moment, ft-kips (kN $\cdot$ m)

M_n = nominal (design) bending moment, ft-kips (kN $\cdot$ m)
$= M_u/\phi$

M_u = ultimate bending moment, ft-kips (kN $\cdot$ m)

n = modular ratio E_s/E_c

N = number of piles

ρ = steel ratio = A_s/bd

ρ_t = total steel ratio = A_{st}/bd

P = any load, kips (kN)

P' = change in pile load due to displacement from critical section, kips (kN)

q = soil pressure, kips/ft^2 (kN/m^2)

ϕR_u = a coefficient for strength design of concrete as given in Table 5.2 (Table 5.3)

V = shear force, kips (kN)

V_c = permissible shear force on concrete, kips (kN)

V_n = nominal shear force, kips (kN) = V_u/ϕ

V_u = ultimate shear force, kips (kN)

W = column load, kips (kN)

W_F = footing weight, kips (kN)

ϕ = capacity reduction factor or resistance factors as specified in the 1983 ACI Code (see Chapter 5; ϕ is *member-type dependent*)

β_c = aspect ratio of column cross section = h/b

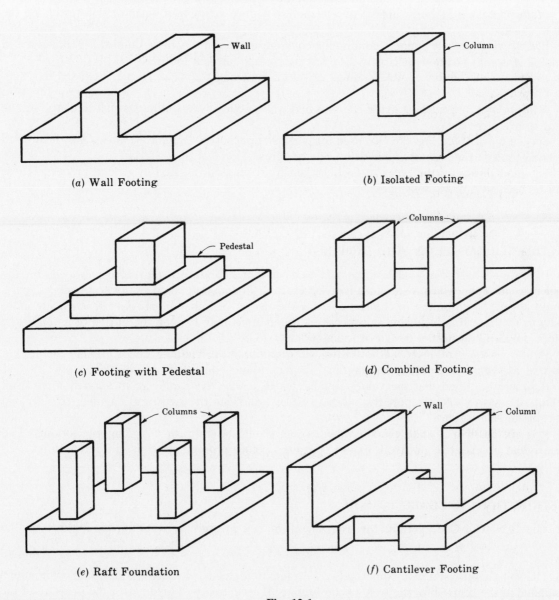

(a) Wall Footing

(b) Isolated Footing

(c) Footing with Pedestal

(d) Combined Footing

(e) Raft Foundation

(f) Cantilever Footing

Fig. 12-1

GENERAL NOTES

Reinforced concrete foundations or *footings* are utilized to support columns and walls composed of a variety of materials, including concrete, steel, masonry and timber.

Spread footings are designed to distribute large loads over a large area of soil near the ground surface in order to reduce the intensity of the force per unit area so that the soil will safely support the structure.

Pile footings are designed to deliver large loads to individual piles. The piles transfer the forces to lower levels by means of *skin-friction* between the soil and the pile surface and *point-bearing* of the pile on a dense soil strata at its base.

Both spread footings and pile footings may be classified into sub-groups such as isolated footings, multiple column footings, wall footings and mat footings.

Isolated footings support the load of a single column. The foundation for a structure may be composed of many isolated footings and, in addition, other types of footings.

Multiple column footings support two or more columns, acting as a beam or slab resting on the soil or piles.

Wall footings usually support continuous concrete or masonry walls around the perimeter of a building. Interior partition walls may also rest on continuous wall footings.

A special application of wall footings exists for retaining walls, which are discussed in Chapter 15.

Raft footings may support many columns and walls, acting as a continuous slab to distribute the loads over a large area.

Special types of footings are also used for particular purposes. *Cantilever footings* may be utilized advantageously near property lines or other structures.

Fig. 12-1 illustrates the types of footings which are used most often in general practice. These may be used with or without piles.

ALLOWABLE LOADS ON SOIL OR PILES

It is always desirable to determine safe bearing values for soil or allowable pile loads for the particular site upon which a structure is to be erected. Foundation engineers or soil testing experts should always be consulted in connection with the design of foundations, since it is usually more costly to rectify errors in judgment concerning foundations than to seek the advice of an authority before proceeding with the design of foundations.

In the absence of more authoritative information, most building codes permit the use of approximate values of safe soil bearing load. Typical values are listed in Table 12.1.

It should be noted that the values listed refer to *service loads and not to ultimate loads*. Soil testing laboratories often report the ultimate values, allowing the structural engineer to select an appropriate safety factor.

It is not possible to state general values which should be used for designing pile footings. Pile tests should be initiated, or a soils expert consulted before proceeding to design pile footings.

DISTRIBUTION OF LOADS TO SOIL

The 1983 ACI Code permits the use of uniform soil pressure beneath spread footings so that

$$q = P/A_F \qquad (12.1)$$

The soil pressure stated in equation (12.1) refers to footings for which the resultant column load is applied at the centroid of the base of the footing. If eccentricity of load exists, the pressure of the soil will vary uniformly, and one of two cases will exist. These are illustrated in Fig. 12-2.

Table 12.1

Soil Composition	Allowable Service-Load Pressure		
	kips/ft^2	kN/m^2	kg/m^2
Soft clay, medium density	3	144	14.6
Medium stiff clay	5	239	24.4
Sand, fine, loose	4	192	19.5
Sand, coarse, loose; compact fine sand; and loose sand-gravel mixture	6	287	29.5
Gravel, loose; and compact coarse sand	8	383	39.1
Sand-gravel mixture, compact	12	575	58.6
Hardpan and exceptionally compacted or partially cemented gravels or sands	20	958	97.6
Sedimentary rocks such as hard shales, sandstones, limestones, silt stones, in sound condition	30	1436	146.5
Foliated rocks such as schist or slate in sound condition	80	3830	390.6
Massive, bed rock, such as granite, diorite, gneiss, trap rock, in sound condition	200	9576	976.4

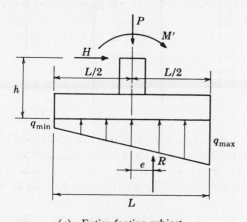

(a) Entire footing subject to soil pressure

Case 1

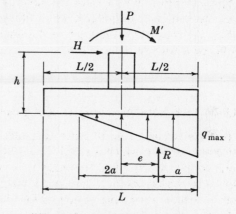

(b) Portion of footing not subject to soil pressure

Case 2

Fig. 12-2

Case 1 of Fig. 12-2 occurs for moderate values of M' or H, and Case 2 occurs for large values of M' or H. The resultant R consists of the applied load P plus the weight of the footing.

If the effects of M' and the moment due to H are combined as M, the maximum and minimum soil pressure can be obtained as follows, considering a constant width of footing B:

For Case 1,
$$q_{max} = R/BL + 6M/BL^2 \qquad (12.2)$$
$$q_{min} = R/BL - 6M/BL^2 \qquad (12.3)$$

For Case 2,
$$q_{max} = 4R/[3B(L - 2e)] \qquad (12.4)$$
$$q_{min} = 0 \qquad (12.5)$$

For both cases, the resultant R is located by determining the eccentricity of load,

$$e = (M' + Hh)/R \tag{12.6}$$

Case 1 will always exist if R lies within the *middle third* of the footing or if

$$e \leqq L/6 \tag{12.7}$$

otherwise, Case 2 will apply. If Case 2 applies,

$$a = L/2 - e \tag{12.8}$$

The foregoing equations for eccentrically loaded footings apply to bending about one axis only and are derived using the equation

$$q = R/A_F \pm Mc/I \tag{12.9}$$

in which A_F = base area of the footing, $M = Re$, $c = L/2$, $I = BL^3/12$, B = width of footing.

When bending occurs about both the x and y axes, and the *entire footing is subjected to pressure*,

$$q_{max} = R/A_F + M_x c_y/I_x + M_y c_x/I_y \tag{12.10}$$
$$q_{min} = R/A_F - M_x c_y/I_x - M_y c_x/I_y \tag{12.11}$$

The values of e_x and e_y are obtained using equation (12.6) first about the x-axis and then about the y-axis. If the resulting point of application of e falls outside of the *kern* of the section (shown shaded in Fig. 12-3), a special case exists and the points of zero pressure must be determined by trial. It should be noted that *tension cannot exist between the soil and the footing*.

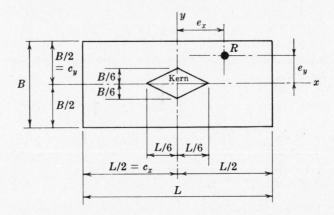

Fig. 12-3   Plan View of Base of Footing with Resultant Outside of Kern

FORCES ON PILES

Vertical Forces

When the resultant R is applied at the *centroid of a pile group* of N piles, each pile receives an identical load, so the pile load P will be

$$P = R/N \tag{12.12}$$

When bending and axial load occur on pile footings, the approach is as follows: Let A_p be the area of each pile, so the average stress q_i in any pile i located at a distance d_i from the pile group centroid will be

$$q_i = R/NA_p \pm Md_i/I$$

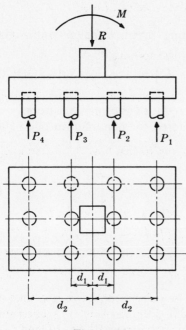

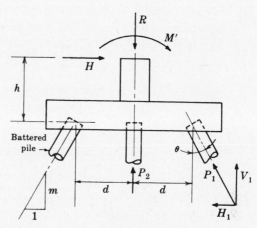

Fig. 12-4   Fig. 12-5

where $I = \Sigma \, (A_p d^2)$; then if each pile has an identical area,

$$q_i = R/NA_p \pm Md_i \Big/ \left(A_p \sum d^2\right)$$

Multiplying both sides of the above equation by A_p and noting that qA_p is the *force* in a pile, the equation for any pile i becomes

$$P_i = R/N \pm Md_i \Big/ \sum d^2 \qquad\qquad (12.13)$$

Piles can assume tension loads if sufficient *negative skin-friction* can be developed, so only one case need be considered for bending about one principal axis. The *positive sign* applies to those piles on the side of the axis which is being rotated downward toward the piles.

When bending occurs about both principal axes, the load in pile i will be

$$P_i = R/N \pm M_x d_{ix} \Big/ \sum d_x^2 \pm m_y d_{iy} \Big/ \sum d_y^2 \qquad\qquad (12.14)$$

In all cases, the x and y principal axes must be directed through the centroid of the *pile group*.

Horizontal Forces

When horizontal loads are applied to spread footings, those loads must be transmitted to the soil by *friction* of concrete on soil or by *passive pressure* of the *side* of the footing (or a key) against the soil. Piles must resist the horizontal forces by shear stress in the piles and passive soil pressure, or battered piles (see Fig. 12-5).

Battered piles are not as effective in resisting vertical loads as are vertical piles. If θ is the batter angle, the efficiency of the battered pile is proportional to $\cos \theta$, and the values of N and I must be adjusted to reflect the loss in efficiency due to the batter.

The vertical component V_1 of the allowable pile load is $P_1 \cos \theta$, and the horizontal component H_1 is $P_1 \sin \theta$. The sum of all of the horizontal components of the pile loads must resist the applied horizontal force H, just as the sum of all the vertical components of the pile loads must resist the applied vertical force W.

WALL FOOTINGS

The equations presented previously for isolated column footings may be used for continuous wall footings if the following factors are utilized:

(1) For spread footings, the width B of the footing is taken as 1 ft (1 m) along the length of the wall.

(2) For *pile footings*, one transverse row of piles is considered for uniform pile patterns, or, one repetition of the pile pattern (or one *panel*) is considered for non-uniform pile patterns. This is illustrated in Fig. 12-6.

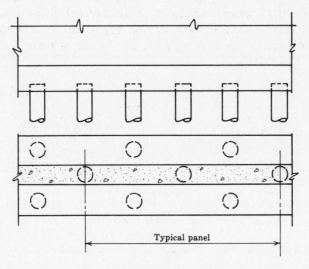

Fig. 12-6

MULTIPLE COLUMN FOOTINGS

The 1983 ACI Code Committee has not chosen to state specific requirements for the design of multiple column footings. Traditionally, such footings have been designed as *inverted beams*, using beam design criteria.

Whether spread footings or pile footings are used, it is necessary to locate the center of gravity of loads at the centroid of the base area of the footing or pile group, in order to spread the load uniformly. *Note*. For pile footings, $R = PN$.

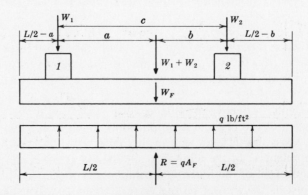

Fig. 12-7

The center of gravity of the total load is determined by summing moments about column *1*, from which

$$a/b = W_2/W_1 \qquad\qquad (12.15)$$

Since the distance c between columns is known and $b = c - a$, it follows that

$$a/c = W_2/(W_1 + W_2) \qquad\qquad (12.16)$$

Whenever possible, the footing width is made constant so that the area of the footing can be obtained in a simple manner.

It is sometimes impossible to use perfect rectangles for combined footings. Clearances with other footings or adjacent structures may prohibit this simplification. An odd shape may be necessary in order to place the center of gravity of the loads at the centroid of the footing base or at the centroid of the pile group. Fig. 12-8 illustrates some of the shapes which are often used.

When multiple column footings are utilized, the footing is treated as a beam on two or more supports in the long direction.

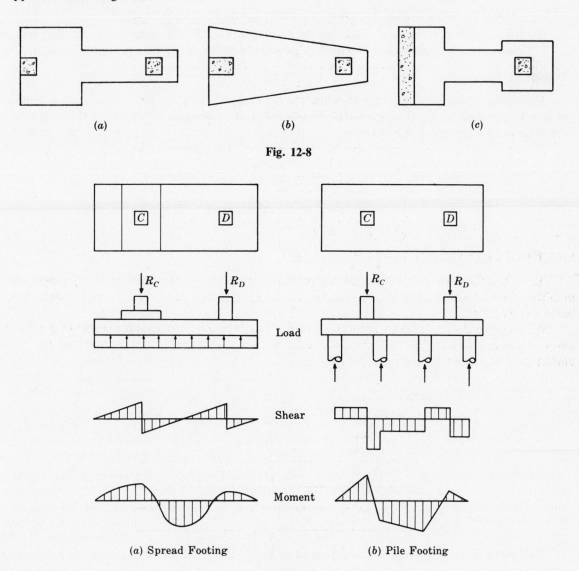

(a)   (b)   (c)

Fig. 12-8

(a) Spread Footing   (b) Pile Footing

Fig. 12-9

Transverse beams are designed to support the *longitudinal beam* and to transmit the soil pressure or pile reactions to the columns. Typical *qualitative shear and moment diagrams* for the longitudinal beam are illustrated in Fig. 12-9.

Several comments must be made regarding Fig. 12-9.

(*a*)   The column loads are actually distributed in some manner over the top of the footing, but the loads are usually considered as *line loads* in the transverse direction at the center of the columns. These loads may also be considered to be uniformly distributed over an area equal to the width of the column in the long direction and the total short width of the footing. In either case the solution is the same, since critical sections do not occur under the column.

(*b*)   The entire weight of a pile cap is considered by some designers to be supported by the soil. Others assume that these loads are delivered to the piles. Either assumption has merit, although the latter is more conservative.

(*c*)   If the pile spacing is not uniform or if the number of piles is not constant for all transverse rows, the concentrated live loads will differ from one row to another.

(*d*)   For spread footings the weight of the footing is used in obtaining the gross soil pressure while the net soil pressure is used in proportioning the concrete and reinforcement.

Transverse Beams

Regardless of the mode of support of the footing (soil or piles), the transverse beam is required to transfer the loads to the column, and *each* pedestal must be designed to distribute the total load of the column (it supports) to the footing.

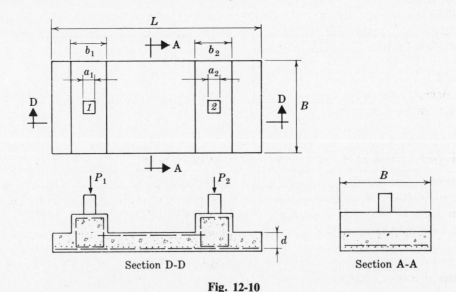

Fig. 12-10

Each *transverse beam* is designed for a uniformly distributed pressure equal to the respective column load divided by the assumed area of the transverse beam. For example, the beam under column *1* has an area equal to $B \times b_1$, and the transverse beam pressure p_T for design is equal to

$$p_T = P_1/Bb_1 \qquad (12.17)$$

The width of the transverse beam has not been standardized in engineering practice. Recommendations range from using the width of the column to using the footing width as the width of the transverse beam.

Using the dimension B as the width of the transverse beam approximates the conditions used in isolated footing design. Using only the width of the column is based on ordinary beam and slab design.

A reasonable width for the transverse beam can be selected considering the usual mode of failure of footings as shown in Fig. 12-11.

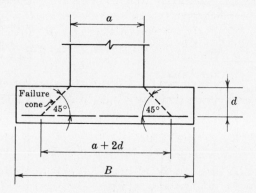

Fig. 12-11

Failure due to shear occurs approximately along 45 degree lines when footings are tested to destruction. This suggests that the width of the transverse beam should be taken as $b = a + 2d$, but not more than B.

As will be shown later, the critical section for shear in slabs and footings occurs at a distance $d/2$ from the face of a support. For this reason, it is quite reasonable to use a transverse beam width equal to $a + 2(d/2)$ or

$$b = a + d \qquad\qquad (12.18)$$

When necessary in order to resist bending or shear forces or to provide sufficient development length, the transverse beam may be deepened to form a *pedestal*.

CRITICAL SECTIONS FOR BENDING AND DEVELOPMENT LENGTH

For *isolated footings* the 1983 ACI Code stipulates that the critical section for bending and flexural bond shall be taken as a section across the full width of the footing *at its top surface* (to account for tapered footings), at planes as shown in Fig. 12-12.

In the case of *combined footings*, the critical sections are used as for isolated footings with respect to bending and development length, as shown in Fig. 12-12. Although no stipulation is contained in the Code with respect to combined footings, it is reasonable to regard such footings as several isolated footings connected together.

The critical sections for bending and development length as shown on the figure may be described as follows:

(a) For footings supporting *concrete columns, pedestals or walls*, the critical section occurs at the *face* of the column, pedestal or wall.

(b) For footings supporting *masonry walls*, the critical section lies midway between the center of the wall and the face of the wall. This is due to the flexibility of a masonry wall.

(c) For footings supporting columns founded on *metal base plates* which are in turn supported on a pier or pedestal, the critical section occurs at the midpoint between the edge of the base plate and the edge of the pedestal.

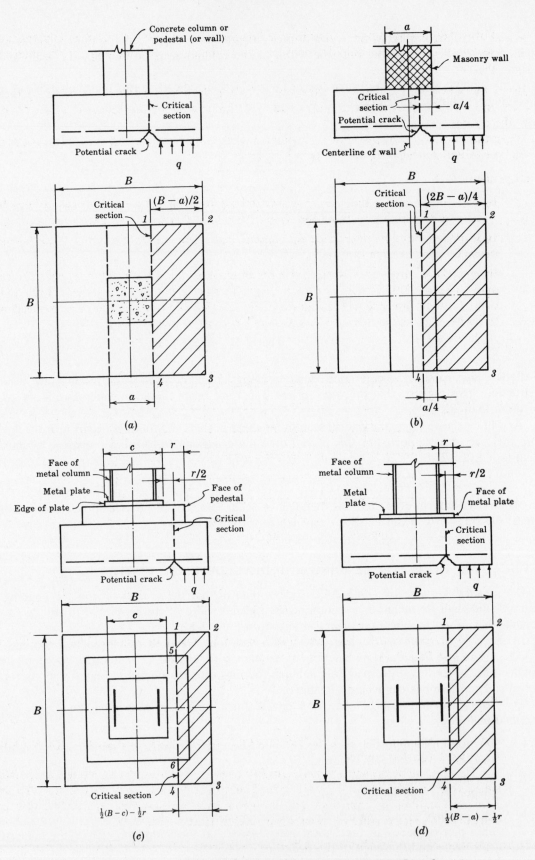

Fig. 12-12   Critical Sections for Bending and Development Length

(*d*)   For footings supporting columns founded on *metal base plates* without pedestals, the critical section occurs at the midpoint between the column face and the edge of the metal base plate.

In all the above cases, the total shear force V_B for flexure is that force exerted by the soil pressure on area *1-2-3-4* as shown in Fig. 12-12. The design bending moment is the moment of V_B about the critical section.

When the footings are *stepped*, or at changes in depth or quantity of reinforcement, the section should be checked for bending and development length.

Reinforcement

(*a*)   In *two-way footings*, the reinforcement should be designed to resist bending and bond stresses at the critical section, and should be uniformly distributed across the footing.

(*b*)   In *square two-way footings*, the steel should be identical in both directions. (Two layers are provided, one in each direction.)

(*c*)   In *rectangular two-way footings*, the steel must be computed separately in each direction, and spaced as follows: (1) The *long bars* should be equally distributed across the footing over the *short dimension B*, as shown in Fig. 12-13. (2) The *short bars* must be distributed in *three band widths* over the long dimension *L*, as shown in Fig. 12-13.

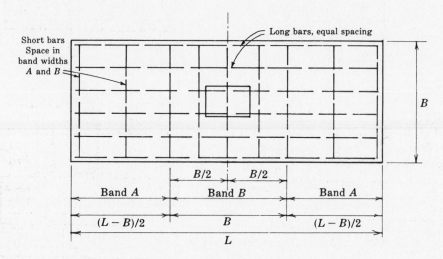

Fig. 12-13

The *area of steel* A_{sB} in band width *B* is obtained from equation (*12.19*), and the steel A_{sA} in band width *A* is obtained from equation (*12.20*):

$$A_{sB} = 2A_{st}/(S+1) \qquad (12.19)$$

and

$$A_{sA} = (A_{st} - A_{sB})/2 \qquad (12.20)$$

where

$$S = L/B \qquad (12.21)$$

and A_{st} is the *total* area of short bars required. Evidently, $A_{st} = A_{sB}$ for a square footing, since $S = 1$ for that condition.

The spacing of the short bars stated above refers *only* to isolated footings, supporting a single column. The short bars are spaced uniformly across the widths of the *transverse beams* in multiple column footings. The short bars are designed separately under each column to resist the bending moments in each transverse beam.

Some quantity of steel is always placed in the segments between the transverse beams, and outside of the transverse beams toward the ends of the footing.

The steel provided *should* be *at least* that which is specified for temperature reinforcement, but the steel should be designed in a manner similar to that for isolated square footings or continuous concrete wall footings, assuming an *imaginary wall face* to exist along the faces of the columns. [*Authors' recommendation*: $\rho_{min} = 200/f_y$ $(1.4f_y)$.]

CRITICAL SECTIONS FOR SHEAR AND DIAGONAL TENSION

Two separate requirements must be considered with respect to shearing forces on *single column footings* and *multiple column footings*, whether square, rectangular or any other shape. The conditions are shown in Fig. 12-14 below.

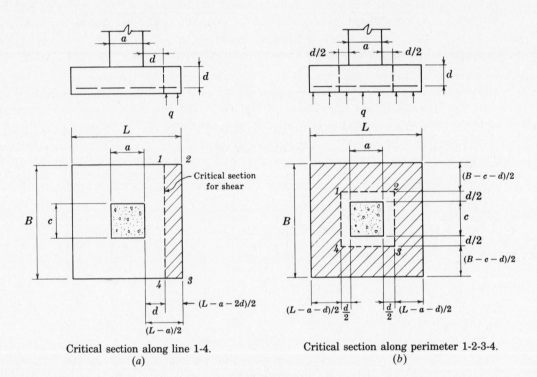

Critical section along line 1-4.
(a)

Critical section along perimeter 1-2-3-4.
(b)

Fig. 12-14

The first condition, shown in Fig. 12-14(*a*), assumes that the section is a wide beam, and all of the provisions for shear in beams apply in this case. The shear force (V or V_u) on Section *1-4* consists of all of the net soil pressure or pile reactions applied on area *1-2-3-4* as shown in Fig. 12-14(*a*). The width *b* of the section resisting shear extends the full width of the footing. In this case, $b = B$. Nominal shear force is computed as

$$V_n = V_u/\phi \qquad (12.22)$$

for strength design, where $\phi = 0.85$.

The permissible shear force for beam (one-way) shear is

$$V_c = 2\sqrt{f_c'}\, bd \quad (\tfrac{1}{6}\sqrt{f_c'}\, bd) \qquad (12.23)$$

The *more detailed analysis* may be used and web reinforcement may be provided as described for shear in beams (see Chapter 6).

The second condition considers perimeter shear or *punching shear* on the critical Section *1-2-3-4* as shown in Fig. 12-14(*b*). The shear force V or V_u consists of all of the net forces on the footing due to soil pressure or pile reactions which exist *outside* of Section *1-2-3-4*. Nominal shear force is given by (12.22); permissible shear force is given by (6.11) and (6.12). In (6.11), β_c is the column aspect ratio and, in terms of Fig. 12-14(b),

$$b_o = 2(a + d) + 2(c + d)$$

For a single circular pile, $b_o = \pi(d_p + d)$.

If the detailed method of analysis is to be used, or if shear reinforcement is to be provided, the appropriate equations discussed in Chapter 6 apply.

ADDITIONAL REQUIREMENTS

(1) Piles are considered as contributing reactions concentrated at the pile center, except in the case of a pile near the critical section. A pile having its center located *at least $d_p/2$ outside of the critical section* delivers a *full pile load* to the critical section. If the center of the pile is located $d_p/2$ *inside of the critical section*, it delivers no load to the critical section. Referring to Fig. 12-15, an equation may be derived for these conditions as

$$P'/P = \begin{cases} 0 & X < -d_p/2 \\ \frac{1}{2} + (X/d_p) & -d_p/2 \leqq X \leqq d_p/2 \\ 1 & X > d_p/2 \end{cases} \qquad (12.24)$$

where P' = The proportion of pile load P delivered to the critical section.

X = The distance of the pile center from the critical section, counted positive when within the critical section, and negative when outside the critical section.

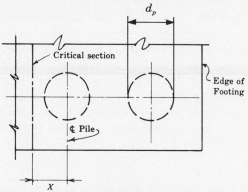

Fig. 12-15

(2) Minimum clearances for footings on piles are shown on Fig. 12-16. Embedment of piles should be 4″ to 6″ (100 to 150 mm) for timber piles, and 6″ to 12″ (150 to 300 mm) for concrete or metal piles. For heavily loaded piles, the larger values apply. Unreinforced concrete may not be used for pile footings.

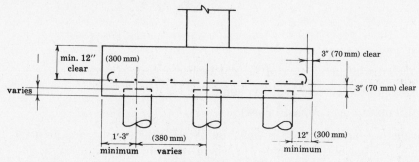

Fig. 12-16

(3) Minimum clearances for footings on soil (spread footings) are shown on Fig. 12-17.

(4) For *round columns* (or pedestals) an equivalent concentric square section having the same area as the round section shall be used for determining the critical sections. This is also practical for use with other symmetrical shapes, i.e. octagonal, hexagonal, cross, etc., and is permitted by the 1983 ACI Code.

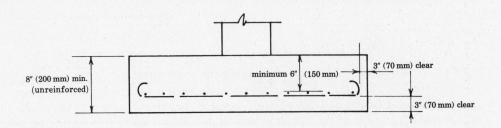

Fig. 12-17

TRANSFER OF FORCE AT THE BASE OF A COLUMN

(1) The forces in the longitudinal steel in any column shall be transferred into its supporting pedestal or footing by extending the vertical bars into the pedestal or footing a sufficient distance to develop the full bar capacity.

(2) In lieu of extending the column bars into the pedestal or footing, the use of *dowels* is a common practice. The dowels must have a cross-sectional area at least equal to that required to transfer the force that is not transferred by the concrete. The minimum area of dowels is $0.05bh$, and at least four dowel bars must be used for each column. The dowel bar diameter may not exceed that of the column bars by more than $\frac{1}{8}''$ (3 mm).

Dowels must be lapped with the column bars in the column for distances equal to the development of the larger bars (see Chapter 7). The dowels must extend into the pedestal or footing a distance at least sufficient to develop the yield strength of the bar.

ALLOWABLE BEARING STRESSES ON THE TOP SURFACES OF REINFORCED CONCRETE PEDESTALS AND FOOTINGS

The compression stress in the concrete at the base of a column shall be assumed to be distributed to the pedestal or footing as *bearing stress*.

The design bearing strength of a column or a pedestal on a footing must satisfy

$$F_b \leqq 0.85 \phi_{br} f'_c \alpha A_1 \qquad (12.25)$$

where

$\phi_{br} = 0.7$ for bearing

A_1 = the cross-sectional area of the column or pedestal (loaded area)

f'_c = strength of concrete in the column

$\alpha = 1.0$ (except as outlined below)

When the supporting surface is wider on all sides than the loaded area (A_1), α may be taken equal to $\sqrt{A_2/A_1}$ but not more than 2.0, where A_2 is the supporting area. See Fig. 12-18 and Fig. 12-19.

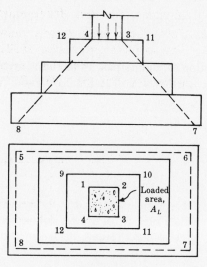

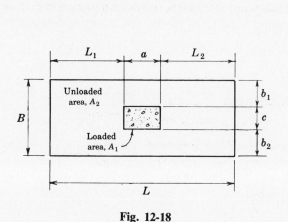

Fig. 12-18   **Fig. 12-19**

BEARING FOR SLOPED OR STEPPED REINFORCED CONCRETE FOOTINGS

For sloped or stepped footings (as shown in Fig. 12-19) the supporting area (A_2) for bearing may be taken as

(1)   the top horizontal surface of the footing (area *9-10-11-12*) or

(2)   the largest base (area *5-6-7-8*) of the frustrum of a cone wholly within the footing, formed by sides sloping from the actual loaded area and extending outward on a slope of 1 vertical to 2 horizontal.

Under certain circumstances, Case (1) will provide the larger area, and under other circumstances Case (2) will provide the larger area.

The area selected shall then be used as the unloaded area as shown in Fig. 12-18. The restrictions shown in that figure are fully applicable here. A_2 and A_1 are used to compute α for (*12.24*).

STRUCTURAL DESIGN OF FOOTINGS

The effective depth required and the steel reinforcement may be determined using either the alternate design method or the strength design method. For both procedures the methods of obtaining the basic elastic shears and moments are identical. The difference lies in the use of safety factors in using one method rather than the other. For the alternate design method, the shears and moments are used as calculated, whereas those values are increased by load factors for the strength design method.

Sections are proportioned by the same methods as used for beams or slabs, in general. For both methods of design, the equations for proportioning can be placed in similar forms. The strength method will be used in the design problems. *The authors recommend* $\rho_{\min} = 200/f_y$ $(1.4/f_y)$.

FLEXURAL DESIGN

The required effective depth to resist compressive stresses is

$$d = \sqrt{M_u/(\phi R_u)b}$$ (*12.26*)

and the required area of reinforcement is

$$A_s = M_u/(\phi a_u)d \qquad (12.27)$$

Attention should be paid to the units involved in (12.26) and (12.27), when these equations are used in conjunction with Table 5.2 or Table 5.3. The reader is referred to the discussion in Chapter 5.

Solved Problems

12.1. Derive an equation for the required effective depth for flexure of a square spread footing which supports a square isolated column as shown in Fig. 12-20. The maximum soil pressure is q. Use the alternate design method.

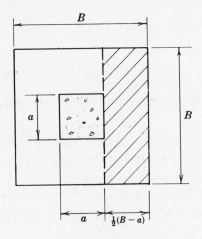

Fig. 12-20

The total shear force V_B on the critical section is $V_B = qB(B - a)/2$, and the moment of that force about the critical section is $M = qB(B - a)^2/8$.

The required depth for flexure is

$$d = \sqrt{M/RB} = \sqrt{qB(B - a)^2/8RB}$$

or

$$d = \tfrac{1}{2}(B - a)\sqrt{q/2R} \qquad (12.28)$$

The equation may be modified for use with strength design by replacing q with q_u and R with ϕR_u.

12.2. Using strength design, derive an equation for the required effective depth of the footing in Problem 12.1 to resist shear as a measure of diagonal tension. Refer to Fig. 12-21 and to (6.11) and (6.12).

For slab shear as shown in Fig. 12-21(a), $V = (B - a - 2d)Bq/2$.
The unit shear is $v = V/bd = V/Bd$, since here $b = B$; then $d = (B - a - 2d)q/2v$.
The allowable shear stress is $\alpha\sqrt{f_c'}$, where $\alpha = 2\phi$ $(\tfrac{1}{6}\phi)$; thus

$$(B - a - 2d)(q/2d) = \alpha\sqrt{f_c'} \qquad (12.29)$$

or

$$\tfrac{1}{2}(B/d - a/d - 2) = \alpha\sqrt{f_c'}/q \qquad (12.30)$$

which can be used effectively for developing design aids since it is dimensionless.

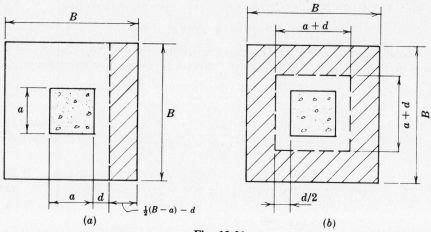

Fig. 12-21

For perimeter shear as shown in Fig. 12-21(b), $V = [B^2 - (a + d)^2]q$; and $v = V/b_o d$ where $b_o = 4(a + d)$. Thus

$$\frac{[B^2 - (a + d)^2]q}{4d(a + d)} = \gamma \sqrt{f_c'} \tag{12.31}$$

where $\gamma = 2 + 4/\beta_c \leqq 4$ $[\frac{1}{6}(1 + 2/\beta_c) \leqq \frac{1}{3}]$. This equation becomes

$$\frac{(B/d)^2 - (a/d + 1)^2}{a/d + 1} = 4\gamma \sqrt{f_c'}/q \tag{12.32}$$

which may be readily utilized to produce design aids since it is dimensionless.

12.3. An 18″ square column is required to support a 214 kip dead load force and a 154 kip live load force. The allowable soil pressure is 4.25 kips/ft². Using $f_c' = 3000$ psi and $f_y = 40,000$ psi, design the footing using strength design. The column contains 10 No. 9 vertical bars.

Since allowable soil pressure is stated rather than ultimate soil pressure, the footing area is calculated using service loads. Thus considering the footing to weigh *approximately* 7.0% of the total applied load, $W_F = (0.070)(214 + 154) = 25.76$ kips (use 26 kips).

The total load is therefore $W = 214 + 154 + 26 = 394$ kips, and the required footing area is $A_F = W/q = 394/4.25 = 92.7$ ft². Thus $B = \sqrt{92.7} = 9.63'$. Try a footing 9′-8″ square for which $A_F = 93.5$ ft².

The *ultimate applied load* will be $P_u = 1.4W_D + 1.7W_L = (1.4)(214) + (1.7)(154) = 561$ kips. The net ultimate soil pressure will then be: net $q_u = P_u/A_F = 561/93.5 = 6.0$ kips/ft².

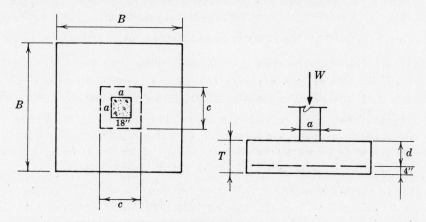

Fig. 12-22

Refer to Fig. 12-22 and check perimeter shear. Assume $d = 19''$. Thus

$$c = a + d = 18'' + 19'' = 37'' = 3.08' \qquad \text{and} \qquad b_o = (4)(37) = 148''$$

The total shear force outside the critical section will be $V_o = 6.0[93.5 - (3.08)^2] = 504$ kips. Since the column aspect ratio is $\beta_c = 18/18 < 2$, the permitted punching shear force is given by (6.12) as

$$V_c = 4\sqrt{3000}\,(148)(19) = 616{,}080 \text{ lb} = 616.08 \text{ kips} > 504 \text{ kips}$$

Hence, the footing is satisfactory for punching shear.

Refer to Fig. 12-23 and check beam shear. The critical section is at D-D', and the total shear force outside of the section is $V_u = (6.0)(30/12)(9.67) = 145$ kips. The beam shear force allowed on the concrete is

$$V_c = 2\sqrt{f_c'}\,bd = 2\sqrt{3000}\,(116)(19) = 241{,}430 \text{ lb} = 241.43 \text{ kips} > 145 \text{ kips}$$

The footing is therefore satisfactory for beam shear.

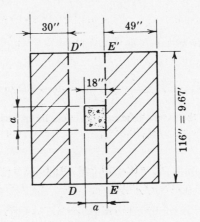

Fig. 12-23

Refer to Fig. 12-23 and design the section for flexure. The critical section is at E-E'. The ultimate moment about E-E' is $M_u = (6.0)(49)^2/[(2)(12)^2] = 50.0$ ft-kips/ft. Now

$$\phi R_u = \frac{M_u}{bd^2} = \frac{50.0 \times 12{,}000}{(12)(19)^2} = 138.5 \text{ psi/ft width}$$

From Table 5.2, for $f_c' = 3000$ psi and $f_y = 40{,}000$ psi, the required steel ratio ρ is 0.004. This is less than $\rho_{min} = 0.005$. However, the 1983 ACI Code permits the use of 1.33 times the required steel ratio when it is less than ρ_{min}. But $(1.33)(0.004) = 0.00532$, which is larger than ρ_{min}; so $\rho = 0.005$ will be used. Thus,

$$A_s = 0.005bd = (0.005)(116)(19) = 11.02 \text{ in.}^2$$

Using No. 6 bars having an area 0.44 in.^{2} each, the number of bars will be $11.02/0.44 = 25$. The distance available for the steel is

$$116.0 - 2(3.0) - 0.75 = 109.25 \text{ in.}$$

The center-to-center distance between bars will be (24 spaces) $109.25/24 = 4.55$ in.

The maximum spacing is the lesser of 18 in. and $3h = 3(19.0 + 0.75 + 3.0) = 3(22.75) = 68.25$ in. The minimum spacing is the greatest of

(a) $2d_b = 2(0.75) = 1.5$ in.

(b) $1.0 + d_b = 1.75$ in.

(c) $(1.33)(\text{maximum aggregate size}) + d_b = (1.33)(1) + 0.75 = 2.08$ in.

It follows that the spacing furnished is satisfactory.

It is necessary to investigate the development length of the bars. The available distance is from the face of the column to the end of the bar,

$$\frac{116.0}{2} - \frac{18.0}{2} - 3.0 = 46.0 \text{ in.}$$

The development length required is the greatest of

(a) $0.04 A_b f_y / \sqrt{f_c'} = (0.04)(0.44)(40,000)/\sqrt{3000} = 12.85$ in.

(b) $0.0004 d_b f_y = (0.0004)(0.75)(40,000) = 12.0$ in.

(c)   12.0 in.

The 46″ furnished is more than satisfactory.

At this point, the design tables in Chapter 5 will be employed. Table 5.2, for $f_c' = 3000$ psi, $f_y = 40,000$ psi, and $\rho = 0.005$, gives $\phi a_u = 2.882$, $\phi j_u = 0.8646$, and $a/d = 0.0784$; furthermore, the area of steel is

$$A_s = \rho b d = (0.005)(12)(19) = 1.14 \text{ in.}^2/\text{ft width}$$

We thus have three equivalent ways of computing the resisting moment:

(a) $M_u = A_s (\phi a_u) d = (1.14)(2.882)(19) = 62.42$ ft-kips/ft

(b) $M_u = A_s f_y (\phi j_u) d = (1.14)(40,000)(0.8646)(19)/12,000 = 62.42$ ft-kips/ft

(c) $M_u = \phi A_s f_y d \left(1 - \dfrac{a/d}{2}\right) = (0.9)(1.14)(40,000)(19)\left(1 - \dfrac{0.0784}{2}\right) \Big/ 12,000 = 62.43$ ft-kips/ft

Transfer of stress at the base of the column must be investigated. The allowable bearing force at the base of the column will be $F_{bc} = \phi_{br}(0.85 f_c') A_1$, where $\phi_{br} = 0.7$ and A_1 is the cross-sectional area of the column; thus,

$$F_{bc} = (0.7)(0.85)(3.0)(18)(18) = 578.3 \text{ kips}$$

On the footing, $F_b = \phi_{br}(0.85 f_c') \sqrt{A_2/A_1}$, provided $\sqrt{A_2/A_1} \leqq 2.0$. But, as seen from Fig. 12-24,

$$\sqrt{A_2/A_1} = \sqrt{(109)(109)/(18)(18)} = 6.06 > 2.0$$

So $F_b = \phi_{br}(2)(0.85)(f_c') A_1 = (0.7)(2)(0.85)(3.0)(18)(18) = 1156.7$ kips

The ultimate load is $(1.4)(200) + (1.7)(145) = 526.5$ kips, which is less than 578.3 kips; so only minimum dowels are required:

$$A_{sd} = 0.005(A_1)_{col} = (0.005)(18)(18) = 1.62 \text{ in.}^2$$

with no fewer than four bars. If No. 6 dowel bars are used, $1.62/0.44 = 3.68$ dowels are needed; use four No. 6 dowels. The development length into the footing required is 12.65 in. which is less than $d = 19$ in.

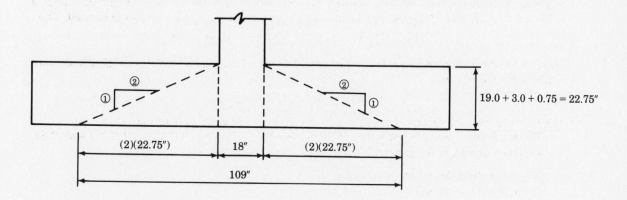

Fig. 12-24

The splice length for the No. 6 dowels up into the column will be based on the larger-size No. 9 bars, which means the greatest of

(a) $(0.04)(1.0)(40,000)/\sqrt{3000} = 29.2$ in.

(b) $(0.0004)(1.128)(40,000) = 18.048$ in.

(c)   12.0 in.

Splice the bars for $29\frac{1}{2}''$ (rounded) up into the column.

The estimated footing weight, 26.0 kips, must also be checked.

$$W_f = (8.67)(8.67)(0.15)(19.0 + 0.75 + 3.0)/12.0 = 21.38 \text{ kips}$$

Hence the design is slightly conservative. The total depth required is 22.75", so 23.0" would be used. The steel consists of twenty-five No. 6 bars in each direction.

12.4.   Design a footing to support a continuous concrete wall as shown in Fig. 12-25. Allowable soil pressure q is 4.75 kips/ft^2. The wall dead load reaction is 12 kips/ft and the live load is 6 kips/ft. Use strength design with $f_c' = 3000$ psi, $f_y = 40,000$ psi. The vertical steel in the wall consists of No. 5 bars at 8" center to center on both faces. The wall width is 12".

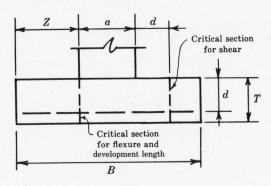

Fig. 12-25

Since the loads are relatively large compared to the allowable soil pressure, the footing will be large. Further, the minimum depth conditions will probably apply. Assume therefore that

$$T = 6'' + \text{bar diameter} + 3'' \text{ cover} = \text{approximately } 10''$$

The *net* service-load soil pressure will be equal to the maximum allowable soil pressure less the weight of the footing. Since the footing weighs 12.5 lbs/ft^2 per inch thickness,

$$\text{net } q = 4.75 - (10)(12.5/1000) = 4.625 \text{ kips/ft}^2$$

and the required area of the footing is $A_F = 18.0/4.625 = 3.89$ ft^2/ft (say $B = 4.0'$).

The factored load will be $P_u = 1.4D + 1.7L = 1.4(12) + 1.7(6) = 27.0$ kips/ft. Hence the net ultimate soil pressure is $q_u = P_u/A_F = 27.0/4.0 = 6.75$ kips/ft^2.

The permissible shear force on the concrete is

$$V_c = 2\sqrt{f_c'}\,bd = 2\sqrt{3000}\,(12)(6) = 7890 \text{ lbs} = 7.89 \text{ kips}$$

The actual ultimate shear force is $V_u = (2.0 - 0.5 - 6.5/12)(6.75) = 6.47$ kips, and

$$V_n = V_u/\phi = 6.47/0.85 = 7.61 \text{ kips} < 7.89 \text{ kips}$$

so the footing is satisfactory for beam shear.

Since $M_u = (6.75)(1.5)(1.5)/2 = 7.59$ ft-kips/ft,

$$\phi R_u = \frac{M_u}{bd^2} = \frac{7.59 \times 12,000}{(12)(6.5)^2} = 179.6 \text{ psi/ft}$$

Table 5.2, for $f'_c = 3000$ psi and $f_y = 40,000$ psi, shows

ρ	ϕR_u
0.0050	173
0.0055	189

Thus, by interpolation, $\rho = 0.0050 + (0.0005)(6.6/16) = 0.0052$. Then $A_s = (0.0052)(12)(6.5) = 0.406$ in.^{2}/ft. One No. 4 bar has an area of 0.20 in.^{2}, so $s = (0.2/0.406)(12) = 5.91''$. Use No. 4 bars at 5.5 inches on centers. The development length for a No. 4 bar is the minimum, 12 in., and the available development length is $18.0'' - 3.0'' = 15.0''$, which is satisfactory.

In the longitudinal direction, shrinkage steel will be provided.

$$A_s = 0.002bd = (0.002)(48)(6.5) = 0.624 \text{ in.}^2$$

Use four No. 4 bars.

The No. 5 bars in the wall are extended into the footing to be developed. The development length of a No. 5 bar in compression must exceed

(a) $0.02 A_b f_y / \sqrt{f'_c} = (0.02)(0.31)(40,000)/\sqrt{3000} = 4.53$ in.

(b) $0.0003 d_b f_y = (0.0003)(0.625)(40,000) = 7.5$ in.

(c)   8 in.

The No. 5 bars in the wall must be developed in the footing at least 8 inches. Add a lug on the bottom of the footing, as shown in Fig. 12-26, to provide for the development length.

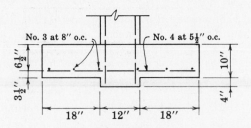

Fig. 12-26

The 1983 ACI Code provides procedures for transfer of stress at the base of a column (see, e.g., Problem 12.3). The procedures for columns might be extrapolated and applied to a wall, considering the wall to be a series of columns side by side. But since it is not clear what width of wall is to be used, and since walls do not contain tie bars, the column procedure might not prove satisfactory.

It should be noted that whenever the top of the footing is below ground, the weight of the soil above the footing must be considered as dead load.

12.5. Referring to Fig. 12-27, design a rectangular footing which supports a 16 in. × 16 in. square column, using $f'_c = 3000$ psi and $f_y = 40,000$ psi. Service loads are $P_D = 210$ kips and $P_L = 150$ kips. The footing width is limited to 6.5 ft. The gross allowable soil pressure at service loads is 5.0 kips/ft^2. Assume No. 8 bars. The base of the footing is 6 ft below grade. Soil density is 100 lbs/ft^3 and concrete density is 150 lbs/ft^3. Use strength design (load factors and understrength or resistance factors). For the column, $f'_c = 5000$ psi, $f_y = 60,000$ psi, and the vertical bars are No. 9.

The solution requires iteration to determine the footing length and the total depth. After several trials, a footing length of 13.0 ft and a total depth of 27 in. are considered to be satisfactory. The final review concerning those dimensions follows.

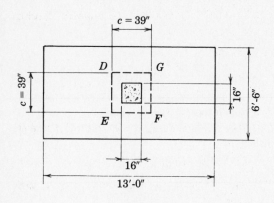

Fig. 12-27

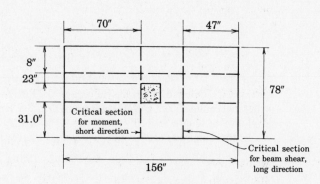

Fig. 12-28

Total service load is $P_D + P_L = 210 + 150 = 360$ kips. The footing area required is therefore $A_F = (P_D + P_L)/q_{net}$, where q_{net} is the allowable net soil pressure after subtracting the weight of the soil overburden and the footing weight, both in kips/ft^2:

$$q_{net} = 5.0 - (27/12)(0.15) - (6.0 - 27/12)(0.10) = 4.827 \text{ kips/ft}^2$$

Thus, $A_F = 360/4.287 = 83.97$ ft^2 and, since $B = 6.5$, $L = 83.97/6.5 = 12.92$ ft. Use

$$L = 13.0 \text{ ft} \qquad \text{and} \qquad A_F = (6.5)(13.0) = 84.5 \text{ ft}^2$$

When the ratio L/B for a rectangular footing approaches 2.0, beam-type shear governs the depth of the footing in the long direction. The effective depth to the top of the lower layer of steel is $d = 27.0 - 3.0 - 1.0 = 23.0$ in. Fig. 12-28 shows that the overhang beyond the critical section for beam shear (d from the face of the support) is 47.0 in. = 3.917 ft. The ultimate soil pressure is

$$(1.4P_D + 1.7P_L)/A_F = [(1.4)(210) + (1.7)(150)]/84.5 = 6.497 \text{ kips/ft}^2$$

Thus, $V_u = 6.497(6.5)(3.917) = 165.4$ kips and $V_n = V_u/\phi = 165.4/0.85 = 194.6$ kips. Allowable is

$$V_c = 2\sqrt{f_c'}\, bd = 2\sqrt{3000}\,(78)(23)/1000 = 196.5 \text{ kips} > 194.6 \text{ kips} \quad \text{(O.K.)}$$

Punching shear will not govern, but the calculations will be illustrated. The perimeter $b_o = 4(16 + 23) = 156$ in., shown as $DEFG$ on Fig. 12-27. The ultimate shear is the total force acting outside of $DEFG$; thus,

$$V_u = 6.497[84.5 - (39/12)^2] = 480.37 \text{ kips}$$

and $V_n = V_u/\phi = 480.37/0.85 = 565.1$ kips. The allowable punching shear around a square column is

$$V_c = 4\sqrt{f_c'}\, b_o d = (4\sqrt{3000})(156)(23)/1000 = 786.1 \text{ kips} > 565.1 \text{ kips} \quad \text{(O.K.)}$$

Beam shear in the short direction is investigated as in the long direction. Because it will never control the design, the calculations are not shown here.

The critical section for bending moment and bar development is at the face of the column. This applies to both long and short directions.

For the long direction, the "cantilever" length of the footing, as shown on Fig. 12-28, is 70.00 in. = 5.833 ft. Hence, the bending moment is

$$M_u = 6.497(6.5)(5.833)(5.833/2.0) = 718.42 \text{ ft-kips}$$

and

$$\phi R_u = M_u/bd^2 = (718.42)(12,000)/[(6.5 \times 12)(23)^2] = 208.93 \text{ psi}$$

Interpolating in Table 5.2, $\rho = 0.006092$, so $A_s = \rho bd = (0.006092)(6.5 \times 12)(23) = 10.93$ in.^{2} Using No. 8 bars, the number required is $10.93/0.79 = 13.8$ (use 14).

The distance available is $(6.5)(12) - 6.0 = 72$ in. and spacing is $72/13 = 5.54$ in. Use 5.5-inch center-to-center spacing.

The development length available is $(13.0 \times 12)/2.0 - 8.0 - 3.0 = 67.0$ in. The required development length is the largest of

(a) $0.04 A_b f_y / \sqrt{f'_c} = 0.04(0.79)(40,000)/\sqrt{3000} = 23.08$ in.

(b) $0.0004 d_b f_y = 0.0004(1.0)(40,000) = 16.0$ in.

(c) 12.0 in.

The governing development length is 23.08 in. < 67.0 in. available   (O.K.).

The cantilever overhang beyond the column face in the short direction is, from Fig. 12-28, 31.0 in. $= 2.583$ ft. The short-direction bending moment is

$$M_u = (6.497)(13.0)(2.583)^2/2.0 = 281.82 \text{ ft-kips}$$

whence $\phi R_u = M_u/bd^2 = (281.82 \times 12,000)/[(13.0 \times 12)(23.0)^2] = 40.98$ psi, which is less than ϕR_u corresponding to $\rho_{min} = 0.0050$. Consequently, $\rho = 0.0050$ and

$$A_s = \rho bd = (0.005)(13.0 \times 12)(23.0) = 17.94 \text{ in.}^2$$

Using No. 8 bars in the short direction, the number of bars required is $17.94/0.79 = 22.7$, or 23 bars, with a total $A_s = 23(0.79) = 18.17$ in.^{2}

The 1983 ACI Code requires that most of the short bars be concentrated over a "band width B," centered on each side of the column; $B = 6.5$ ft in this problem. The amount of steel area for "band width B" is equal to $2.0 A_s/(\beta + 1)$, where $\beta = L/B = 13.0/6.5 = 2.0$. Thus, $\frac{2}{3}$ of the steel must be distributed over 29 in. on each side of the column centerline, and $\frac{1}{6}$ of the steel on each side of "band width B."

The available development length is $(78.0 - 16.0)/2.0 - 3.0 = 28.0$ in. From previous calculations, $L_d = 23.08$ in. required, which is less than 28.0 in. available   (O.K.).

The column forces must be transferred to the footing by bearing and steel dowels. The ultimate axial load from the column is

$$P_u = 1.4(210) + 1.7(150) = 549 \text{ kips}$$

With the column cross-sectional area $A_1 = 16 \times 16 = 256$ in.^{2}, the allowable ultimate axial force is

$$P_u = \phi(0.85)f'_c A_1 = 0.7(0.85)(5.0)(256) = 761.6 \text{ kips} > 549 \text{ kips}$$

so bearing stress on the column is satisfactory. (Note that ϕ for bearing is 0.7.) By (12.24), the bearing force allowed on the footing is

$$(0.85)(0.7)(3000 \times 10^{-3})(2)(256) = 913.9 \text{ kips}$$

which exceeds 549 kips, the ultimate axial load.

The 1983 ACI Code requires minimum area of dowels from the footing into the column, equal to

$$0.005 A_1 = 0.005(256) = 1.28 \text{ in.}^2$$

The minimum number of dowels is four, so each dowel must have an area of $1.28/4 = 0.32$ in.^{2} Using four No. 6 bars, the dowel area is $4(0.44) = 1.76$ in.^{2} The ratio of steel area of dowels required to area of dowels provided is $1.28/1.76 = 0.727$; the required development lengths may be reduced by this factor.

The development length into the footing is based on $f_y = 40,000$ psi and $f'_c = 3000$ psi. The dowels are in compression, so the development length is the greatest of

(a) $0.02 d_b f_y / \sqrt{f'_c} = 0.02(0.75)(40,000)/\sqrt{3000} = 10.95$ in.

(b) $0.003 d_b f_y = 0.003(0.75)(40,000) = 9.0$ in.

(c) 8.0 in.

The governing development length is 10.95 in. However, this may be multiplied by the factor 0.727, making

$$L_d = 10.95(0.727) = 7.96 \text{ in.}$$

With an effective depth of 23.0 in., there is more than sufficient available development length.

The lap of the dowel bars with the vertical bars in the column is based on No. 9 column bars, $f_y = 60,000$ psi, and $f'_c = 5000$ psi. The development length into the column is the greatest of

(a) $0.02d_b f_y / \sqrt{f_c'} = (0.02)(1.128)(60,000) / \sqrt{5000} = 19.14$ in.

(b) $0.0003 d_b f_y = 0.0003(1.128)(60,000) = 20.30$ in.

(c)   8.0 in.

The controlling development length is 20.30 in., but this may be multiplied by the factor 0.727. Therefore, the required development length of the dowels into the column is $0.727(20.3) = 14.76$ in. The dowels should be extended into the column 15 inches.

12.6. Design a pile footing to support an 18″ square column subjected to a live load reaction of 180 kips and a dead load reaction of 160 kips at service loads. The testing laboratory recommends an ultimate pile load of 70 kips per pile, and a service pile load of 42 kips per pile. The vertical steel in the column consists of 12 No. 7 bars. Use $f_c' = 3000$ psi, $f_y = 40,000$ psi, and 12″ diameter piles.

Since the footing weight will be about 3 kips/pile, the net service load per pile is $42.0 - 3.0 = 39.0$ kips/pile. The number of piles required is $N = W/P = 340/39 = 8.7$, or 9 piles. Use a pile pattern as shown in Fig. 12-29.

The net ultimate load is used to design the footing; thus $W_u = (1.4)(160) + (1.7)(180) = 530$ kips, and the load per pile is $P_u = 530/9 = 58.9$, say 59.0 kips/pile, which is less than the maximum ultimate load, 70 kips/pile.

Punching shear around a single pile often governs the footing depth determination, except in cases in which the loads are small. In this case, it will be shown that beam shear governs.

Referring to Fig. 12-30, we calculate the punching shear stress.

After several trials, assume $d = 19.5″$. The shear perimeter is $b_o = \pi(12 + d) = 99.0″$. The permissible shear force around the pile will be

$$V_c = 4\sqrt{f_c'}\, b_o d = 4\sqrt{3000}\,(99)(19.5)/1000 = 423 \text{ kips}$$

Since the actual shear force is the nominal pile reaction, $P_n = P_u/\phi = 59.0/0.85 = 69.4$ kips < 423 kips, the pile will not punch through the pile cap (footing).

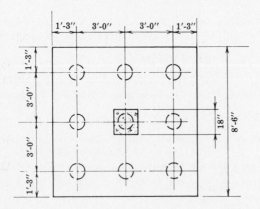

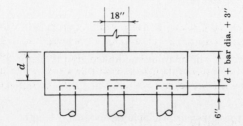

Fig. 12-29

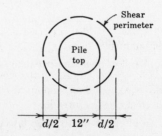

Fig. 12-30

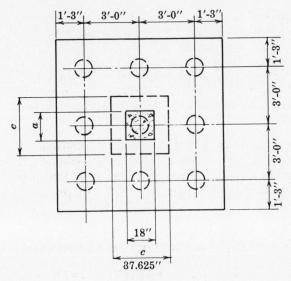

Fig. 12-31   **Fig. 12-32**

Perimeter shear (punching shear) must now be checked around the column in a similar manner. In this case, all of the nominal pile reactions outside of the critical section plus any partial reactions outside of the critical section will contribute to punching shear for the column. Refer to Fig. 12-31.

Assuming No. 6 bars will be used, clearance above the pile butts will be 3″ and embedment of the piles will be 6″. The total depth required will be 28.75″. For practical reasons use 29″; this furnishes an effective depth $d = 19.625″$. Thus $c = a + d = 18.0 + 19.625 = 37.625″$ and $b_o = 4(37.625) = 150.0″$. Hence, $V_{ou} = 472$ kips on 8 piles outside of the critical section as shown on Fig. 12-31. The permissible punching shear force ($\beta_c = 18/18 < 2$) is given by (6.12) as

$$V_c = 4\sqrt{3000}\,(150)(19.625)/1000 = 644.9 \text{ kips}$$

The force to be resisted is $V_n = V_{ou}/\phi = 472/0.85 = 555.3$ kips; therefore the pile cap (footing) is satisfactory for punching shear.

Beam shear must now be checked. Refer to Fig. 12-32. Three piles exist beyond the critical section, so $V_u = (3)(59.0) = 177.0$ kips. Since $b = B = 8'\text{-}6″ = 102″$, the permissible beam shear (one-way shear) force on the critical section is

$$V_c = 2\sqrt{f'_c}\,bd = 2\sqrt{3000}\,(102)(19.625)/1000 = 219.3 \text{ kips}$$

The force to be resisted is the nominal shear force, $V_n = V_u/\phi = 177/0.85 = 208.2$ kips. Hence the footing is satisfactory for beam shear.

The bending moment about the face of the column must now be investigated. Refer to Fig. 12-32.

$$M_u = (177)(27/12) = 398.3 \text{ ft-kips}$$

$$\phi R_u = \frac{M_u}{bd^2} = \frac{398.3 \times 12,000}{(102)(19.625)^2} = 121.67 \text{ psi}$$

Table 5.2, for $f'_c = 3000$ psi and $f_y = 40,000$ psi, discloses the fact that the steel ratio required is less than the minimum steel ratio, $\rho_{\min} = 200/f_y = 0.005$. Further, if the steel ratio required is increased by 1/3, it will still be less than $\rho_{\min}$. It would appear that 4/3 times the required steel ratio would satisfy the 1983 ACI Code. However, the Code does not permit unreinforced (plain concrete) pile caps. Since any section having less than minimum reinforcement is usually considered to be unreinforced, the minimum area of steel will be provided. Thus,

$$A_s = (200/f_y)bd = (200/40,000)(102)(19.625) = 10.0 \text{ in.}^2$$

Use seventeen No. 7 bars ($A_s = 10.2$ in.^{2}).

The 1983 ACI Code is not explicit concerning minimum steel for footings. Hence, some structural engineers use $0.002bh$ for minimum steel area if $f_y = 40,000$ psi and $0.0018bh$ if $f_y = 60,000$ psi. This corresponds to temperature and shrinkage reinforcement requirements.

The assumed footing weight must finally be checked. The total weight is

$$W_F = (8.5)(8.5)(29)(12.5)/1000 = 26.2 \text{ kips}$$

and the weight per pile is $26.2/9 = 2.91$ kips/pile. The assumed weight of 3.0 kips/pile is most satisfactory.

The final details are shown in Fig. 12-33.

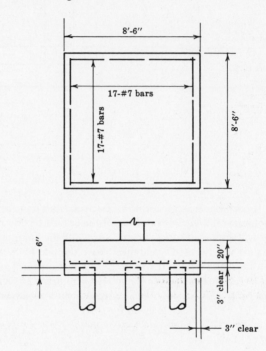

Fig. 12-33

12.7. A column is 18" square and is subjected to a dead load moment of 85 ft-kips and a live load moment of 33 ft-kips, both about the y axis as shown in Fig. 12-34. The axial loads are 143 kips for dead load and 59 kips for live load. The pile butt diameters average 12". Use $f'_c = 3000$ psi, $f_y = 40,000$ psi, and a 25.0 kips/pile service load to design the footing. The column contains 12 No. 7 vertical bars. Suppose that the *test piles* failed under a total load of 50.0 kips per pile. The forces and moments are due to superimposed loads.

Several sets of trial calculations indicate that 12 piles will be required, and a pile pattern as shown in Fig. 12-34 is selected. The total service load is 202 kips, and $M = 118$ ft-kips.

Since the footing is subjected to bending, the *equivalent moment of inertia* of the piles about the centroid of the footing must be calculated:

$$\sum d_y^2 = (2)(3)(1.5)^2 + (2)(3)(4.5)^2 = 135 \text{ ft}^2$$

The load on any pile is obtained using the equation

$$P_i = (R/N) \pm \left(Md_i \Big/ \sum d_y^2 \right) \qquad (1)$$

The maximum P_i will be in row 4 (Fig. 12-35), where $d_i = 4.5$ ft; so

$$P_4 = (202/12) + (118)(4.5)/135 = 20.76 \text{ kips}$$

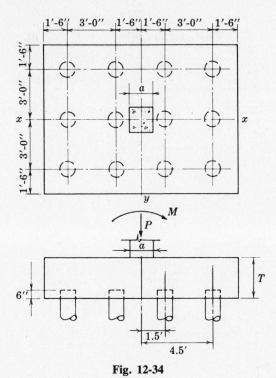

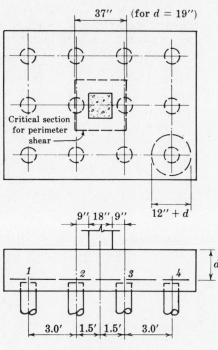

Fig. 12-34   Fig. 12-35

at service loads. Using a safety factor of 2.0 for the piles, a safe service load per pile will be $50/2 = 25$ kips. This allocates $25.0 - 20.76 = 4.24$ kips to each pile as its share in supporting the weight of the pile cap.

The ultimate load on each row of piles can be obtained by substituting the values of P_u and M_u for each row in equation (1). Here,

$$P_u = 1.4D + 1.7L = (1.4)(143) + (1.7)(59) = 300.5 \text{ kips}$$
$$M_u = (1.4)(85) + (1.7)(33) = 175.1 \text{ ft-kips}$$

whence $P_n = P_u/\phi = 300.5/0.7 = 429.3$ kips. Then, the ultimate load (R_n) on each pile can be obtained by substituting the appropriate value of d_i into the equation

$$R_n = (429.3/12) \pm (194.6d_i/135) = 35.8 \pm 1.44d_i \qquad (2)$$

Thus: for row 1, $R_n = 29.32$ kips; for row 2, $R_n = 33.64$ kips; for row 3, $R_n = 37.96$ kips; for row 4, $R_n = 42.28$ kips. Since 42.28 kips plus 1.4 times the weight of the footing per pile will not exceed the ultimate pile load of 50 kips, the pile design is satisfactory.

Investigation of punching shear around a pile, punching shear around the column, beam shear, selection of reinforcement, and transfer of forces at the base of a column proceed as in previous problems. These investigations are left to the reader.

Because a bending moment exists at the base of the column, all the bars should be doweled into the pile cap and appropriately developed.

12.8. Fig. 12-36 shows a multiple column footing which must be designed to support two columns, A and B. For column A, $P_{DL} = 118.0$ kips and $P_{LL} = 84.6$ kips. For column B, $P_{DL} = 203.6$ kips and $P_{LL} = 127.0$ kips. The columns are oriented near the property lines as shown in the figure. The *ultimate* soil pressure is 6.5 kips/ft². Establish the dimensions of the footing and draw the shear and bending moment diagrams for the long direction.

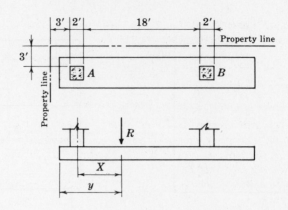

Fig. 12-36

For column A,

$$P_u = 1.4(118.0) + 1.7(84.6) = 309 \text{ kips}$$

For column B,

$$P_u = 1.4(203.6) + 1.7(127.0) = 501 \text{ kips}$$

The total load is 810 kips.

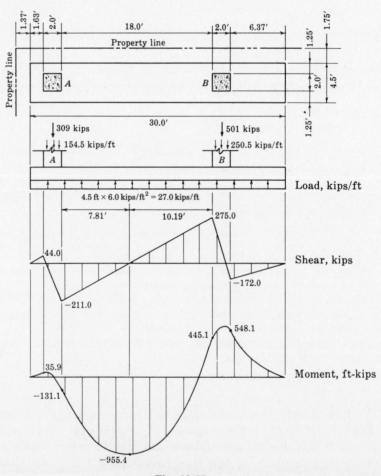

Fig. 12-37

Assuming the soil pressure due to 1.4 times the footing weight will be approximately 0.525 kips/ft^2 (after several trials), the net *ultimate* soil pressure is

$$\text{net } q_u = 6.5 - 0.525 = 5.975 \text{ kips/ft}^2$$

and the required footing area is

$$A_F = 810/5.975 = 135.5 \text{ ft}^2$$

The center of gravity of the loads is located at a distance $X = (501)(20)/810 = 12.37'$ from the center of column A. Considering the overhang at A to be $1.63'$, $y = 12.37 + 1 + 1.63 = 15.0' = L/2$. Hence the length of the footing must be $30'$ in order that the centroid of the footing and the center of gravity of the loads will coincide.

Since the length is $30'$ and the area is 135.5 ft^2, the width of the footing must be $4.5'$, and the net soil pressure is 6.0 kips/ft^2.

The shear and bending moment diagrams for the footing design are shown in Fig. 12-37. Although it is usually considered desirable to make footings as nearly square as possible, it is often necessary to build long narrow footings of the type developed in this problem.

Supplementary Problems

12.9. For the footing shown in Fig. 12-38, $P_{DL} = 150$ kips, $P_{LL} = 200$ kips, $f_c' = 4000$ psi, $f_y = 60,000$ psi. Determine the required effective depth. Use strength design and net $q = 4.92$ kips/ft^2. *Ans.* $d = 15.75''$

12.10. Design the reinforcement for Problem 12.9. (Fig. 12-38) *Ans.* $A_s = 5.414$ in.^{2}

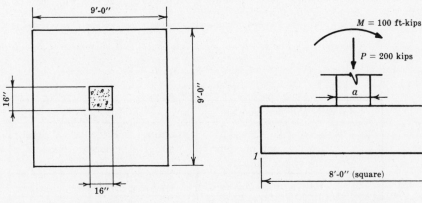

| Fig. 12-38 | Fig. 12-39 |

12.11. Determine the gross soil pressure at points *1* and *2* for the footing shown in Fig. 12-39. Include the weight of the footing. *Ans.* $q_1 = 2.25$ kips/ft^2, $q_2 = 4.59$ kips/ft^2

12.12. Determine the maximum and minimum net pile loads for the footing shown in Fig. 12-40 (not including footing weight). *Ans.* $P_{max} = 36.11$ kips, $P_{min} = 30.55$ kips

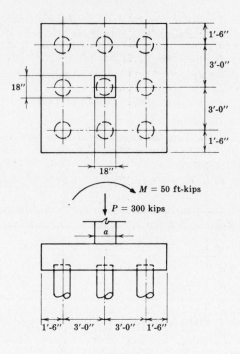

Fig. 12-40

12.13. Verify the designs of square footings indicated in Table 12.2. In each subproblem, $f_c' = 3000$ psi, $f_y = 60,000$ psi, column sizes are 20 in. $\times$ 20 in., and No. 6 bars are assumed. The footing bases are 3.0 ft below grade. Concrete density is 150 lbs/ft^3 and soil density is 100 lbs/ft^3. Subtract the footing weight and soil overburden weight from the given allowable gross soil pressure. Use 3 in. clear cover over steel. Effective depth is to the top of the lower steel layer.

Table 12.2

	GIVEN (Service Loads)			SOLUTIONS		
	P_{DL}, kips	P_{LL}, kips	Allowable soil pressure, kips/ft^2	Footing size, in.	Effective depth, in.	Steel area, in.^{2}
(a)	180	160	3.5	125	17.25	7.187
(b)	210	190	4.0	126	19.25	8.085
(c)	140	215	5.0	105	17.75	6.212
(d)	226	240	6.0	109	20.75	7.539
(e)	240	210	4.5	126	20.75	8.715
(f)	120	150	3.5	112	15.0	5.600
(g)	250	200	5.0	119	20.5	8.132

12.14. (*Hard conversion of Problem 12.13*) Verify the designs of square footings indicated in Table 12.3. In each subproblem, $f'_c = 20$ MPa, $f_y = 400$ MPa, column sizes are 508 mm × 508 mm, and No. 20M bars are assumed. The footing bases are 910 mm below grade. Concrete density is 2400 kg/m^3 and soil density is 1600 kg/m^3. Subtract the footing weight and soil overburden weight (use $g = 9.8$ m/s^2 to convert from mass to weight) from the given allowable gross soil pressure. Use 70 mm clear cover over steel. Effective depth is to the top of the lower steel layer.

Table 12.3

	GIVEN (Service Loads)			SOLUTIONS		
	P_{DL}, kN	P_{LL}, kN	Allowable soil pressure, kN/m^2	Footing size, m	Effective depth, mm	Steel area, mm^2
(a)	800	710	170	3.163	450	4982
(b)	930	845	190	3.225	490	5531
(c)	623	960	240	2.679	450	4219
(d)	1005	1070	290	2.773	530	5144
(e)	1070	935	215	3.205	530	5945
(f)	535	670	170	2.821	380	3752
(g)	1110	890	240	3.014	530	5591

Chapter 13

Two-Way Slabs

1963 ACI CODE METHODS

NOTATION

A = length of clear span in short direction, Method 3

B = length of clear span in long direction, Method 3

B = bending moment coefficient for one-way construction, Method 1

C = factor modifying bending moments prescribed for one-way construction for use in proportioning the slabs and beams in the direction of L of slabs supported on four sides, Method 1

C = moment coefficients for two-way slabs as given in tables. Coefficients have identifying indexes, such as $C_{A\,neg}$, $B_{B\,neg}$, $C_{A\,DL}$, $C_{A\,LL}$, $C_{B\,LL}$, Method 3.

C_s = ratio of the shear at any section of a slab strip distant xL from the support to the total load W on the strip in direction of L, Method 1

C_b = ratio of the shear at any section of a beam distant xL from the support to the total load W on the beam in the direction of L, Method 1

g = ratio of span between lines of inflection to L in the direction of span L, when span L only is loaded, Method 1

g_1 = ratio of span between lines of inflection to L_1 in the direction of span L_1, when span L_1 only is loaded, Method 1

L = length of clear span, Method 1

L_1 = length of clear span in the direction normal to L, Method 1

m = ratio of short span to long span for two-way slabs, Method 2

$r = gL/g_1 L_1$, Method 1

S = length of short span for two-way slabs. The span shall be considered as the center-to-center distance between supports or the clear span plus twice the thickness of slab, whichever value is the smaller, Method 2

w = total uniform load per unit area, kips/ft^2 (kN/m^2)

W = total uniform load between opposite supports on slab strip of any width or total slab load on beam when considered as one-way construction, Method 1

w = uniform load per unit area, kips/ft^2 (kN/m^2). For negative moments and shears, w is the total dead load plus live load. For positive moments, w is to be separated into dead and live loads, Method 3

w_A, w_B = percentages of load w in A and B directions. These shall be used for computations of shear and for loadings on supports, Method 3

x = ratio of distance from support to any section of slab or beam, to span L or L_1, Method 1

INTRODUCTION

The 1963 ACI Building Code defined two-way slabs with supports on four sides: "This construction, reinforced in two directions, includes solid reinforced concrete slabs; concrete joists with fillers of hollow concrete units or clay tile, with or without concrete top slabs; and concrete joists with top slabs placed monolithically with the joists. The slab shall be supported by walls or beams on all sides. . . ."

Fig. 13-1 shows a view of a two-way solid slab supported on all sides by concrete beams. Slabs reinforced in two directions but not supported on four sides will be discussed in Chapter 14. This type slab supported on columns only at the corners of the panel is termed a *flat slab* or *flat plate*.

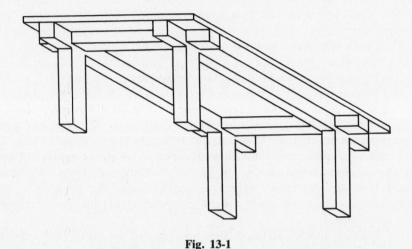

Fig. 13-1

Section 13.3 of the *1983* ACI Code states that: "A slab system may be designed by any procedure satisfying conditions of equilibrium and geometric compatibility if shown that the design strength at every section is at least equal to the required strength considering Sections 9.2 and 9.3 [of the 1983 ACI Code] and that all serviceability conditions, including specified limits on deflections, are met." On the basis of Chapters 4 and 8, it is not difficult to show that the alternate design procedure meets all the above requirements. Therefore, this procedure will be employed in the present chapter.

ANALYSIS OF TWO-WAY SLABS

Two-way slabs are extremely complex and statically indeterminate. Many attempts, analytical and empirical, have been made to determine the division of the moments and shears between the two spans and the distribution of these along the principal axes of the slab.

The slab shown in Fig. 13-2 is subjected to a uniform vertical load. If the slab is supported on all sides by non-yielding supports, the deflection at *e* of a central strip *ab* must be equal to the deflection

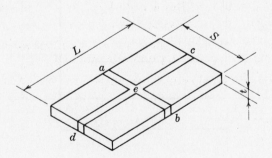

Fig. 13-2

at e of the other central strip cd. The deflection of a simple beam uniformly loaded is $5wl^4/384EI$. If the strips have equal thickness, the deflection of strip ab is $kw_{ab}S^4$ and the deflection of strip cd is $kw_{cd}L^4$, where w_{ab} and w_{cd} are the portions of the load carried by strips ab and cd respectively. If the total load $w = w_{ab} + w_{cd}$, equating deflections of each strip at point e, $w_{ab} = wL^4/(L^4 + S^4)$ and $w_{cd} = wS^4/(L^4 + S^4)$.

If $L = 1.5S$, $w_{ab} = 0.835w$ and $w_{cd} = 0.165w$. If $L = 2.0S$, $w_{ab} = 0.941w$ and $w_{cd} = 0.059w$. Thus the shorter span must resist the major portion of the load. If L/S, the *aspect ratio*, is greater than 2, the portion of the load carried by the long span becomes less significant and the slab is often proportioned as a one-way slab.

Many elastic analysis methods for two-way slabs have been proposed. All have shortcomings. They neglect Poisson's ratio, torsion, changes in stiffness, ultimate capacity, edge restraint, variation of moments and shears along span, and others. As an example: if strip ab in Fig. 13-2 is not a central strip and is moved toward one end of the panel, point e is no longer at the midpoint of strip cd. Hence the constant in the expression for the deflection of cd is not 5/384 and the results of equating deflections would not be the same as obtained before.

As strip ab approaches the edge of the panel, the deflection at e approaches zero. Hence the load sustained by strip ab approaches zero while that sustained by strip cd remains the same. Empirical test results and analytical investigations by Dr. H. M. Westergaard, Joseph Di Stasio, M. P. van Buren and others have demonstrated within reasonable limits the distribution of the bending moments in two-way slabs. However, only on the computer (finite element method) can a highly accurate solution be obtained.

The 1963 ACI Building Code contained three methods of design of two-way slabs. These are based on other than an elastic analysis like that presented herein. Rather than determining the exact distribution of bending moments, the slab is divided into column and middle strips as shown in Fig. 13-3 and the value of the moment is assumed constant across the full width of a strip.

The analysis of two-way slab systems is further complicated by the variation in shear along the panel edge. The methods of analysis proposed by the 1963 ACI Building Code contain different requirements concerning the load distribution to the supporting beams. These requirements will be discussed in the solved problems.

The 1963 ACI Building Code required that the minimum thickness of two-way slabs shall not be less than $3\frac{1}{2}$ inches nor less than the perimeter of the slab divided by 180. (The 1983 Code poses less severe requirements.)

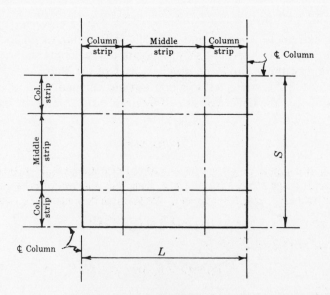

Fig. 13-3

When a two-way slab is subjected to a vertical load, the slab deflects forming a dish-like surface and the corners tend to lift off of the supports. This action may cause distress in the vicinity of the corners. Hence the 1963 ACI Building Code required special reinforcement in these regions. This reinforcement should be provided for a distance from each corner not less than 1/5 of the longest span. The steel area should be equal to that of the maximum positive moment slab steel and be placed in the top and bottom. The top bars must be placed parallel to the diagonal of the panel and the bottom bars are placed perpendicular to the diagonal or parallel to both sides of the slab.

TABLES OF DESIGN COEFFICIENTS

Tables 13.1 and 13.2 contain coefficients for slab moments and shears and beam moments and shears respectively for what was termed Method 1 in the 1963 ACI Code. Tables 13.1 and 13.2 are reproduced with permission of the ACI.

Table 13.3 contains slab moment coefficients for Method 2. Table 13.3 is reproduced with permission of the ACI.

Tables 13.4, 13.5, 13.6 and 13.7 contain coefficients for slab moments and beam shears for Method 3. These tables are reproduced with permission of the ACI.

The use of Tables 13.1 through 13.7 will be demonstrated in the solved problems.

The solved problems have been selected to demonstrate the specific requirements in this chapter. Usually the detailed proportioning and determination of stresses will not be discussed.

Solved Problems

13.1. Determine the central strip bending moments for a simply supported square two-way slab. Assume a concentrated vertical load at the center of the panel. Use the "crossed sticks" method shown in Fig. 13-2.

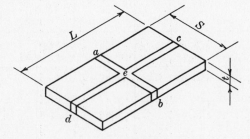

Fig. 13-4

The deflection of a simply supported beam with a concentrated load at the midpoint is $PL^3/48EI$. The deflection at e of strips ab and cd are equal. The load required to deflect ab is P_{ab} and the load required to deflect cd is P_{cd}, and $P = P_{ab} + P_{cd}$. Equating deflections of the two strips,

$$P_{ab}S^3/48EI = P_{cd}L^3/48EI$$

If $L = S$, then $P_{ab}L^3 = P_{cd}L^3$ and

$$P_{ab} = (P - P_{ab})L^3/L^3 = P/2$$

Likewise, $P_{cd} = P/2$.

The maximum simple beam bending moment for a concentrated load at the midpoint is $M = PL/4$. Hence $M_{ab} = M_{cd} = PL/8$.

Table 13.1 Method 1—Slabs

Upper figure: C_s / Lower figure: C_{s1}							C / C_1
r	$r_1 = \dfrac{1}{r}$	0.0	0.1	0.2	0.3	0.4	
0.00	∞	0.50 / 0.00	0.40 / 0.00	0.30 / 0.00	0.20 / 0.00	0.10 / 0.00	1.00 / 0.00
0.50	2.00	0.44 / 0.06	0.36 / 0.03	0.27 / 0.02	0.18 / 0.00	0.09 / 0.00	0.89 / 0.06
0.55	1.82	0.43 / 0.07	0.33 / 0.04	0.23 / 0.02	0.15 / 0.01	0.07 / 0.00	0.79 / 0.08
0.60	1.67	0.41 / 0.09	0.30 / 0.05	0.20 / 0.03	0.12 / 0.01	0.05 / 0.00	0.70 / 0.10
0.65	1.54	0.39 / 0.11	0.28 / 0.06	0.18 / 0.03	0.10 / 0.01	0.04 / 0.00	0.64 / 0.13
0.70	1.43	0.37 / 0.13	0.26 / 0.08	0.16 / 0.04	0.09 / 0.01	0.03 / 0.00	0.58 / 0.15
0.80	1.25	0.33 / 0.17	0.22 / 0.10	0.13 / 0.06	0.07 / 0.02	0.02 / 0.00	0.48 / 0.21
0.90	1.11	0.29 / 0.21	0.19 / 0.13	0.11 / 0.07	0.05 / 0.03	0.01 / 0.01	0.40 / 0.27
1.00	1.00	0.25 / 0.25	0.16 / 0.16	0.09 / 0.09	0.04 / 0.04	0.01 / 0.01	0.33 / 0.33
1.10	0.91	0.21 / 0.29	0.13 / 0.19	0.07 / 0.11	0.03 / 0.05	0.01 / 0.01	0.28 / 0.39
1.20	0.83	0.18 / 0.32	0.11 / 0.21	0.06 / 0.13	0.02 / 0.06	0.00 / 0.02	0.23 / 0.45
1.30	0.77	0.16 / 0.34	0.10 / 0.23	0.05 / 0.14	0.02 / 0.07	0.00 / 0.03	0.19 / 0.51
1.40	0.71	0.13 / 0.37	0.08 / 0.25	0.04 / 0.16	0.02 / 0.09	0.00 / 0.03	0.16 / 0.57
1.50	0.67	0.11 / 0.39	0.07 / 0.27	0.04 / 0.17	0.01 / 0.10	0.00 / 0.04	0.14 / 0.61
1.60	0.63	0.10 / 0.40	0.06 / 0.29	0.03 / 0.19	0.01 / 0.11	0.00 / 0.05	0.12 / 0.66
1.80	0.55	0.07 / 0.43	0.04 / 0.33	0.02 / 0.23	0.01 / 0.15	0.00 / 0.07	0.08 / 0.79
2.00	0.50	0.06 / 0.44	0.03 / 0.36	0.02 / 0.27	0.00 / 0.18	0.00 / 0.09	0.06 / 0.89
∞	0.00	0.00 / 0.50	0.00 / 0.40	0.00 / 0.30	0.00 / 0.20	0.00 / 0.10	0.00 / 1.00

Values of x (column headers 0.0–0.4)

Table 13.2 Method 1—Beams

Upper figure Lower figure		C_b C_{b1}					$\dfrac{1-C}{1-C_1}$
r	$r_1 = \dfrac{1}{r}$	Values of x					
		0.0	0.1	0.2	0.3	0.4	
0.00	∞	0.00 0.50	0.00 0.40	0.00 0.30	0.00 0.20	0.00 0.10	0.00 1.00
0.50	2.00	0.06 0.44	0.04 0.37	0.03 0.28	0.02 0.20	0.01 0.10	0.11 0.94
0.55	1.82	0.07 0.43	0.07 0.36	0.07 0.28	0.05 0.19	0.03 0.10	0.21 0.92
0.60	1.67	0.09 0.41	0.10 0.35	0.10 0.27	0.08 0.19	0.05 0.10	0.30 0.90
0.65	1.54	0.11 0.39	0.12 0.34	0.12 0.27	0.10 0.19	0.06 0.10	0.36 0.87
0.70	1.43	0.13 0.37	0.14 0.32	0.14 0.26	0.11 0.19	0.07 0.10	0.42 0.85
0.80	1.25	0.17 0.33	0.18 0.30	0.17 0.24	0.13 0.18	0.08 0.10	0.52 0.79
0.90	1.11	0.21 0.29	0.21 0.27	0.19 0.23	0.15 0.17	0.09 0.09	0.60 0.73
1.00	1.00	0.25 0.25	0.24 0.24	0.21 0.21	0.16 0.16	0.09 0.09	0.67 0.67
1.10	0.91	0.29 0.21	0.27 0.21	0.23 0.19	0.17 0.15	0.09 0.09	0.72 0.61
1.20	0.83	0.32 0.18	0.29 0.19	0.24 0.17	0.18 0.14	0.10 0.08	0.77 0.55
1.30	0.77	0.34 0.16	0.30 0.17	0.25 0.16	0.18 0.13	0.10 0.07	0.81 0.49
1.40	0.71	0.37 0.13	0.32 0.15	0.26 0.14	0.18 0.11	0.10 0.07	0.84 0.43
1.50	0.67	0.39 0.11	0.33 0.13	0.26 0.13	0.19 0.10	0.10 0.06	0.86 0.39
1.60	0.63	0.40 0.10	0.34 0.11	0.27 0.11	0.19 0.09	0.10 0.05	0.88 0.34
1.80	0.55	0.43 0.07	0.36 0.07	0.28 0.07	0.19 0.05	0.10 0.03	0.92 0.21
2.00	0.50	0.44 0.06	0.37 0.04	0.28 0.03	0.20 0.02	0.10 0.01	0.94 0.11
∞	0.00	0.50 0.00	0.40 0.00	0.30 0.00	0.20 0.00	0.10 0.00	1.00 0.00

Table 13.3   Method 2—Moment Coefficients

Moments	Short span						Long span, all values of m
	Values of m						
	1.0	0.9	0.8	0.7	0.6	0.5 and less	
Case 1—Interior panels							
Negative moment at—							
Continuous edge	0.033	0.040	0.048	0.055	0.063	0.083	0.033
Discontinuous edge	—	—	—	—	—	—	—
Positive moment at midspan	0.025	0.030	0.036	0.041	0.047	0.062	0.025
Case 2—One edge discontinuous							
Negative moment at—							
Continuous edge	0.041	0.048	0.055	0.062	0.069	0.085	0.041
Discontinuous edge	0.021	0.024	0.027	0.031	0.035	0.042	0.021
Positive moment at midspan	0.031	0.036	0.041	0.047	0.052	0.064	0.031
Case 3—Two edges discontinuous							
Negative moment at—							
Continuous edge	0.049	0.057	0.064	0.071	0.078	0.090	0.049
Discontinuous edge	0.025	0.028	0.032	0.036	0.039	0.045	0.025
Positive moment at midspan	0.037	0.043	0.048	0.054	0.059	0.068	0.037
Case 4—Three edges discontinuous							
Negative moment at—							
Continuous edge	0.058	0.066	0.074	0.082	0.090	0.098	0.058
Discontinuous edge	0.029	0.033	0.037	0.041	0.045	0.049	0.029
Positive moment at midspan	0.044	0.050	0.056	0.062	0.068	0.074	0.044
Case 5—Four edges discontinuous							
Negative moment at—							
Continuous edge	—	—	—	—	—	—	—
Discontinuous edge	0.033	0.038	0.043	0.047	0.053	0.055	0.033
Positive moment at midspan	0.050	0.057	0.064	0.072	0.080	0.083	0.050

Table 13.4　Method 3—Coefficients for Negative Moments in Slabs*

$$\left.\begin{array}{l} M_{A\,neg} = C_{A\,neg} \times w \times A^2 \\ M_{B\,neg} = C_{B\,neg} \times w \times B^2 \end{array}\right\} \quad \text{where} \quad w = \text{total uniform dead plus live load}$$

Ratio $m = \dfrac{A}{B}$		Case 1	Case 2	Case 3	Case 4	Case 5	Case 6	Case 7	Case 8	Case 9
1.00	$C_{A\,neg}$		0.045		0.050	0.075	0.071		0.033	0.061
	$C_{B\,neg}$		0.045	0.076	0.050			0.071	0.061	0.033
0.95	$C_{A\,neg}$		0.050		0.055	0.079	0.075		0.038	0.065
	$C_{B\,neg}$		0.041	0.072	0.045			0.067	0.056	0.029
0.90	$C_{A\,neg}$		0.055		0.060	0.080	0.079		0.043	0.068
	$C_{B\,neg}$		0.037	0.070	0.040			0.062	0.052	0.025
0.85	$C_{A\,neg}$		0.060		0.066	0.082	0.083		0.049	0.072
	$C_{B\,neg}$		0.031	0.065	0.034			0.057	0.046	0.021
0.80	$C_{A\,neg}$		0.065		0.071	0.083	0.086		0.055	0.075
	$C_{B\,neg}$		0.027	0.061	0.029			0.051	0.041	0.017
0.75	$C_{A\,neg}$		0.069		0.076	0.085	0.088		0.061	0.078
	$C_{B\,neg}$		0.022	0.056	0.024			0.044	0.036	0.014
0.70	$C_{A\,neg}$		0.074		0.081	0.086	0.091		0.068	0.081
	$C_{B\,neg}$		0.017	0.050	0.019			0.038	0.029	0.011
0.65	$C_{A\,neg}$		0.077		0.085	0.087	0.093		0.074	0.083
	$C_{B\,neg}$		0.014	0.043	0.015			0.031	0.024	0.008
0.60	$C_{A\,neg}$		0.081		0.089	0.088	0.095		0.080	0.085
	$C_{B\,neg}$		0.010	0.035	0.011			0.024	0.018	0.006
0.55	$C_{A\,neg}$		0.084		0.092	0.089	0.096		0.085	0.086
	$C_{B\,neg}$		0.007	0.028	0.008			0.019	0.014	0.005
0.50	$C_{A\,neg}$		0.086		0.094	0.090	0.097		0.089	0.088
	$C_{B\,neg}$		0.006	0.022	0.006			0.014	0.010	0.003

* A cross-hatched edge indicates that the slab continues across or is fixed at the support; an unmarked edge indicates a support at which torsional resistance is negligible.

Table 13.5 Method 3—Coefficients for Dead Load Positive Moments in Slabs*

$$\left.\begin{array}{l} M_{A\ pos\ DL} = C_{A\ DL} \times w \times A^2 \\ M_{B\ pos\ DL} = C_{B\ DL} \times w \times B^2 \end{array}\right\} \quad \text{where} \quad w = \text{total uniform dead load}$$

Ratio $m = \dfrac{A}{B}$		Case 1	Case 2	Case 3	Case 4	Case 5	Case 6	Case 7	Case 8	Case 9
1.00	$C_{A\ DL}$	0.036	0.018	0.018	0.027	0.027	0.033	0.027	0.020	0.023
	$C_{B\ DL}$	0.036	0.018	0.027	0.027	0.018	0.027	0.033	0.023	0.020
0.95	$C_{A\ DL}$	0.040	0.020	0.021	0.030	0.028	0.036	0.031	0.022	0.024
	$C_{B\ DL}$	0.033	0.016	0.025	0.024	0.015	0.024	0.031	0.021	0.017
0.90	$C_{A\ DL}$	0.045	0.022	0.025	0.033	0.029	0.039	0.035	0.025	0.026
	$C_{B\ DL}$	0.029	0.014	0.024	0.022	0.013	0.021	0.028	0.019	0.015
0.85	$C_{A\ DL}$	0.050	0.024	0.029	0.036	0.031	0.042	0.040	0.029	0.028
	$C_{B\ DL}$	0.026	0.012	0.022	0.019	0.011	0.017	0.025	0.017	0.013
0.80	$C_{A\ DL}$	0.056	0.026	0.034	0.039	0.032	0.045	0.045	0.032	0.029
	$C_{B\ DL}$	0.023	0.011	0.020	0.016	0.009	0.015	0.022	0.015	0.010
0.75	$C_{A\ DL}$	0.061	0.028	0.040	0.043	0.033	0.048	0.051	0.036	0.031
	$C_{B\ DL}$	0.019	0.009	0.018	0.013	0.007	0.012	0.020	0.013	0.007
0.70	$C_{A\ DL}$	0.068	0.030	0.046	0.046	0.035	0.051	0.058	0.040	0.033
	$C_{B\ DL}$	0.016	0.007	0.016	0.011	0.005	0.009	0.017	0.011	0.006
0.65	$C_{A\ DL}$	0.074	0.032	0.054	0.050	0.036	0.054	0.065	0.044	0.034
	$C_{B\ DL}$	0.013	0.006	0.014	0.009	0.004	0.007	0.014	0.009	0.005
0.60	$C_{A\ DL}$	0.081	0.034	0.062	0.053	0.037	0.056	0.073	0.048	0.036
	$C_{B\ DL}$	0.010	0.004	0.011	0.007	0.003	0.006	0.012	0.007	0.004
0.55	$C_{A\ DL}$	0.088	0.035	0.071	0.056	0.038	0.058	0.081	0.052	0.037
	$C_{B\ DL}$	0.008	0.003	0.009	0.005	0.002	0.004	0.009	0.005	0.003
0.50	$C_{A\ DL}$	0.095	0.037	0.080	0.059	0.039	0.061	0.089	0.056	0.038
	$C_{B\ DL}$	0.006	0.002	0.007	0.004	0.001	0.003	0.007	0.004	0.002

* A cross-hatched edge indicates that the slab continues across or is fixed at the support; an unmarked edge indicates a support at which torsional resistance is negligible.

Table 13.6 Method 3—Coefficients for Live Load Positive Moments in Slabs*

$$M_{A \text{ pos LL}} = C_{A \text{ LL}} \times w \times A^2$$
$$M_{B \text{ pos LL}} = C_{B \text{ LL}} \times w \times B^2$$
where w = total uniform live load

Ratio $m = \dfrac{A}{B}$		Case 1	Case 2	Case 3	Case 4	Case 5	Case 6	Case 7	Case 8	Case 9
1.00	$C_{A \text{ LL}}$	0.036	0.027	0.027	0.032	0.032	0.035	0.032	0.028	0.030
	$C_{B \text{ LL}}$	0.036	0.027	0.032	0.032	0.027	0.032	0.035	0.030	0.028
0.95	$C_{A \text{ LL}}$	0.040	0.030	0.031	0.035	0.034	0.038	0.036	0.031	0.032
	$C_{B \text{ LL}}$	0.033	0.025	0.029	0.029	0.024	0.029	0.032	0.027	0.025
0.90	$C_{A \text{ LL}}$	0.045	0.034	0.035	0.039	0.037	0.042	0.040	0.035	0.036
	$C_{B \text{ LL}}$	0.029	0.022	0.027	0.026	0.021	0.025	0.029	0.024	0.022
0.85	$C_{A \text{ LL}}$	0.050	0.037	0.040	0.043	0.041	0.046	0.045	0.040	0.039
	$C_{B \text{ LL}}$	0.026	0.019	0.024	0.023	0.019	0.022	0.026	0.022	0.020
0.80	$C_{A \text{ LL}}$	0.056	0.041	0.045	0.048	0.044	0.051	0.051	0.044	0.042
	$C_{B \text{ LL}}$	0.023	0.017	0.022	0.020	0.016	0.019	0.023	0.019	0.017
0.75	$C_{A \text{ LL}}$	0.061	0.045	0.051	0.052	0.047	0.055	0.056	0.049	0.046
	$C_{B \text{ LL}}$	0.019	0.014	0.019	0.016	0.013	0.016	0.020	0.016	0.013
0.70	$C_{A \text{ LL}}$	0.068	0.049	0.057	0.057	0.051	0.060	0.063	0.054	0.050
	$C_{B \text{ LL}}$	0.016	0.012	0.016	0.014	0.011	0.013	0.017	0.014	0.011
0.65	$C_{A \text{ LL}}$	0.074	0.053	0.064	0.062	0.055	0.064	0.070	0.059	0.054
	$C_{B \text{ LL}}$	0.013	0.010	0.014	0.011	0.009	0.010	0.014	0.011	0.009
0.60	$C_{A \text{ LL}}$	0.081	0.058	0.071	0.067	0.059	0.068	0.077	0.065	0.059
	$C_{B \text{ LL}}$	0.010	0.007	0.011	0.009	0.007	0.008	0.011	0.009	0.007
0.55	$C_{A \text{ LL}}$	0.088	0.062	0.080	0.072	0.063	0.073	0.085	0.070	0.063
	$C_{B \text{ LL}}$	0.008	0.006	0.009	0.007	0.005	0.006	0.009	0.007	0.006
0.50	$C_{A \text{ LL}}$	0.095	0.066	0.088	0.077	0.067	0.078	0.092	0.076	0.067
	$C_{B \text{ LL}}$	0.006	0.004	0.007	0.005	0.004	0.005	0.007	0.005	0.004

* A cross-hatched edge indicates that the slab continues across or is fixed at the support; an unmarked edge indicates a support at which torsional resistance is negligible.

Table 13.7   Method 3—Ratio of Load W in A and B Directions for Shear in Slab and Load on Supports*

Ratio $m = \dfrac{A}{B}$		Case 1	Case 2	Case 3	Case 4	Case 5	Case 6	Case 7	Case 8	Case 9
1.00	W_A	0.50	0.50	0.17	0.50	0.83	0.71	0.29	0.33	0.67
	W_B	0.50	0.50	0.83	0.50	0.17	0.29	0.71	0.67	0.33
0.95	W_A	0.55	0.55	0.20	0.55	0.86	0.75	0.33	0.38	0.71
	W_B	0.45	0.45	0.80	0.45	0.14	0.25	0.67	0.62	0.29
0.90	W_A	0.60	0.60	0.23	0.60	0.88	0.79	0.38	0.43	0.75
	W_B	0.40	0.40	0.77	0.40	0.12	0.21	0.62	0.57	0.25
0.85	W_A	0.66	0.66	0.28	0.66	0.90	0.83	0.43	0.49	0.79
	W_B	0.34	0.34	0.72	0.34	0.10	0.17	0.57	0.51	0.21
0.80	W_A	0.71	0.71	0.33	0.71	0.92	0.86	0.49	0.55	0.83
	W_B	0.29	0.29	0.67	0.29	0.08	0.14	0.51	0.45	0.17
0.75	W_A	0.76	0.76	0.39	0.76	0.94	0.88	0.56	0.61	0.86
	W_B	0.24	0.24	0.61	0.24	0.06	0.12	0.44	0.39	0.14
0.70	W_A	0.81	0.81	0.45	0.81	0.95	0.91	0.62	0.68	0.89
	W_B	0.19	0.19	0.55	0.19	0.05	0.09	0.38	0.32	0.11
0.65	W_A	0.85	0.85	0.53	0.85	0.96	0.93	0.69	0.74	0.92
	W_B	0.15	0.15	0.47	0.15	0.04	0.07	0.31	0.26	0.08
0.60	W_A	0.89	0.89	0.61	0.89	0.97	0.95	0.76	0.80	0.94
	W_B	0.11	0.11	0.39	0.11	0.03	0.05	0.24	0.20	0.06
0.55	W_A	0.92	0.92	0.69	0.92	0.98	0.96	0.81	0.85	0.95
	W_B	0.08	0.08	0.31	0.08	0.02	0.04	0.19	0.15	0.05
0.50	W_A	0.94	0.94	0.76	0.94	0.99	0.97	0.86	0.89	0.97
	W_B	0.06	0.06	0.24	0.06	0.01	0.03	0.14	0.11	0.03

* A cross-hatched edge indicates that the slab continues across or is fixed at the support; an unmarked edge indicates a support at which torsional resistance is negligible.

13.2. Repeat Problem 13.1 and assume $L = 2.0S$.

As before, $P_{ab}S^3/48EI = P_{cd}L^3/48EI$ and $P_{ab} = (P - P_{ab})L^3/S^3 = 8(P - P_{ab})S^3/S^3 = 8P/9$. Likewise, $P_{cd} = P/9$.

The maximum bending moments for the strips are

$$M_{ab} = (8P/9)(S/4) = 2PS/9 \qquad M_{cd} = (P/9)(L/4) = PL/36 = PS/18$$

13.3. An interior two-way slab panel 25'-0" by 25'-0" must carry a live load of 300 lb/ft². The slab is $8\frac{1}{2}$" thick and is supported on masonry walls. Determine the principal bending moments and shears in the slab.

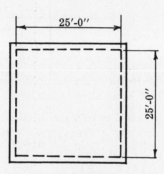

25'-0"

25'-0"

Fig. 13-5

The 1963 ACI Building Code requires that in the design of two-way slabs according to Method 2 the supports must be built monolithically with the slab. Hence Method 1 or 3 must be used. Method 1 will be used in this solution.

For a continuous slab, $g = g_1 = 0.76$. If $L = L_1$, then $r = gL/gL = 1.00$. From Table 13.1 for $r = 1.00$, $C = C_1 = 0.33$. For a continuous one-way slab $B = 1/16$ and $1/11$ for positive and negative moments respectively.

The bending moments in the slab are $M = CBWL$ and $M_1 = C_1BW_1L_1$, where W is the total uniform load between opposite supports on the slab strip.

The loads are

$$\begin{aligned} \text{D.L.—}8\tfrac{1}{2}\text{" slab} &= 106 \text{ lb/ft}^2 \\ \text{L.L.} &= \underline{300 \text{ lb/ft}^2} \\ \text{Total Load} &= 406 \text{ lb/ft}^2 \end{aligned}$$

Hence $W = 25.0(406) = 10{,}100$ lb/ft.

Because the panel is square, $M = M_1 = CBWL$. The maximum moments are

$$+M = \frac{0.33(10{,}100)(25.0)}{1000(16)} = 5.22 \text{ ft-kips/ft}, \qquad -M = \frac{0.33(10{,}100)(25.0)}{1000(11)} = 7.60 \text{ ft-kips/ft}$$

The shear in the slab is $V = C_sW$ and $V_1 = C_{s1}W_1$, where the coefficients C_s and C_{s1} are taken from Table 13.1. At $x = 0$ for $r = 1.0$, $C_s = C_{s1} = 0.25$. Hence at the support $V = V_1 = 0.25(10{,}100)/1000 = 2.53$ kips/ft.

13.4. Assume that the slab in Problem 13.3 is supported on concrete beams 18" wide by 36" deep. Determine the principal beam shears and bending moments. Use Method 1. (See also Chapter 2, page 26).

The bending moments in the beams due to the slab load are $M = (1 - C)BWL$ and $M_1 = (1 - C_1)BW_1L_1$.

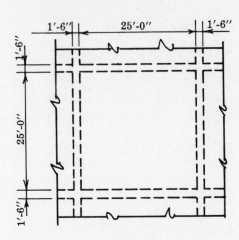

Fig. 13-6

The total load on the beam is part of the slab load plus the dead and live loads on the beam itself. The moments and shears of both these effects must be added. The loads on the beam itself are

$$\text{D.L.} = 18(36)(150)/144 = 674 \text{ lb/ft}$$
$$\text{L.L.} = 300(18)/12 \quad = \underline{450 \text{ lb/ft}}$$
$$\text{Total Load} \quad\quad = 1124 \text{ lb/ft}$$

The maximum moments in the beam due to dead and live loads on the beam are

$$+M = \frac{wl^2}{16} = \frac{1124(25.0)^2}{16(1000)} = 44.0 \qquad -M = \frac{wl^2}{11} = \frac{1124(25.0)^2}{11(1000)} = 63.9 \text{ ft-kips}$$

The maximum moments in the beam due to slab loads are

$$+M = (1-C)BWL = \frac{0.67(10{,}100)(25.0)^2}{16(1000)} = 264 \text{ ft-kips}$$

$$-M = (1-C)BWL = \frac{0.67(10{,}100)(25.0)^2}{11(1000)} = 385 \text{ ft-kips}$$

Summing the total beam moments,

$$+M = 264 + 44.0 = 308 \text{ ft-kips at midspan}$$
$$-M = 385 + 63.9 = 449 \text{ ft-kips at support}$$

The shears in the beams due to the slab load are $V = C_b W$ and $V_1 = C_{b1}W_1$. The maximum shear at the face of the support due to the dead and live loads on the beam is approximately $V = wl/2 = 14.0$ kips. The shear at the face of the support or at $x = 0$ due to the slab loads is $V = V_1 = C_b W = 0.25(10{,}100)(25.0)/1000 = 63.1$ kips. C_b is obtained from Table 13.2 for $r = 1.0$. Summing the total shears,

$$\text{Shear due to slab load} \quad = 63.1 \text{ kips}$$
$$\text{Shear due to beam load} = \underline{14.0 \text{ kips}}$$
$$\text{Total maximum shear} \quad = 77.1 \text{ kips}$$

13.5. Assume the slab in Problem 13.3 is supported on monolithically cast concrete beams. Determine the principal bending moments and shears in the slab and beams. Use Method 2.

(*a*) The loads on the slab are

$$\text{D.L.}—8\tfrac{1}{2}'' \text{ slab} = 106 \text{ lb/ft}^2$$
$$\text{L.L.} \quad\quad = \underline{300 \text{ lb/ft}^2}$$
$$\text{Total Load} \quad = 406 \text{ lb/ft}^2$$

(b) Method 2 defines the span length as the distance center-to-center of supports or the clear span plus twice the slab thickness, whichever is smaller. Assuming 18″ wide supporting beams, $25.0 + 1.5 =$ 26.5 ft and $25.0 + 2(8.5)/12 = 26.4$ ft. Hence $S = 26.4$ ft. Because the panel is square, $m = 1.0$.

From Table 13.3 the moment coefficients for a panel with four edges continuous are 0.033 and 0.025 for negative and positive moments respectively.

The bending moments in the middle strip are $M = CwS^2$ or

$$+M = \frac{0.025(406)(26.4)^2}{1000} = 7.1 \text{ ft-kips/ft} \qquad -M = \frac{0.033(406)(26.4)^2}{1000} = 9.3 \text{ ft-kips/ft}$$

The bending moments in the column strips are 2/3 the above values.

(c) The shear in the slab at the supports is computed assuming that the load on the supporting beams is that within an area bounded by the 45 degree lines from the panel corners and the centerline of the long span. This assumption conforms with the *yield-line theory* of slab and plate analysis, the application of which is permitted by the 1983 ACI Code in this case, though not in general. See Fig. 13-7. (See also Problems 2.17 and 2.18.)

The maximum shear in the slab is $V = 406(25.0)/(2)(1000) = 5.06$ kips/ft. The average shear in the slab is one-half this value or 2.53 kips/ft.

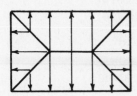

Fig. 13-7

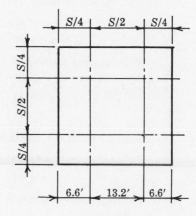

Fig. 13-8

(d) When $2S > L$, the middle strip is one-half panel wide symmetrical about the slab centerline. See Fig. 13-8.

The total panel moment in the slab is approximately

$$+M = 7.1(13.2) + 0.67(7.1)(13.2) = 156 \text{ ft-kips}$$
$$-M = 9.3(13.2) + 0.67(9.3)(13.2) = \underline{205 \text{ ft-kips}}$$
$$\text{Total panel moment in slab} \qquad = 361 \text{ ft-kips}$$

(e) The equivalent uniform load on the beams used to determine moments may be approximated by $wS/3$ for each panel supported. Hence the moments in the supporting beams due to slab loads are

$$+M = \left(\frac{2wS}{3}\right)\frac{S^2}{16} = \frac{2(406)(26.4)^3}{3(16)(1000)} = 312 \text{ ft-kips} \qquad -M = \left(\frac{2wS}{3}\right)\frac{S^2}{11} = 452 \text{ ft-kips}$$

(f) The total moment in panel slab and beams in one direction is

$$\begin{aligned} \text{Slab moments, positive and negative} &= 361 \text{ ft-kips} \\ \text{Beam moments, positive and negative} &= \underline{764 \text{ ft-kips}} \\ \text{Total panel moment} &= 1125 \text{ ft-kips} \end{aligned}$$

From Problem 13.3(c) and 13.4(b), the total panel moment due to slab loads obtained by Method 1 is

$$\text{Slab moment} = 25.0(5.22 + 7.60) = 320 \text{ ft-kips}$$
$$\text{Beam moment} = 264 + 385 \qquad = \underline{649 \text{ ft-kips}}$$
$$\text{Total panel moment} \qquad\qquad = 969 \text{ ft-kips}$$

The difference in the above answers is due to the difference in effective span lengths used in Methods 1 and 2. In Problems 13.3 and 13.4, the span length was 25.0 ft; so $(969)(26.4/25.0)^3 = 1140$, which is close to 1125.

13.6. If an interior two-way slab has a width to length ratio 0.7, compare the total panel moments determined by Methods 1 and 2. Assume $L < L_1$.

(a) For Method 1, $g = g_1 = 0.76$ and $r = gL/g_1L_1 = 0.7$. From Table 13.1, $C = 0.58$ and $C_1 = 0.15$. If the coefficients for positive and negative bending moments are assumed to be 1/16 and 1/11 respectively, the slab bending moments in the long direction are

$$+M_1 = C_1BW_1L_1 = 0.15wLL_1^2/16 = 0.00938wLL_1^2$$
$$-M_1 = C_1BW_1L_1 = 0.15wLL_1^2/11 = 0.0136wLL_1^2$$

Likewise, the slab bending moments in the short direction are

$$+M = CBWL = 0.58wL^2L_1/16 = 0.0363wL^2L_1$$
$$-M = CBWL = 0.58wL^2L_1/11 = 0.0527wL^2L_1$$

With the same coefficients 1/11 and 1/16 the bending moments in the supporting beams in the long direction are

$$+M_1 = (1 - C_1)BW_1L_1 = 0.85wLL_1^2/16 = 0.0531wLL_1^2$$
$$-M_1 = (1 - C_1)BW_1L_1 = 0.85wLL_1^2/11 = 0.0773wLL_1^2$$

Likewise, the beam bending moments in the short direction are

$$+M = (1 - C)BWL = 0.42wL^2L_1/16 = 0.0263wL^2L_1$$
$$-M = (1 - C)BWL = 0.42wL^2L_1/11 = 0.0382wL^2L_1$$

The total panel moment in the long direction is

$$M_1 = (0.00938 + 0.0136 + 0.0531 + 0.0773)wLL_1^2 = 0.1534wLL_1^2$$

The total panel moment in the short direction is

$$M = (0.0363 + 0.0527 + 0.0263 + 0.0382)wL^2L_1 = 0.1535wL^2L_1$$

(b) For Method 2 the coefficients for bending moments are taken from Table 13.3. $C = 0.055$ and 0.041 for negative and positive moments respectively in the short directions and $C = 0.033$ and 0.025 for negative and positive moments respectively in the long direction. The slab bending moments in the long direction are

$$+M_1 = CwL^3/2 + CwL^3/3 = 5(0.025)wL^3/6 = 0.0208wL^3$$
$$-M_1 = CwL^3/2 + CwL^3/3 = 5(0.033)wL^3/6 = 0.0275wL^3$$

Likewise, the slab bending moments in the short direction are

$$+M = CwL^2L_1/2 + CwL^2L_1/3 = 5(0.041)wL^2L_1/6 = 0.0342wL^2L_1$$
$$-M = CwL^2L_1/2 + CwL^2L_1/3 = 5(0.055)wL^2L_1/6 = 0.0458wL^2L_1$$

The equivalent loadings on the supporting beams are $2(0.33)wL$ and $2(wL/3)(3 - m^2/2) = 2(0.418)wL$ for the short and long spans respectively. Assuming moment coefficients of 1/16 and 1/11, the bending moments in the supporting beam in the long direction are

$$+M_1 = 2(0.418)wLL_1^2/16 = 0.0522wLL_1^2$$
$$-M_1 = 2(0.418)wLL_1^2/11 = 0.0760wLL_1^2$$

Likewise, the beam bending moments in the short direction are

$$+M = 2(0.33)wL^3/16 = 0.0412wL^3$$
$$-M = 2(0.33)wL^3/11 = 0.0600wL^3$$

The total panel moment in the long direction is

$$M_1 = (0.0208 + 0.0275)wL^3 + (0.0552 + 0.0760)wLL_1^2 = 0.1549wLL_1^2$$

The total panel moment in the short direction is

$$M = (0.0342 + 0.0458)wL^2L_1 + (0.0412 + 0.0600)wL^3 = 0.1508wL^2L_1$$

(c)   Methods 1 and 2 yield practically the same results. This close correlation will be true for most two-way slabs of usual length to width ratios. Exterior and corner panels do not check as near as the above. It should be noted that the same effective span lengths were used in both methods.

13.7. Determine the principal slab and beam bending moments and shears in the two-way slab in Problem 13.3 by use of Method 3.

(a)   From Problem 13.3, D.L. = 106 lb/ft^2, L.L. = 300 lb/ft^2, $A = B = 25.0$ ft and $m = 1.0$.

The bending moment in the middle strip of the slab is $M = CwA^2$ or CwB^2. For Case 2 in Table 13.4, $C_{A\,neg} = C_{B\,neg} = 0.045$. For w = total uniform load = 406 lb/ft^2, $M_{A\,neg} = M_{B\,neg} = CwA^2 = 0.045(406)(25.0)^2/1000 = 11.4$ ft-kips/ft. For Case 2 in Tables 13.5 and 13.6, $C_{A\,DL} = C_{B\,DL} = 0.018$ and $C_{A\,LL} = C_{B\,LL} = 0.027$. Hence

$$M_{A\,pos} = M_{B\,pos} = C_{A\,DL}w_{DL}A^2 + C_{A\,LL}w_{LL}A^2$$
$$= 0.018(106)(25.0)^2/1000 + 0.027(300)(25.0)^2/1000 = 6.25 \text{ ft-kips/ft}$$

The average bending moments in the column strips are 2/3 of the above values.

(b)   For Case 2 in Table 13.7, the ratio of total panel load to load resisted in A and B directions is 0.50. Hence,

$$V_A = V_B = \frac{0.50wAB}{2A} = \frac{0.50(406)(25.0)^2}{2(25.0)(1000)} = 2.53 \text{ kips/ft}$$

(c)   The load on the supporting beams is determined by the shear in the slab. However, the total load on the short span beam must not be less than that of an area bounded by the 45 degree lines from the panel corners. The equivalent uniform load on the beams used to determine moments is $wA/3$ for each panel supported. If the coefficients 1/11 and 1/16 are assumed, the beam moments due to slab loads are

$$+M_A = +M_B = \left(\frac{2wA}{3}\right)\frac{A^2}{16} = \frac{2(406)(25)^3}{3(16)(1000)} = 264 \text{ ft-kips}$$

$$-M_A = -M_b = \left(\frac{2wA}{3}\right)\frac{A^2}{11} = \frac{2(406)(25)^3}{3(11)(1000)} = 385 \text{ ft-kips}$$

(d)   The maximum beam shears at the supports are

$$V = \frac{2(2.53)A}{2} = \frac{2(2.53)(25.0)}{2(1000)} = 63.1 \text{ kips}$$

13.8. An apartment building is designed using 20'-0" by 20'-0" two-way slabs. The live load is 40 lb/ft^2, the partition load is 20 lb/ft^2 and the floor finish is 5 lb/ft^2. Design, using the alternate design method of the 1983 ACI Code, a typical exterior, interior and corner panel. Assume $f_c' = 4000$ psi, $f_s = 20,000$ psi and, as usual, $w_c = 150$ lb/ft^3. The columns are 12" by 12" and the supporting beams are 12" wide.

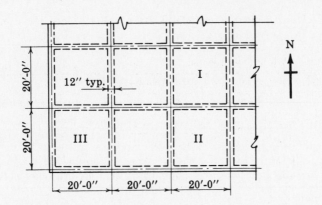

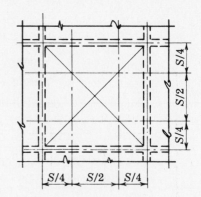

Fig. 13-9   Fig. 13-10

(a) The slab is cast monolithically with the supporting beams. Hence Method 2 will be used.

(b) The 1963 ACI Building Code required that the slab thickness h_f must be greater than $3\frac{1}{2}''$ and the slab perimeter divided by 180. The perimeter is approximately 76 feet. Hence $76/180 = 0.422$ ft $= 5.06''$. Try a slab thickness of $5.5''$. The loads on the slab are

$$
\begin{array}{lr}
\text{D.L.—Slab} & = \quad 69 \, \text{lb/ft}^2 \\
\text{Floor finish} = & 5 \, \text{lb/ft}^2 \\
\text{Partitions} & = \quad 20 \, \text{lb/ft}^2 \\
\text{Total D.L.} & = \quad 94 \, \text{lb/ft}^2 \\
\text{L.L.} & = \quad 40 \, \text{lb/ft}^2 \\
\text{Total Load} & = 134 \, \text{lb/ft}^2
\end{array}
$$

(c) By Method 2 the panel is divided into column and middle strips as shown in Fig. 13-10 above.

The 1963 ACI Building Code specified that the length of the short span S shall be taken as center-to-center of supports or the clear span plus two times the slab thickness, whichever is the smaller. The clear span is $19'-0''$. Hence $19'0'' + 2(5.5'') = 19'-11'' < 20'-0''$ or $S = 19.92$ ft.

(d) Before computing the slab moments and shears, it is advisable to check the approximate slab thickness required for flexural and shearing stresses.

In Table 13.3 the maximum moment coefficient for $m = 1.0$ is $C = 0.049$ for the corner panel. Hence the maximum moment is $M = CwS^2 = 0.049(134)(19.92)^2/1000 = 2.61$ ft-kips/ft. The maximum shear in the slab is approximately $V = wS/2 = 134(19.92)/2(1000) = 1.33$ kips/ft.

From Table 4.2, a slab with a $4''$ effective depth can resist a moment of 5.19 ft-kips/ft > 2.61.

The unit shear stress $v = V/bd = 1.33(1000)/12(4) = 27.7$ psi. From Table 4.1, the allowable shear is 70 psi > 27.7. This check is somewhat conservative because the actual shear stress at a distance d from the face of the support (cf. the 1983 ACI Code) would be less than the above.

In the above, the value of d was assumed as the average depth to steel. See Fig. 13-11. This is common practice in structural engineering design offices. However, some engineers use the actual effective depths to the individual layers of steel. When this is done, it is important to determine that the steel layers are properly placed during construction. Using the plane where the two layers intersect to calculate the effective depth minimizes this problem.

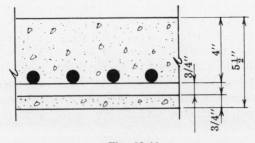

Fig. 13-11

(e) Perhaps the most efficient way to execute the slab design is by use of a table. The following table is self-explanatory and is complete.

Two-Way Slab Design

Direction of Span	Strip		Panel I $wS^2 = 53.2$ ft-kips/ft		Panel II $wS^2 = 53.2$ ft-kips/ft			Panel III $wS^2 = 53.2$ ft-kips/ft		
			$+M$	$-M$	$-M$	$+M$	$-M$	$-M$	$+M$	$-M$
NORTH-SOUTH	Middle Strip	C	0.025	0.033	0.021	0.031	0.041	0.025	0.037	0.049
		M	1.33	1.76	1.12	1.65	2.18	1.33	1.97	2.61
		d	4.0	4.0	4.0	4.0	4.0	4.0	4.0	4.0
		A_s/ft	0.23	0.31	0.19	0.29	0.38	0.23	0.34	0.45
		A_s/Strip	2.29	3.09	1.89	2.89	3.78	1.29	3.39	4.48
		Reinf.	9-#5	10-#5	9-#5	10-#5	13-#5	9-#5	11-#5	15-#5
	Column Strip	C								
		M								
		d								
		A_s/ft								
		A_s/Strip	0.76	1.03	0.63	0.96	1.26	0.76	1.13	1.49
		Reinf.	5-#4	6-#4	5-#4	5-#4	7-#4	5-#4	6-#4	8-#4
EAST-WEST	Middle Strip	C								
		M								
		d								
		A_s/ft								
		A_s/Strip								
		Reinf.	9-#5	10-#5	9-#5	10-#5	13-#5	9-#5	11-#5	15-#5
	Column Strip	C								
		M								
		d								
		A_s/ft								
		A_s/Strip								
		Reinf.	5-#4	6-#4	5-#4	5-#4	7-#4	5-#4	6-#4	8-#4

The top and bottom portions of the table are the same because the panels are square.

The 1963 ACI Building Code specified that the minimum reinforcement in two-way slabs be $0.0020bh_f = 0.0020(12)(5.5) = 0.132$ in^2/ft. For a middle strip, minimum $A_s = 0.132(19.92)/2 = 1.31$ in^2. For a column strip, minimum $A_s = 0.132(19.92)/4 = 0.62$ in^2.

The maximum bar spacing is $3h_f = 3(5.5) = 16.5''$; and $10'-0''/1'-4.5'' = 8$ spaces or 9 bars.

(f) The reinforcement in the column strips is two-thirds of the corresponding reinforcement in the middle strips. If trussed bars are used, it is sometimes convenient to select the same bar size in the table for positive and negative moment regions. This practice coupled with the maximum spacing requirement often necessitates the furnishing of an excess of reinforcement. As an example, 8-#5's would be sufficient for the positive steel requirement in Panel I. However, 9 bars are furnished in order to comply with the maximum spacing of $3h_f$.

The column strip steel spacing may be varied from the spacing in the middle strip to a value three times this at the edge of the panel. In this example, such a practice cannot be used because of the $3h_f$ maximum.

(g) In the table, the average depth to the reinforcing steel was used for d. Some designers prefer to use the minimum depth in order to be conservative. Other designers prefer to use the two different depths, indicating which steel is to go in the upper layer and which in the lower.

(h) For the details of reinforcement that must be investigated at this point, the reader is referred to Chapters 7 (development lengths) and 8 (crack control).

(*i*) In continuous beams and slabs, the longitudinal reinforcement is frequently bent up or trussed. See Fig. 13-12. Theoretically, the bars should be bent up or down at points where they are no longer needed to resist flexural stresses. These points would be points of inflection. Due to the practice of maximizing positive and negative moments in a span by alternate loading of the live load, the points of inflection are located at different positions for the various loading conditions.

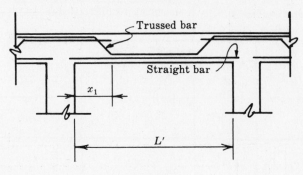

Fig. 13-12

If positive moment coefficients of 1/14 and 1/16 and negative moment coefficients of 1/10 and 1/11 are used, the various moment diagrams and points of inflection are as shown in Figs. 13-13 and 13-14.

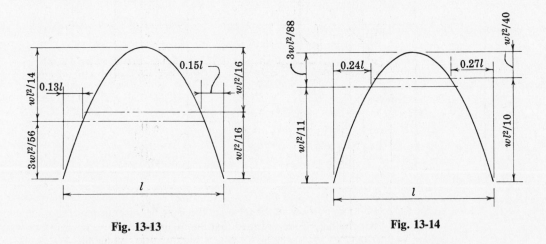

Fig. 13-13   **Fig. 13-14**

The 1983 ACI Building Code requires that in continuous beams at least 1/4 of the positive reinforcement must be extended into the support at least 6''. If in Fig. 13-12 $x_1 = L'/4$ for an interior span and $x_1 = L'/7$ for the exterior end of an exterior span, the bottom bar extension will provide adequate positive reinforcement for the various loading conditions.

The 1963 ACI Building Code required that at least 1/3 of the negative reinforcement must be extended past the extreme interior point of inflection not less than $L'/16$ or d, whichever is greater. In addition, the anchorage bond requirements must be met. Two general rules have been used to provide adequate negative reinforcement for the various loading conditions: (1) Extend all bars to $L'/4$, or (2) Extend half of the bars to $L'/3$ and remainder to $L'/6$.

The general rules for bending bars are illustrated in Fig. 13-15. It should be noted that these bendup and cutoff points are based on moment coefficients and are approximate only, and they should be verified before adoption as a standard practice in designing a structure.

(*j*) Fig. 13-16 is part plan showing the reinforcement for the three panels designed. The reinforcement in both directions is the same.

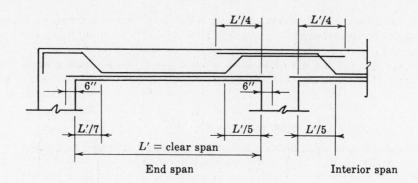

Fig. 13-15

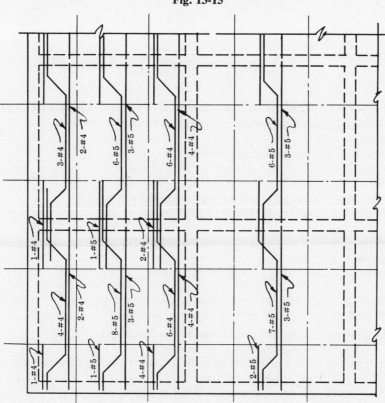

Fig. 13-16

(*k*) The load on the supporting beams is the slab load within the area shown in Fig. 13-7. If the clear span of 19′-0″ is used, the total slab load on an interior beam is $W = 2(19.0)^2(134)/4(1000) = 24.2$ kips. In addition there is dead and live load on the beam itself. If the total depth of the beam is assumed to be 19″, the dead and live loads on the beam are $W' = 19.0(5 + 20 + 40 + 237) = 5730$ lb $= 5.73$ kips.

The shear at the support is approximately $V = (24.2 + 5.73)/2 = 15.0$ kips.

For computing moments an equivalent uniform load may be used, $w' = 2wS/3 = 2(134)(19.92)/3(1000) = 1.78$ kips/ft. The total equivalent uniform load acting on the beam is approximately $w'' = 1.78 + 0.30 = 2.08$ kips/ft. Using the moment coefficients 1/16 and 1/11, the beam moments are approximately

$$+M = 2.08(19.0)^2/16 = 46.9 \qquad -M = 2.08(19.0)^2/11 = 68.2 \text{ ft-kips}$$

(*l*)   If $b = 12''$ and $d = 17.5''$, the unit shear stress in the beam at a distance d from the face of the support is approximately

$$v = \frac{V}{bd} = \frac{15.0 - 1.46(0.30) - (1.46)^2(0.134)/2}{12(17.5)} = 69 \text{ psi}$$

The allowable shear is $v_{\text{All}} = 1.1\sqrt{f'_c} = 70$ psi > 69. Hence no web reinforcement is required.

(*m*)   For $f_c = 1800$ psi and $f_s = 20{,}000$ psi, $R = 324$. The resisting moment of the beams is $M = Rbd^2 = 324(12)(17.5)^2/12{,}000 = 99.0$ ft-kips > 68.2.

(*n*)   The selection of the reinforcing size and details would be similar to that illustrated earlier and will not be done here. The design of the other supporting beams would be the same as above except that the spandrel beams carry less load and the moment and shear coefficients for end spans will not be the same as those in (*l*).

Supplementary Problems

13.9.   Using the "crossed sticks" method, develop a curve showing the ratio of bending moments in each direction as a function of the ratio of the width to length of a two-way slab.

13.10.   Repeat Problem 13.3 but use strength design.
          Ans. $+M = 9.0$, $-M = 13.05$ ft-kips/ft, $V = 4.35$ kips/ft

13.11.   Repeat Problem 13.6 but assume the width to length ratio is 0.85.
          Ans.   Total panel moment $= 0.15wL^2L_1$ or $0.15wLL_1^2$

13.12.   Design a typical interior two-way slab panel that is $25'-0''$ by $25'-0''$. Assume $f'_c = 3000$ psi, $f_s = 20{,}000$ psi, superimposed D.L. $= 10$ lb/ft^2, L.L. $= 75$ lb/ft^2. The supporting beams are $14''$ wide. Use the alternate design method (Chapter 4).   *Partial Ans.* $h_f = 6''$

13.13.   Repeat Problem 13.12 using strength design and $f_y = 50{,}000$ psi.

13.14.   Using the data given in Problem 13.12, design a corner panel.

13.15.   An interior two-way slab panel is $20'-0''$ by $20'-0''$. Determine by Method 3 the slab bending moments and shears if $h_f = 7''$ and the live load 150 lb/ft^2.
          Partial Ans.   Maximum negative moment $= 4.76$ ft-kips/ft

Chapter 14

The Equivalent Frame Method

NOTATION

A_c = area of concrete for transfer of unbalanced bending moment by direct shear stress (see Appendix A-10)

c_1 = size of rectangular or equivalent rectangular column, capital, or bracket measured in the direction of the span for which moments are being determined

c_2 = size of rectangular or equivalent rectangular column, capital, or bracket measured transverse to the direction of the span for which moments are being determined

C = cross-sectional constant to define torsional properties

d = distance from extreme compression fiber to centroid of tension reinforcement

E_{cb} = modulus of elasticity of beam concrete

E_{cc} = modulus of elasticity of column concrete

E_{cs} = modulus of elasticity of slab concrete

h = overall thickness of member

I_b = moment of inertia about centroidal axis of gross section of beam

I_c = moment of inertia of gross section of column

I_s = moment of inertia about centroidal axis of gross section of slab

  = $h^3/12$ times width of slab defined in notations α and β_t

J = equivalent polar moment of inertia of rectangular cross section

J/c = modulus of critical section for transfer of bending moment by torsional shear stress (see Appendix A-10)

k = stiffness factor occurring in $K = kEI/L$

K_b = flexural stiffness of beam; moment per unit rotation

K_c = flexural stiffness of column; moment per unit rotation

K_{ec} = stiffness of equivalent column at a joint

K_s = flexural stiffness of slab; moment per unit rotation

K_t = torsional stiffness of torsional member; moment per unit rotation

L_c = height of column

L_n = length of clear span in direction that moments are being determined, measured face-to-face of supports

L_1 = length of span in direction that moments are being determined, measured center-to-center of supports

L_2 = length of span transverse to L_1, measured center-to-center of adjacent slab panels

M_o = total factored static moment

v_c = permissible combined direct and torsional shear stress at critical section

v_u = actual combined direct and torsional shear stress at critical section

w_D = factored dead load per unit area

w_L = factored live load per unit area

W_u = factored load per unit area

x = shorter overall dimension of rectangular part of cross section

y = longer overall dimension of rectangular part of cross section

α = ratio of flexural stiffness of beam section to flexural stiffness of a width of slab bounded laterally by centerlines of adjacent panels (if any) on each side of the beam = $E_{cb}I_b/E_{cs}I_s$

α_c = ratio of flexural stiffness of columns above and below the slab to combined flexural stiffness of the slabs and beams at a joint taken in the direction of the span for which moments are being determined = $\Sigma K_c/\Sigma (K_s + K_b)$

α_{min} = minimum α_c

α_1 = α in direction of L_1

α_2 = α in direction of L_2

β_a = ratio of dead load per unit area to live load per unit area (in each case without load factors)

β_t = ratio of torsional stiffness of edge beam section to flexural stiffness of a width of slab equal to span length of beam, center-to-center of supports = $E_{cb}C/2E_{cs}I_s$

γ_f = fraction of unbalanced moment transferred by flexure

$$= \frac{1}{1 + \frac{2}{3}\sqrt{(c_1 + d)/(c_2 + d)}}$$

γ_v = fraction of unbalanced moment transferred by eccentricity of shear = $1 - \gamma_f$

THE EQUIVALENT FRAME

The 1983 ACI Code permits the use of an *equivalent frame* like that shown in Fig. 14-1. Because of the many possibilities of live load variation, the Code requires the application of *pattern loads* (also called *checkerboard loads* and *skip loads*) to the frame in order to determine the maximum bending moments and shear forces in the beams and/or slabs, and the maximum shear forces, axial forces and bending moments in the columns. The patterns that must be investigated involve full factored dead load on all spans, plus (i) 3/4 of factored live load on spans *AB* and *CD* only; (ii) 3/4 of factored live load on span *BC* only; (iii) 3/4 of factored live load on spans *AB* and *BC* only; (iv) 3/4 of factored live load on spans *BC* and *CD* only; (v) full factored live load on all spans. (Refer to Chapter 5 for the values of the various load factors.) Out of these five analyses, the "worst case" for a given member is chosen to represent the maximum moment or force at critical locations.

Because the slab-beam system acts as a *diaphragm*, the horizontal deflections of all columns are assumed identical at each floor level. Hence, a three-dimensional frame can be modeled as a series of

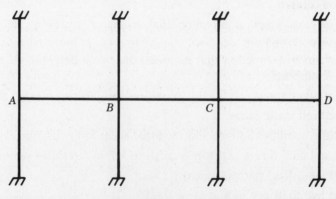

Fig. 14-1

plane frames linked together for lateral loads (e.g. winds, earthquakes). For gravity loads (dead load and live load), the plane frames can be considered individually, each with its hand-calculated moment distribution. Columns are assumed to be fixed at the levels above and below the floor under consideration. When a computer program is utilized, it is common practice to use the entire frame in determining axial forces, shears and bending moments, both for gravity loads and lateral loads.

The 1983 ACI Code provides the following requirements relative to the equivalent frame method:

Design of slab systems by the Equivalent Frame Method shall be based on assumptions given in Sections 13.7.2 through 13.7.6 of the Code, and all cross sections of slabs and supporting members shall be proportioned for moments and shears thus obtained.

The structure shall be considered to be made up of equivalent frames on column lines taken longitudinally and transversely through the building. Each frame shall consist of a row of columns or supports and slab-beam strips, bounded laterally by the centerline of panel on each side of the centerline of columns or supports. Columns or supports shall be assumed to be attached to slab-beam strips by *torsional members* transverse to the direction of the span for which moments are being determined and extending to bounding lateral panel center-lines on each side of a column.

Frames adjacent and parallel to an edge shall be bounded by that edge and the centerline of the adjacent panel. Each equivalent frame may be analyzed in its entirety, or for gravity loading, each floor and the roof (slab-beams) may be analyzed separately with far ends of columns considered fixed. Where slab-beams are analyzed separately, it may be assumed in determining moment at a given support that the slab-beam is fixed at any support two panels distant therefrom, provided the slab continues beyond that point.

Slab-Beams and Columns

Moment of inertia of slab-beams at any cross section outside of joints or column capitals may be based on the gross area of concrete. Variation in moment of inertia along axis of slab-beams shall be taken into account. Moment of inertia of slab-beams from center of column to face of column, bracket, or capital shall be assumed equal to the moment of inertia of the slab-beam at face of column, bracket, or capital divided by the quantity $(1 - c_2/L_2)^2$, where c_2 and L_2 are measured transverse to the direction of the span for which moments are being determined. [This applies also to *flat plate* construction, where there are no beams. In all cases, the moment of inertia of the columns is assumed to be infinite from the lower ends of beams to the centerline of the slab above, and from the slab surface to the centerline of the slab at the lower end of the column.]

Torsional Members

Torsional members shall be assumed to have a constant cross section throughout their length consisting of the larger of: (*a*) a portion of slab having a width equal to that of the column, bracket, or capital in the direction of the span for which moments are being determined; (*b*) for monolithic or fully composite construction, the portion of slab specified in (*a*) plus that part of the transverse beam above and below the slab; (*c*) transverse beam torsional resistance shall be included.

Stiffness K_t of the torsional members shall be calculated from

$$K_t = \sum \frac{9 E_{cs} C}{L_2 \left(1 - \dfrac{c_2}{L_2}\right)^3} \qquad (14.1)$$

where c_2 and L_2 relate to the transverse spans on each side of the column.

The constant C in (14.1) may be evaluated by dividing the cross section into separate rectangular parts and carrying out the following summation:

$$C = \sum \left(1 - 0.63\,\frac{x}{y}\right) \frac{x^3 y}{3} \qquad\qquad (14.2)$$

Where beams frame into columns in the direction of the span for which moments are being determined, the value of K_t from (14.1) and (14.2) shall be multiplied by the ratio of the moment of inertia of slab with such beam to moment of inertia of slab without such beam.

Arrangement of Live Load

When the loading pattern is known, the equivalent frame shall be analyzed for that load.

When live load is variable but does not exceed three-quarters of the dead load, or the nature of live load is such that all panels will be loaded simultaneously, maximum factored moments may be assumed to occur at all sections with *full factored live load* on the entire slab system.

For loading conditions other than those defined above, maximum positive factored moment near midspan of a panel may be assumed to occur with *three-quarters of factored live load* on the panel and on alternate panels; and maximum negative factored moment in the slab at a support may be assumed to occur with *three-quarters of factored live load* on adjacent panels only.

Factored moments shall be taken not less than those occurring with full factored live load on all panels.

Factored Moments

At interior supports, the critical section for negative factored moment (in both column and middle strips) shall be taken at the face of rectilinear supports, but not greater than $0.175L_1$ from the center of a column.

At exterior supports provided with brackets or capitals, the critical section for negative factored moment in the span perpendicular to an edge shall be taken at a distance from the face of the supporting element not greater than one-half the projection of the bracket or capital beyond the face of the supporting element.

Circular or regular polygonal supports shall be treated as square supports of the same area for location of the critical section for negative design moment.

SLAB-BEAMS AND COLUMNS

Figure 14-2 shows the members at a joint in the equivalent frame. Figure 14-3 gives the parameters involved in the calculation of the fixed-end moments, the stiffness factors (k_1, k_2, k_3), and the carry-over factors $(C_{AB} = k_2/k_1, C_{BA} = k_2/k_3)$. The calculations themselves are best carried out by computer: suitable programs, in ANSI FORTRAN-77, immediately precede the Solved Problems of this chapter. Figures 14-4 and 14-5 are diagrams for computing the stiffnesses K_{sb} of various slab-beam systems, while Fig. 14-6 applies to the calculation of column stiffness K_c. The above-mentioned computer programs can be used on most microcomputers that have ANSI FORTRAN-77 compilers; job control statements are not included, because they differ from computer to computer. An illustration of the hand calculation of k is given in Problem 14.2.

TORSIONAL RESISTANCE; THE EQUIVALENT COLUMN

In modeling a three-dimensional structure, it is necessary to include the torsional resistance of slabs and/or beams that are perpendicular to the plane frame being analyzed. Figures 14-7 and 14-8 illustrate the various torsional members. The torsional resistances are included with the rotational stiffnesses of the columns to provide an *equivalent column*, of *flexibility*

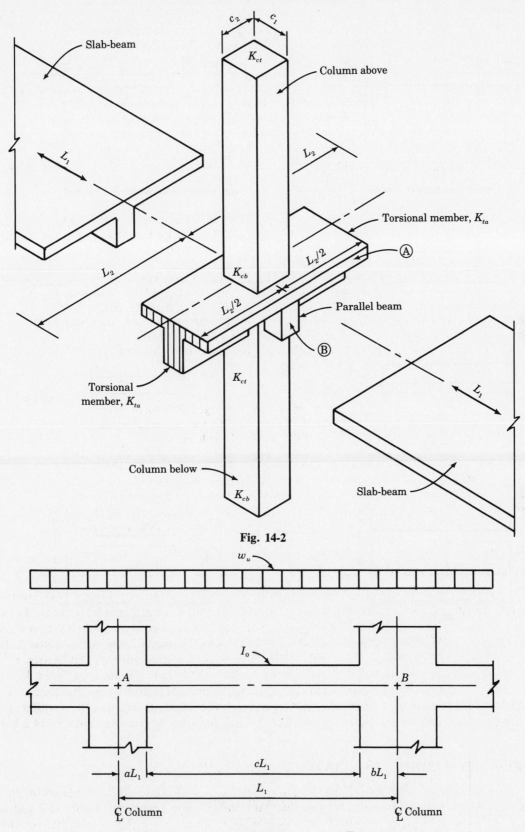

Fig. 14-2

Fig. 14-3

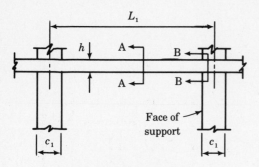

Slab system without beams

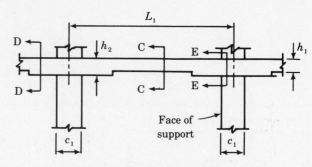

Slab system with drop panels

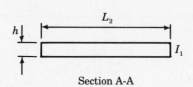

Section A-A

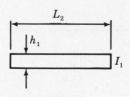

Section C-C

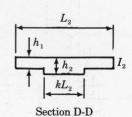

Section D-D

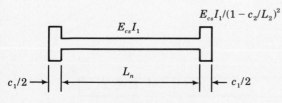

Section B-B

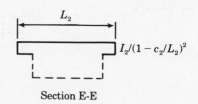

Section E-E

Equivalent slab-beam stiffness diagram

Equivalent slab-beam stiffness diagram

Fig. 14-4

$$\frac{1}{K_{ec}} = \frac{1}{\sum K_c} + \frac{1}{\sum K_t} \tag{14.3}$$

or, equivalently, of *stiffness*

$$K_{ec} = \frac{\left(\sum K_c\right)\left(\sum K_t\right)}{\sum K_c + \sum K_t} \tag{14.4}$$

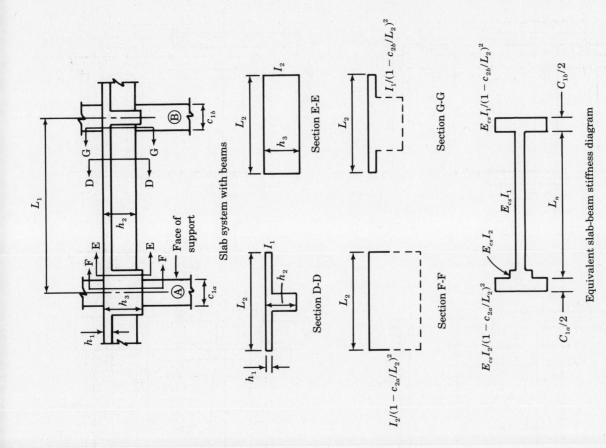

Fig. 14-5

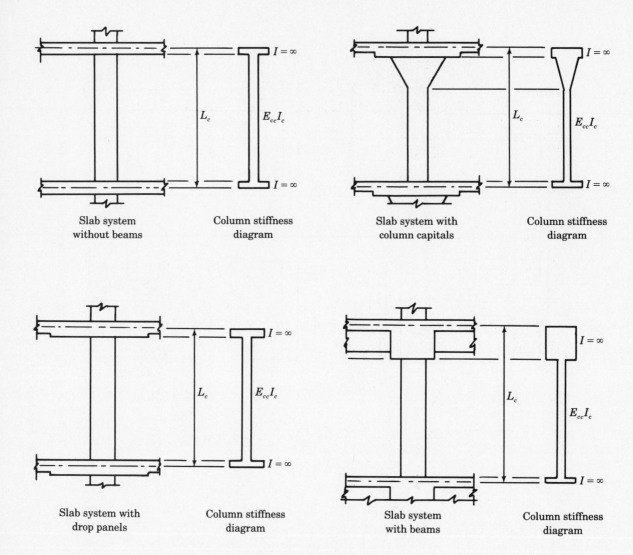

Fig. 14-6

LOADING CONDITIONS

It is important to note that when the equivalent frame method is used with lateral loads, the entire frame must be considered. Also, the moment of inertia of the slab-beam system must be reduced to account for the reduced stiffness caused by cracking of the concrete. The 1983 ACI Code does not provide requirements concerning the reduction, but the Commentary for the Code recommends using from 1/4 to 1/2 of the gross cross-sectional moment of inertia of the slab.

Figure 14-9 shows the loading conditions that must be considered in the analysis of an equivalent frame for gravity loads (live load plus dead load). In the loading patterns, loads are uniformly distributed over the entire area of the slab. The slab may also be subjected to *line loads* due to walls supported by the slab or beam webs beneath the slab. Furthermore, concentrated loads of various types can—and usually do—occur within the spans.

The partial frame analysis indicated in Fig. 14-9 must be followed by a second analysis, in which, along with the vertical loads, a portion of the slab moments is assigned to the beam cross sections.

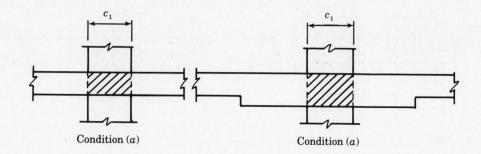

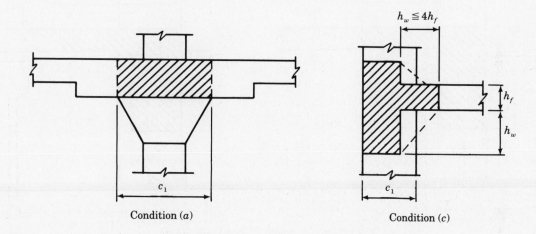

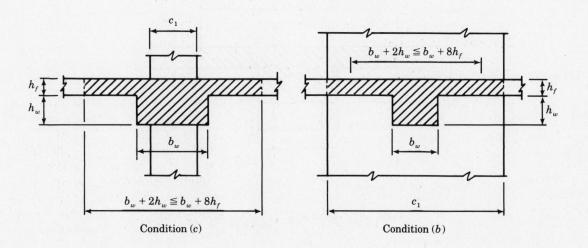

Fig. 14-7.   Torsional Members

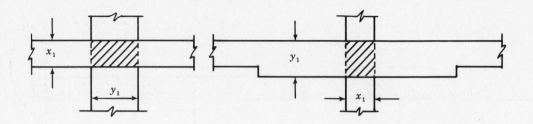

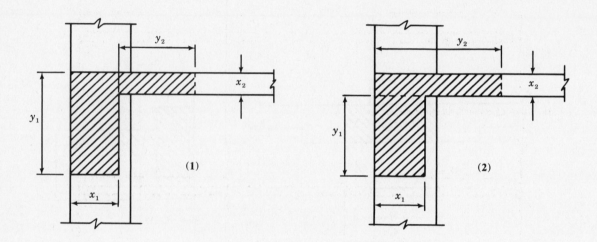

Use larger value of C computed from (1) or (2)

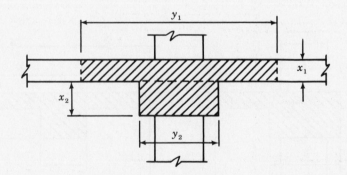

$$C = \sum \left[\left(1 - 0.63\frac{x_1}{y_1}\right)\frac{x_1^3 y_1}{3}\right] + \left[\left(1 - 0.63\frac{x_2}{y_2}\right)\frac{x_2^3 y_2}{3}\right]$$

Fig. 14-8. Torsional Constant C

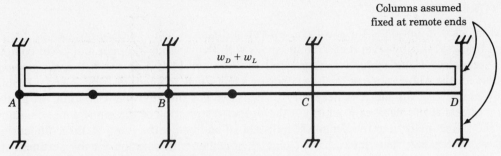

(1)   Loading pattern for design moments in all spans with $L \leq \frac{3}{4}D$

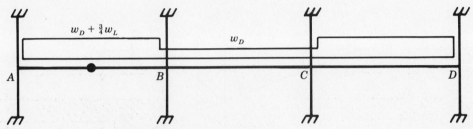

(2)   Loading pattern for positive design moment in span AB

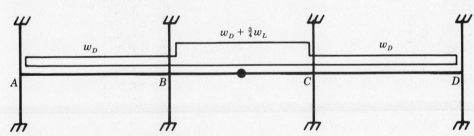

(3)   Loading pattern for positive design moment in span BC

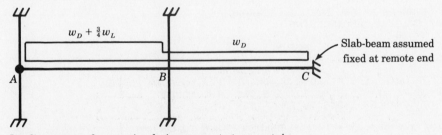

(4)   Loading pattern for negative design moment at support A

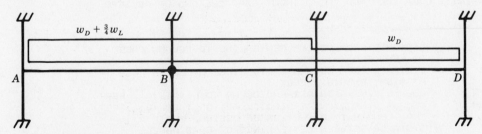

(5)   Loading pattern for negative design moment at support B

Fig. 14-9

FACTORED MOMENTS

Moment distribution is the most convenient and simple hand calculation method for analyzing partial frames involving several continuous spans with the far ends of upper and lower columns fixed. The mechanics of the method is illustrated in Problem 14.1. For the *two-cycle method*, which accounts for the whole range of loading conditions in a single moment distribution, see Problem 14.4. In any moment technique it is necessary to consider (1) the equivalent column, so as to include torsion effects at the slab-beam joint; (2) the proper distribution of the equivalent column moments from the frame analysis to the actual columns above and below the joint. A three-dimensional view of the members that frame into a joint (joint 2), together with notation for the pertinent stiffnesses ($K = kEI/L$), is given in Fig. 14-10.

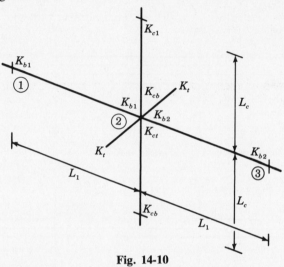

Fig. 14-10

COMPUTER PROGRAM LISTINGS

```
C----------------------------------------------------------------C
C----------------------------------------------------------------C
C
C       PROGRAM "FEMWFP" BY DR. NOEL J. EVERARD
C       PROFESSOR OF CIVIL ENGINEERING
C       P.O. BOX 19308
C       UNIVERSITY OF TEXAS AT ARLINGTON
C       ARLINGTON, TEXAS 76019
C
C       PROGRAM FOR OBTAINING FIXED-END MOMENTS FOR UNIFORM LOADS
C       FOR SLAB-BEAMS WITH FINITE JOINT SIZES. THE MOMENT
C       OF INERTIA FOR SLAB-BEAMS IS TAKEN AS:
C
C       ISB = I1/((1.0-C2/L2)**2) FROM COLUMN FACES TO COLUMN
C       CENTERLINES, WHERE I1 IS THE MOMENT OF INERTIA OF THE
C       SLAB-BEAM AT COLUMN FACES. IF MOMENT OF INERTIA IS TAKEN AS
C       INFINITY OVER "AL" AND "BL", INPUT C2A, C2B AND L2 AS ZERO.
C
C       DEFINITIONS OF INPUT VALUES:
C
C       AL1     = DISTANCE FROM COLUMN CENTERLINE TO COLUMN FACE
C                 AT LEFT END (A), FEET
C       BL1     = DISTANCE FROM COLUMN CENTERLINE TO COLUMN FACE
C                 AT RIGHT END (B), FEET
C       C2A     = COLUMN DIMENSION PERPENDICULAR TO L1 AT (A), FEET
C       C2B     = COLUMN DIMENSION PERPENDICULAR TO L1 AT (B), FEET
C       L1      = SPAN LENGTH IN DIRECTION OF DESIGN, FEET
C       L2      = DISTANCE BETWEEN CENTERLINES OF SLAB PANELS
C                 PERPENDICULAR TO L1, FEET
C       WU      = ULTIMATE UNIT LOAD, KIPS/FOOT (1.4 WD + 1.7 WL)
C
C----------------------------------------------------------------C
C----------------------------------------------------------------C
```

```
C
      REAL L,L1,L2, INA, INB, INERTA
      DIMENSION X(110),BM(110),AMEI(110)
      DIMENSION BEAM (4)
      KOUNT = 0
      NOPAGE = 1
      NSEGC = 100
      J2 = NSEGC
      J3 = J2 + 1
      ASEGC = NSEGC
      WRITE (6,81) NOPAGE
   81 FORMAT (50X,'PAGE',I4,//)
  450 FORMAT (2X,4A2,7F10.3)
    1 FORMAT (13X,'BEAM',6X,'SPAN',7X,'AL',8X,'BL',
     & 8X,'WU',6X,'FEM(A)',4X,'FEM(B)')
    2 FORMAT (24X,'FT.',7X,'FT.',7X,'FT.',5X,'KIPS/FT.',
     & 2X,'FT-KIPS',3X,'FT-KIPS',/)
    3 FORMAT (8X,72('-'),/)
      WRITE (6,1)
      WRITE (6,2)
      WRITE (6,3)
 1000 READ (5,450,END=999)(BEAM(K),K=1,4),L,AL,BL,WU,C2A,C2B,L2
  420 FORMAT (11X,4A2,F9.3,4X,F6.3,4X,F6.3,F9.3,F11.3,
     &    F10.3)
?
      CL = L - AL - BL
      DX = CL/ASEGC
      X(1) = AL
      RL = WU * L/2.0
      RR = WU * L/2.0
      BM(1) = RL * AL -WU*(AL**2)/2.0
      SUMA = 0.0
      SUMOM = 0.0
      DO 400 K = 2,J2
      XK = K - 1
      X(K) = XK * DX + AL
      BM(K) = RL * X(K) - WU * (X(K) **2)/ 2.0
      AMEI (K) = BM(K) * DX
      SUMA = SUMA + AMEI (K)
      SUMOM = SUMOM + AMEI(K) * X(K)
  400 CONTINUE
      IF (L2 .LT. 0.1) FACAL = 1.0
      IF (L2 .LT. 0.1) FACBL = 1.0
      IF (L2 .LT. 0.1) GO TO 65
      FACAL = 1.0/((1.0-C2A/L2)**2)
      FACBL = 1.0/((1.0-C2B/L2)**2)
   65 AM1A = (BM(1) * AL/2.0)/FACAL
      IF (L2 .LT. 0.1) AM1A = 0.0
      AM2A = (BM(1)* DX/2.0)
      SUMA = SUMA + AM1A + AM2A
      SUMOM = SUMOM + (AM1A*2.0*AL/3.0) + AM2A * (AL + DX/3.0)
      BM(J3) = RR * BL - WU * (BL**2)/ 2.0
      X(J3) = X(J2) + DX
      AM1B = BM(J3) * DX/2.0
      AM2B = (BM(J3) * BL/2.0) / FACBL
      IF (L2 .LT. 0.1) AM2B = 0.0
      SUMA = SUMA + AM1B + AM2B
      SUMOM = SUMOM + AM1B*(X(J2)+2.0*DX/3.)
     &    + AM2B * (AL + CL + BL/3.0)
      XTOP = SUMOM/SUMA
      IF (L2 .LT. 0.1) GO TO 92
C     FLAT PLATE TYPE
      AREA = CL + AL/FACAL + BL/FACBL
      ENUM = (AL/FACAL)*AL/2. + CL*(AL+CL/2.) + (BL/FACBL)*(L-BL/2.)
      CA = ENUM/AREA
      CB = L - CA
      E = XTOP - CA
      INERTA = (CL**3)/12. + (CL*(CA-AL-(CL/2.))**2) + (AL/FACAL)*
     &    (AL**3/12.) + (AL/FACAL)*((CA-AL/2.)**2) + (BL/FACBL)*
     &    (BL**3/12.)   + (BL/FACBL) * ((C-BL/2.)**2)
      GO TO 93
```

```
C      INFINITE MOMENT OF INERTIA OVER "AL" AND "BL"
    92 CA = AL + CL/2.
       CB = BL + CL/2.
       AREA = CL
       E = XTOP - CA
       INERTA = (CL**3) / 12.0
?
    93 CONTINUE
C      RESULTANT OF M/EI ASSUMED TO RIGHT OF CENTERLINE OF SPAN.
C      I.E. ECCENTRICITY OF ELASTIC LOAD IS ASSUMED TO BE POSITIVE
       FEML = SUMA/AREA - SUMA * E * CA/INERTA
       FEMR = -(SUMA/AREA + SUMA * E * CB / INERTA)
C      CLOCKWISE MOMENTS ARE CONSIDERED TO BE IN NEGATIVE DIRECTION.
       WRITE (6,420) (BEAM(M), M=1,4),L,AL,BL,WU,FEML,FEMR
       KOUNT = KOUNT + 1
       IF (KOUNT .LT. 55) GO TO 1000
       KOUNT = 0
       NOPAGE = NOPAGE + 1
       WRITE (6,81) NOPAGE
       WRITE (6,1)
       WRITE (6,2)
       WRITE (6,3)
       GO TO 1000
   999 STOP
       END
```

DATA FOLLOWS

(1)	18.0	0.67	0.67	3.06	1.33	1.33	14.0
(2)	22.0	1.0	2.0	1.0	0.0	0.0	0.0
(3)	20.0	1.0	1.0	2.0	0.0	0.0	0.0

SOLUTION:

BEAM	SPAN FT.	AL FT.	BL FT.	WU KIPS/FT.	FEM(A) FT-KIPS	FEM(B) FT-KIP
(1) *	18.000	0.670	0.670	3.060	83.612	-83.611
(2)	22.000	1.000	2.000	1.000	40.080	-51.080
(3)	20.000	1.000	1.000	2.000	72.994	-72.994

* FIXED-END MOMENT SOLUTION FOR PROBLEM 14.1

```
C-----------------------------------------------------------------C
C-----------------------------------------------------------------C
C
C
C     PROGRAM "CLBMSTIF" BY DR. NOEL J. EVERARD
C     PROFESSOR OF CIVIL ENGINEERING
C     P.O. BOX 19308
C     UNIVERSITY OF TEXAS AT ARLINGTON
C     ARLINGTON, TEXAS 76019
C
C     PROGRAM FOR OBTAINING STIFFNESS FACTORS K1, K2 AND K3
C     FOR BEAMS AND COLUMNS WITH FINITE JOINT SIZES. FOR
C     BEAMS, THE MOMENT OF INERTIA IS CONSIDERED TO BE INFINITE
C     FROM COLUMN CENTERLINES TO COLUMN FACES. FOR COLUMNS,
C     THE MOMENT OF INERTIA IS CONSIDERED TO BE INFINITE FROM
C     THE CENTERLINE OF THE SLAB ABOVE TO THE BOTTOM OF THE
C     BEAM ABOVE THE COLUMN, AND FROM THE SURFACE OF THE SLAB
C     BELOW THE COLUMN TO THE CENTERLINE OF THE SLAB BELOW THE
C     COLUMN. FOR FLAT PLATE SLABS USE PROGRAM "SLBMSTIF".
C
C
C-----------------------------------------------------------------C
C-----------------------------------------------------------------C
C
      COMMON LINES,NOPAGE
C      NOBM IS THE MEMBER NUMBER, 5 DIGITS OR LESS
      REAL L, INERTA, I, K1, K2, K3, K1AB, K2AB, K2BA, K3BA
      CALL TITLE
C     TAKE ACTUAL EI AS 1.0
   20 READ (5,10,END=999) L,AL,BL,NOBM
      EI = 1.0
      NOPAGE = 0
      LINES = 0
      IBLANK = 0
      CL = L - AL - BL
      XL = AL + ( 0.5 * CL)
      AREA = CL/EI
      INERTA = (1.0/EI) * (CL **3 ) /12.0
      K1AB = (1.0/AREA) + 1.0 * XL * XL/INERTA
      K1 = K1AB * L
      K2AB =      ((1.)/AREA) - (1.0* XL * (L-XL))/ INERTA
      K3BA = (1.0/AREA) + 1.0 * (L-XL) * (L-XL)/INERTA
      K3 = K3BA * L
      K2BA =      ( 1.0/AREA) - ( 1.0 * (L-XL) * XL) / INERTA
C     AVERAGE THE K2 VALUES
      K2 = 0.5 * (K2AB + K2BA)  * L * (-1.0)
      WRITE (6,30) NOBM,L,AL,BL,K1,K2,K3
      LINES = LINES + 1
      IBLANK = IBLANK + 1
      IF (IBLANK .GT. 4 ) WRITE (6,40)
   40 FORMAT (/)
   10 FORMAT (3F10.5,2I5)
      IF (IBLANK .GT. 4 ) IBLANK = 0
   30 FORMAT (5X,I5,6F10.3)
      IF (LINES .GT. 50 ) WRITE (6,99)
   99 FORMAT (1H1)
      IF ( LINES .GT. 50 ) CALL TITLE
      GO TO 20
  999 STOP
      END

      SUBROUTINE TITLE
      COMMON LINES,NOPAGE
      LINES = 0
      WRITE (6,10)
```

```
 10 FORMAT ('1')
    NOPAGE = NOPAGE + 1
    WRITE (6,42) NOPAGE
 42 FORMAT (50X,'PAGE',I5,///)
    WRITE (6,15)
 15 FORMAT (20X,'STIFFNESS FACTORS',/)
    WRITE (6,16)
    WRITE (6,17)
 16 FORMAT (5X,'MEMBER',3X,'LENGTH',7X,'AL',8X,'BL',
   &          7X,'K1',8X,'K2',9X,'K3')
 17 FORMAT (5X,'NUMBER',5X,'(L)',//)
    RETURN
    END

    DATA FOLLOWS

20.0        0.0        0.0            1
30.0        6.0        6.0            2
10.0        2.0        0.0            3
```

SOLUTIONS   PAGE   1

                        STIFFNESS FACTORS

MEMBER NUMBER	LENGTH (L)	AL	BL	K1	K2	K3
1	20.000	0.000	0.000	4.000	2.000	4.000
2	30.000	6.000	6.000	15.556	12.222	15.556
3	10.000	2.000	0.000	9.688	4.375	5.000

```
C--------------------------------------------------------------C
C--------------------------------------------------------------C
C
C
C     PROGRAM "SLBMSTIF" BY DR. NOEL J. EVERARD
C     PROFESSOR OF CIVIL ENGINEERING
C     P.O. BOX 19308
C     UNIVERSITY OF TEXAS AT ARLINGTON
C     ARLINGTON, TEXAS 76019
C
C     PROGRAM FOR OBTAINING STIFFNESS FACTORS K1, K2 AND K3
C     FOR SLAB-BEAMS WITH FINITE JOINT SIZES. THE MOMENT
C     OF INERTIA FOR SLAB-BEAMS IS TAKEN AS:
C
C     ISB = I1/((1.0-C2/L2)**2) FROM COLUMN FACES TO COLUMN
C     CENTERLINES, WHERE I1 IS THE MOMENT OF INERTIA OF THE
C     SLAB-BEAM AT COLUMN FACES, C2 IS THE COLUMN DIMENSION
C     PERPENDICULAR TO THE DIRECTION OF BENDING, AND
C     L2 IS THE DISTANCE FROM SLAB CENTERLINE TO SLAB
C     CENTERLINE ON BOTH SIDES, PERPENDICULAR TO THE DIRECTION
C     OF BENDING. IF THERE IS AN INTERSECTING PERPENDICULAR
C     BEAM, THE MOMENT OF INERTIA IS CONSIDERED TO BE
C     INFINITE FROM THE COLUMN FACE TO THE COLUMN
C     CENTERLINE. (USE PROGRAM CLBMSTIF)
C
C
C
C--------------------------------------------------------------C
C--------------------------------------------------------------C
```

```
C
      COMMON LINES,NOPAGE
C       NOBM IS THE MEMBER DESIGNATION, 8 CHARACTERS OR LESS
      REAL I1,IA1,IA2,IA3,K1,K2,K3,K2L,K2R,L1,L2
      DIMENSION NOBM (4)
      WRITE (6,44)
   44 FORMAT (2X,'MEMBER',9X,'K1',8X,'K2',8X,'K3',//)
C     TAKE ACTUAL EI AS 1.0 FOR SLAB-BEAM AT COLUMN FACES
      I1 = 1.0
   20 READ (5,300,END=999) L1,AL1,BL1,C2L,C2R,L2,
     &           (NOBM(KK), KK = 1,4)
  300 FORMAT (6F10.2,2X,4A2)
      CL1 = L1 - AL1 - BL1
      EILL = I1/((1.0-C2L/L2)**2)
      EILR = I1/((1.0-C2R/L2)**2)
      A1 = (1.0/EILL) * AL1
      A2 = (1.0/EILR) *BL1
      A3 = (1.0/I1) * CL1
      AREA = A1 + A2 + A3
      XBARL = (A1*AL1/2.0 + A2*(L1-BL1/2.0)
     &        + A3*(CL1/2.0 + AL1))/AREA
      XBARR = L1-XBARL
      XR = XBARR
      XL = XBARL
      IA1 = (1.0/EILL) * (AL1 **3)/12.0
     &        + A1 * (XL-AL1/2.0)**2
      IA2 = (1.0/EILR) * (BL1**3)/12.0
     &        + A2 * (XR - BL1/2.0) **2
      IA3 = (1.0/I1) * (CL1**3)/12.0
     &        + A3 * (L1/2.0 - XL) **2
      Q = IA1 + IA2 + IA3

      K3 = (1.0/AREA) + (1.0 * XR * XR/Q)
      K2R = (1.0/AREA) - (1.0 * XR * XR/Q)
      K1 = (1.0/AREA) + (1.0 * XL * XL/Q)
      K2L = (1.0/AREA) - (1.0 * XL * XL/Q)
C     USE AVERAGE K2 FROM BOTH ENDS OF THE MEMBER
      K2 = 0.5 * (K2R + K2L)
      K1 = K1 * L1
      K2 = K2 * L1
      K3 = K3 * L1
      WRITE (6,100) (NOBM(KK), KK = 1,4), K1,K2,K3
  100 FORMAT (2X,4A2,3F10.3)
      GO TO 20
  999 STOP
      END
```

DATA FOLLOWS

```
20.0        1.0         1.0         2.0        2.0        20.0       BEAM1
20.0        0.0         0.0         0.0        0.0        20.0       BEAM2
```

SOLUTION

MEMBER	K1	K2	K3
BEAM1	4.182	-2.143	4.182
BEAM2	4.000	-2.000	4.000

Solved Problems

14.1. Referring to the slab plan, Fig. 14-11, use the equivalent frame method to determine the design moments for the slab in the transverse direction for an intermediate floor. (*This problem is provided by courtesy of the Portland Cement Association.*)

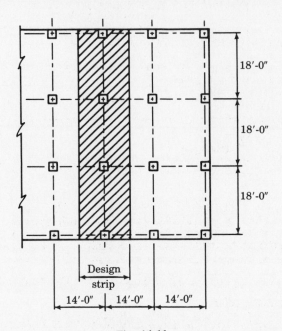

Fig. 14-11

Data are: story height $= 9$ ft; columns are 16 in. $\times$ 16 in. square; lateral loads are to be resisted by shear walls and there are no spandrel beams; partition weight $= 20$ lbs/ft^2; service live load $= 40$ lbs/ft^2. $f'_c = 3000$ psi (for slab); $f'_c = 5000$ psi (for column); $f_y = 60,000$ psi.

Preliminary design for slab thickness

(*a*) *Control of deflections.* For slab systems without beams, the minimum overall thickness h is governed by equation (9-13) of the 1983 ACI Code. Using Grade 60 reinforcement, with $f_y = 60,000$ psi,

$$h = \frac{L_n(800 + 0.005f_y)}{36,000} = \frac{(216 - 16)[800 + 0.005(60,000)]}{36,000} = 6.11 \text{ in.}$$

and this exceeds the 5-in. minimum for slabs without beams or drop panels. For slab systems without edge beams (spandrels), the panel having a discontinuous edge must be increased by 10 percent:

$$h = 6.11 \times 1.10 = 6.72 \text{ in.}$$

Use $h = 7$ in. for all panels; density $= 87.5$ lbs/ft^2.

(*b*) *Shear strength.* Use an average effective depth d of about 5.75 in. for $\frac{3}{4}$ in. cover plus No. 4 bar. The total factored load is

$$W_u = 1.4(87.5 + 20) + 1.7(40) = 218.5 \text{ lbs/ft}^2$$

Wide-beam action is investigated for a 12-in.-wide strip at distance d from the face of support in the long direction (see Fig. 14-12). Thus, the actual and allowable shear forces are

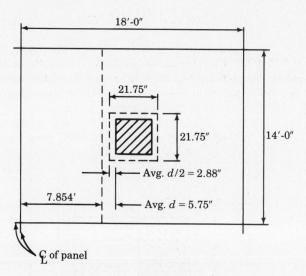

Fig. 14-12

$$V_u = (218.5)(7.854)/1000 = 1.72 \text{ kips}$$
$$\phi V_c = 2\phi\sqrt{f_c'}\,b_w d = 2(0.85)\sqrt{3000}\,(12)(5.75)/1000 = 6.43 \text{ kips}$$

$V_u < \phi V_c$ (O.K.). As for two-way action, since there are no shear forces at the centerlines of adjacent panels, the punching shear strength at distance $d/2$ around the support is

$$V_u = (218.5)[(18)(14) - (1.81)^2]/1000 = 54.3 \text{ kips}$$

while, for a square column,

$$\phi V_c = 4\phi\sqrt{f_f'}\,b_o d = 4(0.85)\sqrt{3000}\,(4 \times 21.75)(5.75)/1000 = 93.2 \text{ kips}$$

Again, $V_u < \phi V_c$ (O.K.).

The conclusion is that a 7-in. overall slab thickness is adequate for control of deflections and shear strength.

Members of equivalent frame

Determine moment distribution constants and fixed-end moments for the equivalent frame members. The column analogy procedure of Problems 14.2 and 14.3 yields the stiffness factors $k_1 = k_3 = 4.13$; carry-over factors COF $= k_2/k_1 = k_2/k_3 = 0.509$ (since $k_2 = 2.10$); and the fixed-end moment factor 0.0843, for the slab-beams. For the columns, $k_1 = k_3 = 4.74$; COF $= 0.549$ (so $k_2 = 2.60$).

(a) *Slab-beams*. From

$$I_s = \frac{L_2 h^3}{12} = \frac{(168)(7)^3}{12} = 4802 \text{ in.}^4$$
$$E_{cs} = 57{,}000\sqrt{3000} = 3.12 \times 10^6 \text{ psi}$$
$$L_1 = 18'\text{-}0'' = 216 \text{ in.}$$

it follows that the flexural stiffness at either end is

$$K_{sb} = 4.13\,\frac{E_{cs}I_s}{L_1} = 4.13\,\frac{(3.12 \times 10^6)(4802)}{216} = 286 \times 10^6 \text{ in.-lb}$$

(b) *Columns*. From

$$I_c = \frac{c^4}{12} = \frac{(16)^4}{12} = 5461 \text{ in.}^4$$
$$E_{cc} = 57{,}000\sqrt{5000} = 4.03 \times 10^6 \text{ psi}$$
$$L_c = 9'\text{-}0'' = 108 \text{ in.}$$

it follows that the flexural stiffness at either end is (flat plate, no beams)

$$K_c = 4.74 \frac{E_{cc}I_c}{L_c} + 4.74 \frac{(4.03 \times 10^6)(5461)}{108} = 966 \times 10^6 \text{ in.-lb}$$

(c) *Torsional members*. Since (Fig. 14-13) $c_2 = 16$ in. and $L_2 = 14'\text{-}0'' = 168$ in., equations (14.2) and (14.1) give

$$C = \left(1 - 0.63 \frac{x}{y}\right) \frac{x^3 y}{3} = \left(1 - 0.63 \frac{7}{16}\right) \frac{(7)^3(16)}{3} = 1325 \text{ in.}^4$$

$$K_t = \frac{9E_{cs}C}{L_2(1 - c_2/L_2)^3} = \frac{9(3.12 \times 10^6)(1325)}{168(1 - 16/168)^3} = 299 \times 10^6 \text{ in.-lb}$$

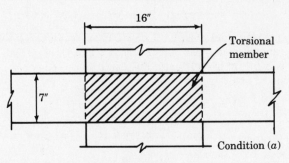

Fig. 14-13

The equivalent column stiffness is given by (14.4) as

$$K_{ec} = \left(\sum K_c \times \sum K_t\right) \Big/ \left(\sum K_c + \sum K_t\right) = (2 \times 966)(2 \times 299) / [(2 \times 966) + (2 \times 299)]$$

$$= 457 \times 10^6 \text{ in.-lb}$$

where ΣK_t is for two torsional members, one on each side of the column, and ΣK_c is for the upper and lower columns at the slab-beams joint of an intermediate floor (see Fig. 14-14). By use of Fig. 14-15, with the known values of K_{sb} and K_{ec}, the slab-beam joint distribution factors are calculated as

$$\textit{exterior joint} \qquad \text{DF} = \frac{286}{286 + 457} = 0.385$$

$$\textit{interior joint} \qquad \text{DF} = \frac{286}{286 + 286 + 457} = 0.278$$

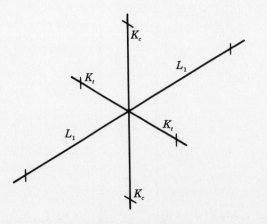

Fig. 14-14

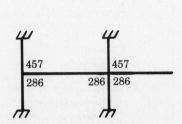

Fig. 14-15

Partial analysis of the equivalent frame

Determine maximum negative and positive moments for the slab-beams using the moment distribution method. Since $L/D = 40/(87.5 + 20) = 0.37 < 3/4$, design moments are assumed to occur at all critical sections with full factored live load on all spans.

Factored dead load per unit area: $w_D = 1.4(87.5 + 20) = 150.5$ lbs/ft^2

Factored live load per unit area: $w_L = 1.7(40) = 68$ lbs/ft^2

Factored load per unit area: $w_D + w_L = 218.5$ lbs/ft^2

FEM due to $w_D + w_L$: $0.0843wL_2L_1^2 = 0.0843(218.5/1000)(14)(18)^2 = 83.6$ ft-kips

The moment distribution is shown in Table 14.1 and the accompanying Fig. 14-16; all moments are in ft-kips. Positive span moments are determined from

$$M(\text{midspan}) = M_s - \tfrac{1}{2}(|M_L| + |M_R|)$$

where M_s is the moment at midspan for a simple beam with uniform load. When the end moments are not equal, the maximum moment in the span does not occur at midspan; but its value is close to that at midspan, and it is common to use the positive moment at midspan in practical design cases. Thus, the positive moment in span 1-2 is

$$+M = (0.2185)(14)(18)^2/8 - \tfrac{1}{2}(86.4 + 86.4) = 37.5 \text{ ft-kips}$$

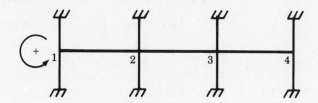

Fig. 14-16

Table 14.1

Joint	1	2		3		4
Member	1–2	2–1	2–3	3–2	3–4	4–3
DF	0.385	0.278	0.278	0.278	0.278	0.385
COF	0.509	0.509	0.509	0.509	0.509	0.509
FEM	+83.6	−83.6	+83.6	−83.6	+83.6	−83.6
COM*	0	−16.4	0	0	+16.4	0
COM*	+2.3	0	−2.3	+2.3	0	2.3
COM*	+0.3	−0.5	−0.3	+0.3	+0.5	−0.3
Σ	+86.2	−100.5	+81.0	−81.0	+100.5	−86.2
DM**	−33.2	+5.4	+5.4	−5.4	−5.4	+33.2
Neg. M	+53.0	−95.1	+86.4	−86.4	+95.1	−53.0
***	+53.6	−95.8	+87.0	−87.0	+95.8	−53.6
M at C.L.	49.8		37.5		49.8	
***	50.1		37.8		50.1	

  * Carry-over moment, COM, is the negative product of the distribution factor, the carry-over factor, and the unbalanced joint moment carried over to the opposite end of the span.

 ** Distributed moment, DM, is the negative product of the distribution factor and the unbalanced joint moment.

*** Moments obtained from a computer program for the analysis and design of slab systems by the Portland Cement Association.

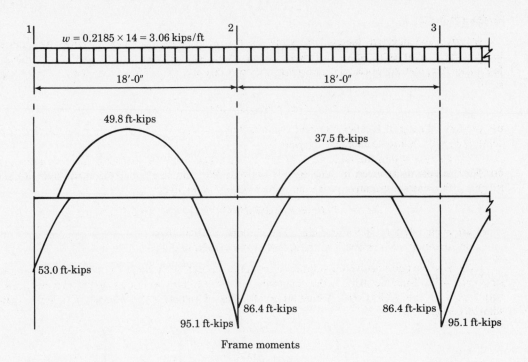

Frame moments

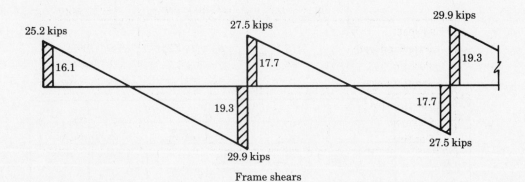

Frame shears

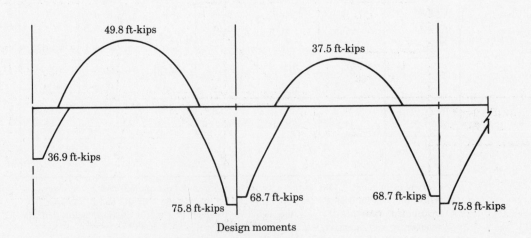

Design moments

Fig. 14-17

Design moments

Positive and negative factored moments for the slab system (all spans carrying full factored live load) in the transverse direction are shown in Fig. 14-17. The negative design moments are taken at the face of rectangular supports, but not greater than a distance of $0.175L_1$ from the center of supports; since

$$0.67 \text{ ft} < 0.175(18) = 3.15 \text{ ft}$$

use the face-of-support location. Shaded areas in Fig. 14-17 are areas under the moment diagrams from centerlines of columns to faces of columns.

Slab systems within the limitations of Section 13.6.1 of the 1983 ACI Code may have the resulting analytical moments reduced in such proportion that the numerical sum of the positive and absolute values of the average negative moments need not be greater than:

$$M_o = wL_2L_n^2/8 = (0.2185)(14)(16.67)^2/8 = 106.3 \text{ ft-kips}$$

End span:　$49.8 + (36.9 + 75.8)/2 = 106.2 \text{ ft-kips} \approx 106.3 \text{ ft-kips}$
Interior span:　$37.5 + (68.7)/2 = 106.2 \text{ ft-kips} \approx 106.3 \text{ ft-kips}$

The negative and positive factored moments at the critical sections may be distributed to the column strip and two half-middle strips of the slab-beam according to the proportions specified in Sections 13.6.4 and 13.6.6 of the ACI Code. Distribution of factored moments (in ft-kips) at critical sections is summarized in Table 14.2.

Table 14.2

	Factored moment	To column strip	To two half-middle strips
End span:			
Exterior negative	36.9	36.9 (100%)	0.0 (0%)
Positive	49.8	29.9 (60%)	19.9 (40%)
Interior negative	75.8	56.9 (75%)	18.9 (25%)
Interior span:			
Negative	68.7	51.5 (75%)	17.2 (25%)
Positive	37.5	22.5 (60%)	15.0 (40%)

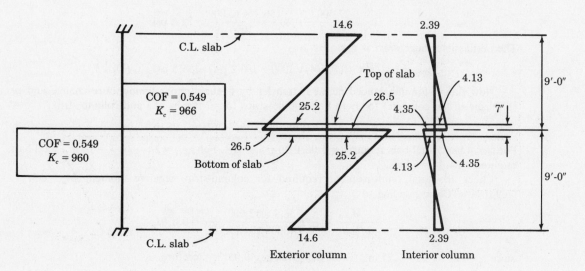

Fig. 14-18

The unbalanced moments from the slab-beams at the supports of the equivalent frame are distributed to the actual columns above and below the slab-beam in proportion to the relative stiffnesses of the actual columns. Referring to Table 14.1, the unbalanced moments are $+53.0$ ft-kips at joint 1 and $-95.1 + 86.4 = -8.7$ ft-kips at joint 2. The stiffness and carry-over factors of the actual columns and the distribution of the unbalanced moment (in ft-kips) to the exterior and interior columns are shown in Fig. 14-18. The design moments for the columns may be taken at the juncture of the face of the column and the slab. Summarizing: design moment in exterior column = 25.2 ft-kips; design moment in interior column = 4.13 ft-kips.

Transfer of gravity-load shear and moment at exterior columns

(a) *Factored shear force transfer.*

$$V_u = w_u L_1 L_2 / 2 = (0.2185)(14)(18)/2 = 27.5 \text{ kips}$$

(b) *Unbalanced moment transfer.* From Table 14.2, the unbalanced moment at exterior columns is $M_u = 36.9$ ft-kips. For an edge column with bending perpendicular to the edge, the calculation of the actual combined shear *stress* v_u at the inside face of the critical transfer section makes use of Appendix A-10:

$$a = c_1 + 2d/2 = 16 + 5.75/2 = 18.88 \text{ in.}$$
$$b = c_2 + d = 16 + 5.75 = 21.75 \text{ in.}$$
$$c = a^2/(2a + b) = (18.88)^2/[2(18.88) + 21.75] = 5.99 \text{ in.}$$
$$A_c = (2a + b)d = [2(18.88) + 21.75](5.75) = 342.2 \text{ in.}^2$$
$$J/c = [2ad(a + 2b) + d^3(2a + b)]/6$$
$$= \{2(18.88)(5.75)[18.88 + (2)(21.75)] + (21.75)^3[(2)(18.88) + 21.75]\}/6.0 = 2358 \text{ in.}^3$$

$$\gamma_f = \cfrac{1}{1 + \cfrac{2}{3}\sqrt{\cfrac{c_1 + d}{c_2 + d}}} = \cfrac{1}{1 + 0.667\sqrt{\cfrac{16.0 + 5.75}{16.0 + 5.75}}} = 0.6$$

$$\gamma_v = 1.0 - \gamma_f = 0.4$$

and finally

$$v_u = \frac{V_u}{A_c} + \gamma_v \frac{M_u}{J/c} = \frac{27,500}{342.2} + 0.40 \frac{36.9 \times 12,000}{2358} = 155.5 \text{ psi}$$

Similarly, the combined shear *stress* at the outside face of the critical transfer section is

$$v_u = \frac{27,500}{342.2} - 0.40 \frac{36.9 \times 12,000}{1096} = 81.2 \text{ psi}$$

The permissible shear *stress* is given by

$$\phi v_c = \phi 4\sqrt{f_c'} = (0.85)(4)\sqrt{3000} = 186.2 \text{ psi} > 155.5 \text{ psi} \qquad \text{(O.K.)}$$

Now design for unbalanced moment transfer by flexure (under temperature change and/or shrinkage of the concrete) for both half-middle strips ($b = 7$ ft = 84 in.) and column strip:

$$A_{s(\min)} = 0.0018bh = 0.0018(84)(7) = 1.06 \text{ in.}^2$$

For No. 4 bars, total bars required = $1.06/0.20 = 5.3$ bars. For $s_{\max} = 2h = 2 \times 7 = 14$ in., total bars required = $84/14 = 6$ bars.

Check the total reinforcement required for column-strip negative moment $M_u = \phi M_n = 36.9$ ft-kips. Corresponding to

$$\frac{M_u}{\phi f_c' b d^2} = \frac{36.9 \times 12,000}{(0.9)(3000)(84)(5.75)^2} = 0.0591$$

since $d = 7.0 - 1.25 = 5.75$ in., Table 5.1 gives $\omega = 0.0613$; therefore,

$$A_s = \omega f_c' b d / f_y = (0.0613)(3000)(84)(5.75)/60,000 = 1.48 \text{ in.}^2$$

For No. 4 bars, total bars required $= 1.48/0.20 = 7.4$ bars. Use six No. 4 bars at 14-inch spacing in the middle strip and two No. 4 bars in the portion of the column strip outside of the unbalanced moment transfer section,

$$c + 2(1.5h) = 16 + 2(1.5 \times 7) = 37 \text{ in.}$$

Additional reinforcement required over the column within an effective slab width of 37 in., to resist the fraction of the unbalanced moment transferred by flexure, is computed from Eq. (13-1) of the 1983 ACI Code. For a square column,

$$\gamma_f M_u = (0.60)(36.9) = 22.1 \text{ ft-kips}$$

must be transferred within the effective slab width of 37 in. Try two additional bars over the column, making a total of four No. 4 bars. Checking moment strength for four No. 4 bars within a 37-in. width,

$$A_s = 4(0.20) = 0.80 \text{ in.}^2 \quad \text{and} \quad \omega = \frac{A_s f_y}{f'_c b d} = \frac{(0.80)(60,000)}{(3000)(37)(5.75)} = 0.0752$$

Enter Table 5.1 at $\omega = 0.0752$ to obtain $M_n/f'_c b d^2 = 0.0719$, whence

$$M_n = (0.0719)(3000)(37)(5.75)^2/12,000 = 22.0 \text{ ft-kips}$$

and so $\phi M_n = (0.9)(22.0) = 19.8 \text{ ft-kips} < 22.1 \text{ ft-kips}$ (N.G.). Try, then, three additional bars, for a total of five No. 4 bars. For five No. 4 bars within a 37-in. width:

$$A_s = 5(0.20) = 1.00 \text{ in.}^2 \quad \omega = \frac{(1.00)(60,000)}{(3000)(37)(5.75)} = 0.0928$$

From Table 5.1, $M_n/f'_c b d^2 = 0.0877$, and thus

$$\phi M_n = (0.9)(0.0877)(3000)(37)(5.75)^2/12,000 = 24.1 \text{ ft-kips}$$

which, exceeding 22.1 ft-kips, is satisfactory.

14.2. Use the *column analogy method* to obtain the stiffness factors (k) for the beam shown in Fig. 14-19. Consider the moment of inertia to be infinite over the distances aL and bL, from the column centerlines to the faces of the columns.

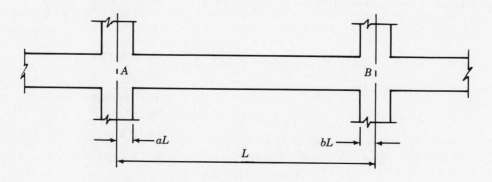

Fig. 14-19

In the solution, k_1 and k_3 will be the stiffness coefficients in $K_{AB} = k_1 EI/L$ and $K_{BA} = k_3 EI/L$. The stiffness factor k_2 in $K_2 = k_2 EI/L$ relates to the member stiffness matrix. For moment distribution, the carry-over factors are defined as $C_{AB} = k_2/k_1$ and $C_{BA} = k_2/k_3$. Use $L = 20$ ft and $aL = bL = 1.0$ ft (symmetric case).

In the column analogy method, the beam is replaced by an *analogous column* (or *equivalent column*), so defined that, at any location along the beam, the column is of width $1/EI$, where EI is the beam stiffness at the given location. In general, the cross-sectional moment of inertia I will be variable, so that the analogous column will have variable width. Figure 14-20 shows a cross section of the analogous column for the present problem; because of the assumption made in the problem statement, $c = 1 - a - b$.

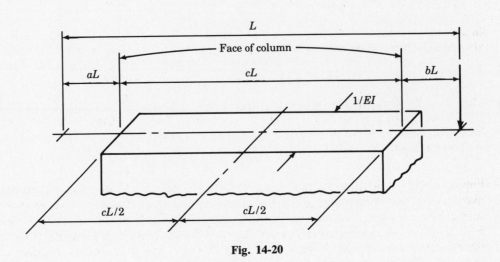

Fig. 14-20

Under the analogy, stresses in the equivalent column represent moments in the actual beam. Specifically, let end A of the beam be held fixed and let end B undergo a unit rotation, $\phi = 1$ rad. Then, at distance x from the centroidal axis of the analogous column,

beam   column

$$K_{xA} \leftrightarrow \frac{P}{A^*} + \frac{Pex}{Q}$$

where P is a unit rotational load; A^* is the cross-sectional area of the analogous column $[A^* = (1/EI)(cL)$ if EI is constant]; e is the eccentricity of the elastic load $(e = L/2)$; and Q is the cross-sectional moment of inertia of the analogous column $[Q = (1/EI)(cL)^3/12$ if EI is constant]. This gives for the flexural stiffness at joint B ($x = L/2 = 10$ ft):

$$K_{BA} = \frac{1}{(1/EI)(18)} + \frac{(1)(10)(10)}{(1/EI)(18)^3/12} = 0.26133EI$$

from which the stiffness factor at joint B is

$$k_3 = \frac{L}{EI} K_{BA} = (20)(0.26133) = 5.227$$

By symmetry, $K_{AB} = K_{BA}$, and so $k_1 = k_3 = 5.227$.

The carry-over stiffness is obtained by using $x = -L/2 = -10$ ft in the analogy:

$$K_2 = \frac{1}{(1/EI)(18)} - \frac{(1)(10)(10)}{(1/EI)(18)^3/12} = -0.1502EI$$

$$k_2 = 20(+0.1502) = 3.004$$

The carry-over stiffness factor k_2 has been taken positive in accordance with the modern practice of distributing *directional* moments. Because of *Maxwell's law of reciprocity*, k_2 is the stiffness factor both for carry-over from end B to end A and for carry-over from end A to end B.

Use of the computer program SLBMSTIF leads to the same k-values as found above; see Problem 14.5(p).

14.3. For the beam of Fig. 14-19, determine the fixed-end moments at A and B by the column-analogy method. Assume that $a = b = 0$; $L = 22$ ft; EI is constant over the span, which carries a uniform load of $w = 1.0$ kips/ft.

The moment diagram is parabolic, with a maximum simple beam moment $wL^2/8$ at the center of the span. The elastic load N on the analogous column is equal to the area under the M/EI diagram (also parabolic):

$$N = \frac{2}{3}\left(\frac{wL^2/8}{EI}\right)L = \frac{2}{3}\frac{(1.0)(22)^3}{8EI} = \frac{887.33}{EI}$$

In this case, because of symmetry, there is no eccentricity of the elastic load, so the fixed-end moments at both ends of the beam are simply

$$\text{FEM}_{AB} = \text{FEM}_{BA} = \frac{N}{A^*} = \frac{887.33/EI}{(1/EI)(22)} = 40.33 \text{ ft-kips}$$

The moment is counterclockwise at A and clockwise at B.

The computer program FEMWFP provides solutions $\text{FEM}_{AB} = \text{FEM}_{BA} = 40.23$ ft-kips; see Problem 14.7(a). The computer program uses finite elements to calculate the fixed-end moments, this being the best approach when a and b are nonzero and/or unequal.

14.4. Figure 14-21 shows the loading conditions that must be investigated to determine the maximum negative moments at the supports and the maximum positive moments at the centers of the spans. Use the *two-cycle moment distribution method* to obtain a solution that *includes all seven loading conditions*, with only one distribution of moments. (The results of the two-cycle moment distribution method agree reasonably well with the maximum negative and positive moments obtained by distributing moments seven times, once for each loading condition, and finding the maximum moment at all points.) For simplicity assume that members have constant EI and that column stiffnesses are one-half of beam stiffnesses.

(a) Calculate distribution factors $\text{DF} = K/\Sigma K$ at the joints:

For members AB and DE, $\text{DF} = 1/[1 + (2)(\frac{1}{2})] = \frac{1}{2}$

For all other ends of members, $\text{DF} = 1/[1 + 1 + (2)(\frac{1}{2})] = \frac{1}{3}$

For uniformly distributed loads, the fixed-end moments are $\text{MF} = wL^2/12$. Hence, with *dead load* of 0.72 kips/ft:

For AB, BA, BC and CB, $\text{MF} = 072(20)^2/12 = 24.0$ ft-kips

For CD and DC, $\text{MF} = 0.72(23.3)^2/12 = 32.6$ ft-kips

For DF and ED, $\text{MF} = 0.72(25)^2/12 = 37.6$ ft-kips

With *live load* of 1.08 kips/ft:

For AB, BA, BC and CB, $\text{MF} = 1.08(20)^2/12 = 36.0$ ft-kips

For CD and DC, $\text{MF} = 1.08(23.3)^2/12 = 48.9$ ft-kips

For DE and ED, $\text{MF} = 1.08(25)^2/12 = 56.3$ kips

(b) Calculate the products of the distribution factors and carry-over factors, $\text{DF} \times \text{COF}$, at the ends of each member. In this problem, each $\text{COF} = \frac{1}{2}$, by assumption of uniform beam stiffness. In Table 14.3, record DF, then COF, and then $\text{DF} \times \text{COF}$. Next, list the dead-load fixed-end moments and the total-load fixed-end moments (total load = dead load + live load). The moments are considered to be positive if clockwise and negative if counterclockwise.

(c) Start with distribution item D1, going diagonally down across each joint from right to left. Obtain the algebraic sum of DLFEM and TLFEM. Multiply the sum by ($\text{DF} \times \text{COF}$), and record on the left end of each beam. Do not include joint A, since there is no beam to the left of it.

(d) Follow with distribution item D2, going diagonally down from left to right. Obtain the algebraic sum of DLFEM and TLFEM. Multiply the sum by ($\text{DF} \times \text{COF}$) and record on the right side of each member. Do not include joint E, since there is no member to the right of it.

(e) Calculate and record on the left side of all beams the term $X = \text{TLFEM} + \text{D1}$.

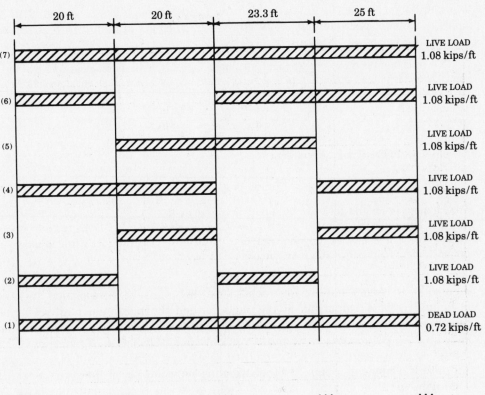

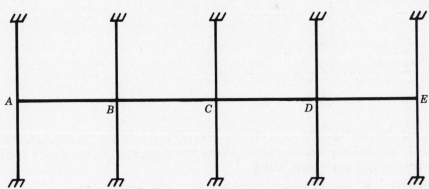

Fig. 14-21

(*f*) Calculate and record on the right side of all beams the term $Y = \text{TLFEM} + \text{D2}$.

(*g*) Record zero (0.0) for X on the right side of all beams and for Y on the left side of all beams.

(*h*) Calculate $Z = -\Sigma\,[(X + Y) \times \text{DF}]$ where the summation is around the joint, at both sides of all joints and record.

(*i*) Add algebraically and obtain the sums $(X + Y + Z)$ at both sides of all joints. These are the *final directional moments* at the ends of all beams.

(*j*) Transform directional moments to *bending moments*. Counterclockwise moments are negative bending moments on the left sides of beams, and clockwise moments are negative bending moments on the right sides of beams. In general, the end moments on all beams will be *negative bending moments*.

Table 14.3

	A		B		C		D		E
DF	1/2	1/3	1/3	1/3	1/3	1/3	1/3	1/2	
COF	1/2	1/2	1/2	1/2	1/2	1/2	1/2	1/2	
DF × COF	1/4	1/6	1.6	1/6	1/6	1/6	1/6	1/6	
DLFEM	—	24	−24	24	−32.6	32.6	−37.6	—	
TLFEM	−60	60	−60	60	−81.5	81.5	−94	94	
−(TLFEM + DLFEM) × (DF × COF) = D1	−(60 − 24)(1/6)		−(60 − 32.6)(1/6)		−(81.5 − 37.6)(1/6)		−(94 − 0)(1/4)(6)		
	−6		−4.56		−7.31		−23.5		
−(TLFEM + DLFEM) × (DF × COF) = D2	−(60 − 0)(1/4)		−(−60 + 24)(1/6)		−(−81.5 + 24)(1/6)		−(−94 + 32.6)(1/6)		
		15		6		9.58		10.23	
X = TLFEM + D1	−66	0	−64.56	0	−88.81	0	−117.5	0	
Y = TLFEM + D2	0	75	0	66	0	91.08	0	104.23	
Z = −Σ [(X + Y) × DF]	33	−3.48	−3.48	7.6	7.6	8.81	8.81	−52.1	
SUM = (X + Y + Z) = DIRECTION M	−33	71.52	−68.04	73.6	−81.21	99.89	−108.69	52.13	
BENDING MOMENT	−33	−71.52	−68.04	−73.6	−81.21	−99.89	−108.69	−52.13	
CLMFE	30		30		40.8		47		
P = −D1(1 + DF)/2	4.5		3.03		4.86		15.6		
Q = D2(1 + DF)/2	9.98		3.99		6.37		7.67		
SUM = CLMFE + P + Q = Final Moment at Center of Beam	44.48		37.02		52.03		70.27		

(k) Calculate the moments at the centers of all beams due to total load (dead load + live load) under fixed-end conditions. For uniformly distributed loads, the centerline moment is CLMFE = $wL^2/24$. Record at the centers of all beams.

(l) Calculate $(-D1)(1 + DF)/2$ for all beams and record at the centerlines of the beams as item P. Use DF for the left sides of beams.

(m) Calculate $(D2)(1 + DF)/2$ for all beams and record at the centerlines of the beams as item Q. Use DF for the right sides of beams.

(n) Obtain the sum CLMFE + P + Q and record this as the *final positive moments* at the centers of all beams.

(o) Record the solution for negative and positive moments as in Fig. 14-22.

It should be noted that the positive moment at the center of a beam (CLMFE) is obtained by calculating the *simply supported beam moment* for the particular loading condition ($wL^2/8$, for uniformly distributed loads), and subtracting the average of the absolute moments due to fixed-end conditions (cf. Problem 14.1). The factor k in $kEI/L = K$ is 4.0 for beams of constant EI, and is larger than 4.0 for variable moment inertia. Likewise, the carry-over factor is greater than 1/2 for beams having variable moment of inertia.

For comparison with the solution yielded by the two-cycle moment distribution method, Fig. 14-23 shows the results of using (a) seven separate loading conditions and seven separate moment distributions, (b) the 1963 ACI Code moment coefficients (see Table 2.1).

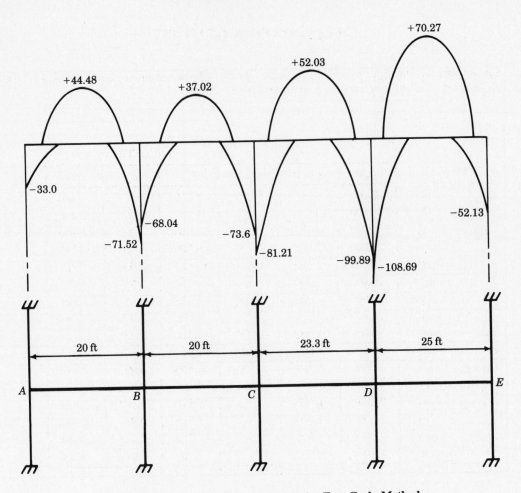

Fig. 14-22.   Final Moments, in ft-kips, by Two-Cycle Method

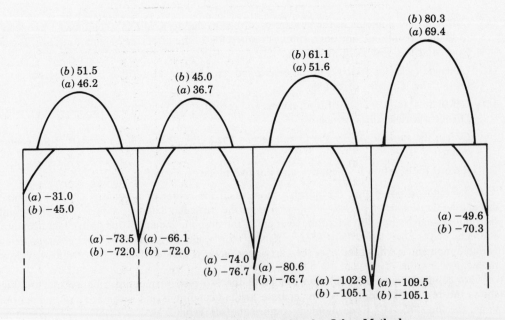

Fig. 14-23.   Final Moments, in ft-kips, by Other Methods

Supplementary Problems

14.5. Subproblems (a) through (p) in Table 14.4 have to do with the continuous beam of Fig. 14-19. Confirm the solutions yielded by computer program CLBMSTIF.

Table 14.4

	GIVEN			SOLUTIONS		
	L, ft	aL, ft	bL, ft	k_1	k_2	k_3
(a)	20.000	0.000	0.000	4.000	2.000	4.000
(b)	30.000	6.000	6.000	15.556	12.222	15.556
(c)	10.000	2.000	0.000	9.688	4.375	5.000
(d)	22.000	2.000	2.000	6.700	4.255	6.700
(e)	24.000	1.500	1.500	5.621	3.335	5.621
(f)	18.000	1.500	1.500	6.384	3.984	6.384
(g)	20.000	2.000	1.500	6.825	4.127	6.291
(h)	26.000	1.500	2.000	5.608	3.472	5.964
(i)	16.000	1.000	1.000	5.621	3.335	5.621
(j)	15.000	1.000	1.500	6.044	3.978	6.735
(k)	17.000	1.250	1.500	6.138	3.899	6.437
(l)	19.000	1.000	1.500	5.494	3.426	5.976
(m)	24.000	1.500	1.250	5.542	3.192	5.362
(n)	36.000	2.000	1.500	5.299	2.969	5.073
(o)	23.000	1.500	1.500	5.713	3.413	5.713
(p)	20.000	1.000	1.000	5.226	3.004	5.226

Table 14.5

	GIVEN			SOLUTIONS		
	L, m	aL, m	bL, m	k_1	k_2	k_3
(a)	6.100	0.000	0.000	4.000	2.000	4.000
(b)	6.700	0.600	0.300	6.203	3.460	5.375
(c)	6.700	0.300	0.600	5.375	3.460	6.203
(d)	7.600	0.600	0.300	5.865	3.237	5.174
(e)	7.600	0.300	0.600	5.174	3.237	5.865
(f)	7.300	0.400	0.400	5.373	3.127	5.373
(g)	7.300	0.300	0.350	5.012	2.871	5.121
(h)	9.000	0.460	0.460	5.260	3.032	5.260
(i)	8.000	0.500	0.500	5.621	3.335	5.621
(j)	8.700	0.400	0.300	5.035	2.770	4.858
(k)	9.800	0.500	0.500	5.237	3.030	5.257
(l)	10.000	0.400	0.400	4.940	2.766	4.940
(m)	7.300	0.300	0.400	5.055	2.953	5.277
(n)	8.800	0.400	0.400	5.093	2.893	5.093
(o)	10.400	0.300	0.300	4.647	1.524	4.647
(p)	10.000	0.400	0.400	4.940	2.766	4.940

14.6. (SI analog of Problem 14.5) Subproblems (a) through (p) in Table 14.5 have to do with the continuous beam of Fig. 14-19. Confirm the solutions yielded by computer program CLBMSTIF.

14.7. Subproblems (a) through (t) in Table 14.6 have to do with the continuous beam of Fig. 14-19. Confirm the solutions yielded by computer program FEMWFP.

Table 14.6

| | GIVEN | | | | SOLUTIONS | |
	L, ft	aL, ft	bL, ft	w, kips/ft	FEM(A), ft-kips	FEM(B), ft-kips
(a)	22.00	0.00	0.00	1.00	40.33	−40.33
(b)	22.00	2.00	1.00	1.00	51.08	−40.08
(c)	22.00	1.00	2.00	1.00	40.08	−51.08
(d)	25.00	2.00	1.00	1.00	64.33	−51.83
(e)	25.00	1.00	2.00	1.00	51.83	−64.33
(f)	24.00	1.00	1.50	0.80	39.81	−44.61
(g)	24.00	1.50	1.00	0.80	44.61	−39.81
(h)	26.00	2.00	1.00	1.20	82.89	−67.29
(i)	26.00	1.00	2.00	1.20	67.29	−82.89
(j)	30.00	2.00	1.00	1.40	125.64	−104.64
(k)	30.00	1.00	2.00	1.40	104.64	−125.64
(l)	28.00	2.00	1.00	1.20	94.89	−78.09
(m)	28.00	1.00	2.00	1.20	78.09	−94.89
(n)	32.00	2.00	1.00	1.40	141.51	−119.11
(o)	32.00	1.00	2.00	1.40	119.11	−141.51
(p)	28.00	2.00	1.00	1.30	102.80	−84.60
(q)	28.00	1.00	2.00	1.30	84.60	−102.80
(r)	26.00	2.50	1.50	1.00	70.95	−57.95
(s)	26.00	1.50	2.50	1.00	57.95	−70.95
(t)	24.00	2.00	1.50	1.00	57.52	−51.52

14.8. (SI analog of Problem 14.7) Subproblems (a) through (p) in Table 14.7 have to do with the continuous beam of Fig. 14-19. Confirm the solutions yielded by computer program FEMWFP.

14.9. Refer to Fig. 14-21. Find the bending moments at the supports and at the centerlines of the spans. Distribution factors and carry-over factors are the same as for Problem 14.5. Span lengths and uniformly distributed loads are given in Table 14.8.
Ans. See Table 14.9; moments are symmetrical about structure centerline.

14.10. (Soft conversion of Problem 14.9) Refer to Fig. 14-21. Find the bending moments at the supports and at the centerlines of the spans. Distribution factors and carry-over factors are the same as for Problem 14.5. Span lengths and uniformly distributed loads are given in Table 14.10.
Ans. See Table 14.11; moments are symmetrical about structure centerline.

Table 14.7

	GIVEN				SOLUTIONS	
	L, m	aL, m	bL, m	w, kN/m	FEM(A), kN·m	FEM(B), kN·m
(a)	6.71	0.00	0.00	4.59	54.68	−54.68
(b)	6.71	0.61	0.30	4.59	69.25	−54.34
(c)	6.71	0.30	0.61	4.59	54.34	−69.35
(d)	7.62	0.61	0.30	4.59	87.22	−70.27
(e)	7.62	0.30	0.61	4.59	70.27	−87.22
(f)	7.32	0.30	0.46	3.68	53.98	−60.49
(g)	7.32	0.46	0.30	3.68	60.49	−53.98
(h)	7.92	0.61	0.30	5.51	112.39	−91.24
(i)	7.92	0.30	0.61	5.51	91.24	−112.39
(j)	9.14	0.61	0.30	6.43	170.34	−141.87
(k)	9.14	0.30	0.61	6.43	141.87	−170.34
(l)	8.53	0.61	0.30	5.51	128.66	−105.88
(m)	8.53	0.30	0.61	5.51	105.88	−128.66
(n)	9.75	0.61	0.30	6.43	191.85	−161.48
(o)	9.75	0.30	0.61	6.43	161.48	−191.85
(p)	8.53	0.61	0.30	5.97	139.38	−114.70

Table 14.8

	Span Length, ft		w_D, kips/ft	w_L, kips/ft
	AB and DE	BC and CD		
(a)	17.0	20.0	0.90	1.5
(b)	18.0	22.0	1.0	1.5
(c)	21.0	25.0	1.2	1.8
(d)	18.0	22.0	1.3	1.6
(e)	16.0	20.0	1.2	1.2
(f)	20.0	15.0	1.0	1.5
(g)	22.0	22.0	1.4	1.8

Table 14.9

	Negative Moments, ft-kips				Positive Moments, ft-kips	
	AB and ED	BA and DE	BC and DC	CB and CD	AB and DE	BC and CD
(a)	−37.3	−83.4	−85.9	−89.3	47.5	51.7
(b)	−36.0	93.2	−102.1	−113.1	48.4	65.3
(c)	−59.1	−149.2	−160.5	−174.9	79.5	101.0
(d)	−41.3	−107.8	−117.8	−130.6	55.4	74.7
(e)	−26.5	−71.5	−79.1	−89.1	35.5	50.5
(f)	−47.0	86.5	69.1	−49.1	63.6	28.1
(g)	−70.6	−154.6	−147.9	−141.2	95.1	80.6

Table 14.10

	Span Length, m		w_D, kN/m	w_L, kN/m
	AB and DE	BC and CD		
(a)	5.18	6.10	13.13	21.89
(b)	5.49	6.71	14.59	21.89
(c)	6.40	7.62	17.51	26.27
(d)	5.49	6.71	18.97	23.35
(e)	4.88	6.10	17.51	17.51
(f)	6.10	4.57	14.59	21.89
(g)	6.71	6.71	20.43	26.27

Table 14.11

	Negative Moments, kN · m				Positive Moments, kN · m	
	AB and ED	BA and DE	BC and DC	CB and CD	AB and DE	BC and CD
(a)	−50.6	−113.1	−116.5	−121.1	64.4	70.1
(b)	−48.8	−126.4	−138.4	−153.3	65.6	88.5
(c)	−80.1	−202.3	−217.6	−237.1	107.8	136.9
(d)	−56.0	−146.2	−159.7	−177.1	75.1	101.5
(e)	−35.9	−96.9	−107.2	−120.8	48.1	68.5
(f)	−63.7	−117.3	−93.7	−66.6	86.2	38.1
(g)	−95.7	−209.6	−200.5	−191.4	128.9	109.3

Chapter 15

Retaining Walls

NOTATION

c = unit cohesion for clay soils, $\mathrm{lb/ft^2}$ $(\mathrm{kN/m^2})$

e = eccentricity of resultant of vertical loads, ft (m)

f = friction coefficient

h = total thickness of a member, in. (mm); also, height of wall stem, ft (m)

h_s = surcharge height in equivalent soil, ft (m)

h_w = height of wall from base of footing, ft (m)

H = horizontal force, kips (kN)

H_a = active pressure horizontal force, kips (kN)

H_p = passive pressure horizontal force, kips (kN)

H_s = total pressure force due to surcharge, kips (kN)

k_a = coefficient of active soil pressure

k_p = coefficient of passive soil pressure

L = length of base of a footing, ft (m)

M = bending moment, ft-kips $(\mathrm{kN \cdot m})$

M_o = overturning moment, ft-kips $(\mathrm{kN \cdot m})$

M_r = resisting moment, ft-kips $(\mathrm{kN \cdot m})$

N = normal force, kips (kN)

p = unit pressure, $\mathrm{lb/ft^2}$ $(\mathrm{kN/m^2})$

p_a = active soil pressure, $\mathrm{lb/ft^2}$ $(\mathrm{kN/m^2})$

p_p = passive soil pressure, $\mathrm{lb/ft^2}$ $(\mathrm{kN/m^2})$

p_s = surcharge pressure, $\mathrm{lb/ft^2}$ $(\mathrm{kN/m^2})$

P_a = sloping active pressure force, kips (kN) (May also be horizontal)

P_p = sloping passive pressure force, kips (kN) (May also be horizontal)

R = resultant vertical force, kips (kN)

SF = safety factor

T = surcharge load on a footing, lb/ft (kN/m)

V = shear force, kips (kN)

w = density of soil, $\mathrm{lb/ft^3}$ $(\mathrm{kg/m^3})$

w_s = pressure of surcharge, $\mathrm{lb/ft^2}$ $(\mathrm{kg/m^2})$

W = vertical load due to soil or concrete, kips (kN)

α = natural angle of repose of soil, degrees (radians)

ϕ = angle of internal friction of soil, degrees (radians); also, capacity reduction factor (resistance factor)

USES OF RETAINING WALLS

Very often in the construction of buildings or bridges it is necessary to retain earth in a relatively vertical position. Whenever embankments are involved in the construction, retaining walls are usually necessary. In the construction of buildings having basements, retaining walls are mandatory. Fig. 15-1 indicates some of the various types of retaining walls which are commonly utilized.

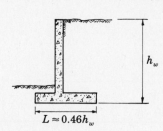

(a) T-shaped cantilever wall, horizontal fill

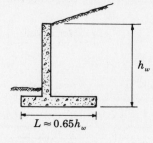

(b) T-shaped cantilever wall, sloping fill

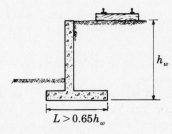

(c) T-shaped cantilever wall, with surcharge

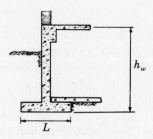

(d) Basement wall

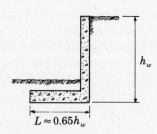

(e) L-shaped retaining wall, horizontal fill

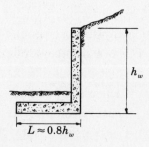

(f) L-shaped retaining wall, sloping fill

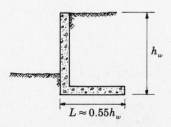

(g) Reversed L-shaped retaining wall, horizontal fill

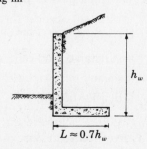

(h) Reversed L-shaped retaining wall, sloping fill

Fig. 15-1

FORCES ON RETAINING WALLS

The usual gravity loads due to the weights of the materials do not present great problems with respect to retaining walls. Actually, the *horizontal pressures due to the retained soil* will present the greatest problems.

If we construct a box having a sliding wall, as shown in Fig. 15-2, fill the box with sand, and then suddenly lift the wall, the sand will *slide along a shear plane* and form a slope as shown in Fig. 15-3.

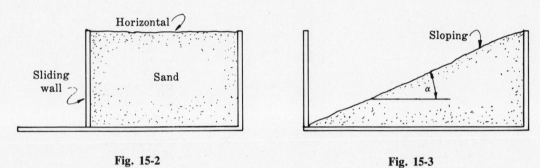

Fig. 15-2   Fig. 15-3

The angle α formed by the sand is called the *natural angle of repose* of the material. Different materials exhibit widely varying slopes of repose. Further, the *moisture content* of the material is an important factor with respect to the slope of repose.

If the moisture content of the sand in Fig. 15-2 would be at an *optimum value*, the sand might stand vertically for a short time.

Granular materials such as sand or gravel behave differently than *cohesive materials* such as clay, when retained in some manner. Materials which contain combinations of the two types of soil will act similarly to the predominant material. Since the percentages of cohesive and non-cohesive materials vary extensively in nature, it is necessary to conduct tests in order to determine the properties of a natural deposit of soil.

Although a soil will assume its natural angle of repose when not confined, it is improper to use the angle of repose in design for confined material. Under confinement, the soil has a tendency to slide in a manner similar to that discussed with respect to Figs. 15-2 and 15-3, but somewhat modified. The sliding surface will be more like those shown in Figs. 15-4 and 15-5.

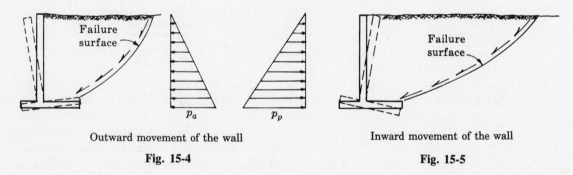

Outward movement of the wall   Inward movement of the wall

Fig. 15-4   Fig. 15-5

If the wall is absolutely rigid, *earth pressure at rest* will develop. If the wall should deflect or move a very small amount away from the earth, *active earth pressure* will develop, as shown in Fig. 15-4. If the wall moves toward the earth, *passive earth pressure* will develop, as shown in Fig. 15-5. The magnitude of earth pressure at rest lies at some value between active and passive earth pressure.

Under normal conditions earth pressure at rest is so intense that the wall deflects, relieving itself of this type of pressure, and active pressure results. For this reason, most retaining walls are designed for active pressure due to the retained soil.

Although the actual pressure intensity diagram is very complex, it is usual to assume a *linear pressure distribution* due to active or passive pressure. The intensity is assumed to increase with depth as a function of the weight of the soil, so that the horizontal pressure of the earth against the wall is often called *equivalent fluid pressure*.

Fig. 15-6 shows the usual types of pressure diagrams assumed in designing retaining walls.

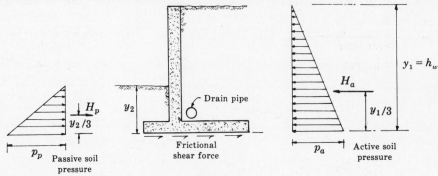

Fig. 15-6

For granular, non-cohesive materials which are dry, the assumed linear pressure diagram is nearly correct. Cohesive soils do not behave in this manner, and it is not good practice to *backfill* a wall with cohesive soils. Granular materials are used for backfill whenever possible to provide a linear pressure distribution, and also to provide for *drainage* to release water pressure from behind the wall.

When the materials used behind a wall are such that a linear pressure distribution may be assumed, the pressure intensity at any depth y may be stated as

$$p_a = k_a w y \qquad (15.1)$$

for active pressure, and as

$$p_p = k_p w y \qquad (15.2)$$

for passive pressure, where k_a = coefficient of active pressure, k_p = coefficient of passive pressure, w = density of the soil in lb/ft^3 (kg/m^3), y = distance from the surface to the plane in question in ft (m).

Fundamentally, $k_a w$ and $k_p w$ are *equivalent liquid pressures* to be used in determining the total horizontal force acting against the wall at any given depth.

The coefficients k_a and k_p are usually only approximate, and should never be treated as though they are scientifically exact.

Experience has proven that walls can be safely designed using the approximate coefficients obtained from mathematical theories such as those of *Coulomb* and *Rankine*. *The angle of internal friction* ϕ is obtained experimentally or approximated from experience for use with those theories.

Figs. 15-7 and 15-8 illustrate results of test data for cohesionless soils and cohesive soils, respectively. It is important to note from the figures that the friction angle line passes through the origin for cohesionless soils, and for cohesive soils the line springs from a non-zero value. The latter value is called the *cohesion c* and represents the greatest portion of the shear strength of the cohesive soil.

The general case concerning the forces acting on the back face of a retaining wall is shown in Fig. 15-9.

It is significant to note that Coulomb's theory accounts for the frictional force between the wall and the soil while Rankine's theory neglects this force. Since it is rather difficult to accurately determine the frictional force on the back face of the wall, Coulomb's theory is not widely used in retaining wall design.

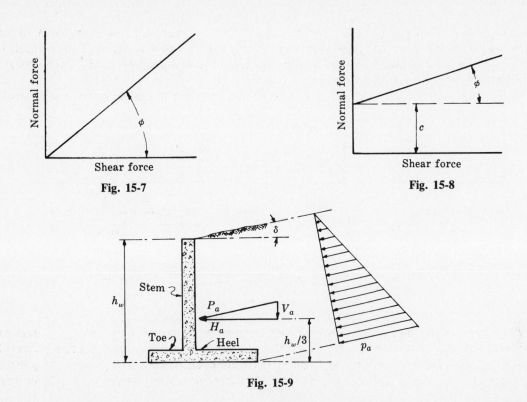

Fig. 15-7

Fig. 15-8

Fig. 15-9

For a cohesionless soil a triangular pressure diagram is assumed to develop. This assumption is applied to both passive and active pressure. The nature of the triangular pressure diagram shown in Fig. 15-6 is such that the total force which develops at a depth y from the surface will be

$$H_a = (k_a w) y^2 / 2 \qquad (15.3)$$

for active pressure and

$$H_p = (k_p w) y^2 / 2 \qquad (15.4)$$

for passive pressure.

Although it is an approximation, for sloping backfill as shown in Fig. 15-9 it is usually assumed that the force P_a is parallel to the surface of the fill so the components of the force are easy to determine. In such cases equations (15.3) and (15.4) are usually applied in practice.

A more detailed study of the Rankine theory indicates that the force is not only dependent on the slope of the surface of the fill, but is also dependent on the angle between the face of the wall and the fill surface. For active pressure the angle between the back face of the wall and the horizontal is used.

The more accurate equations are shown in Fig. 15-10 for the Rankine theory. When the wall is vertical and the soil surface is horizontal the detailed equations are identical to the equations previously stated for active pressure. Thus

$$C = k_a = (1 - \sin \phi) / (1 + \sin \phi) \qquad (15.5)$$

which is often replaced by the identity

$$k_a = \tan^2 (45° - \phi/2) \qquad (15.6)$$

Passive pressure occurs at the front of a retaining wall and on *shear keys*, as shown in Fig. 15-11. The passive pressure on the front of the wall is usually neglected, since it is possible that this soil will erode away or that shearing will occur, thus eliminating this passive pressure. It is always safe to neglect this passive pressure.

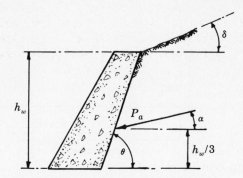
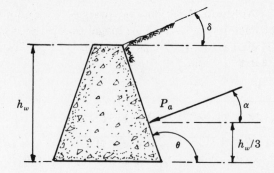

Case I. Wall leaning toward fill ($\theta < 90°$)   **Case II.** Wall leaning away from fill ($\theta > 90°$)

$\beta = \phi$ = angle of internal $\beta = \delta$ = slope of fill surface
friction for the soil

$$\tan \alpha = \frac{C \sin \delta - \cos \theta}{C \cos \delta}$$

$$C = \cos \delta \left\{ \frac{\cos \delta - \sqrt{\cos^2 \delta - \cos^2 \phi}}{\cos \delta + \sqrt{\cos^2 \delta - \cos^2 \phi}} \right\} [1 - (\cot \theta)(\tan \beta)]$$

$$P_a = \frac{C w h_w^2}{2} \left\{ \frac{\cos \delta}{\cos \alpha} [1 - (\cot \theta)(\tan \beta)] \right\}$$

For $\theta = 90°$ and $\delta = 0$,

$$\alpha = 0 \qquad \text{and} \qquad P_a = H_a = \frac{w h_w^2}{2} \left[\frac{1 - \sin \phi}{1 + \sin \phi} \right]$$

Fig. 15-10. Rankine Soil Pressure and Force for Cohesionless Soils

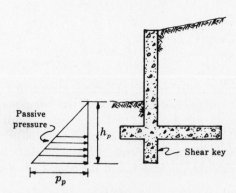

Fig. 15-11

The active horizontal soil pressure must be resisted by opposite forces so that the wall will not slide. Those forces are provided by *friction of the footing base on the soil* and by passive forces on the shear key when such is provided.

The unit passive pressure at any depth h is

$$p_p = k_p w h_p \tag{15.7}$$

where $$k_p = (1 + \sin \phi)/(1 - \sin \phi) \tag{15.8}$$

which may also be written as

$$k_p = \tan^2 (45° + \phi/2) \tag{15.9}$$

Although it is common practice to ignore the erodable soil above the toe of the wall, it is satisfactory to use the least magnitude of the pressure on the toe in front of the key as a *surcharge* in calculating the passive pressure on the shear key. This will be discussed in detail in connection with shear key design.

ANGLE OF INTERNAL FRICTION, ϕ

True values of the angle of internal friction can only be obtained by tests of the soil. In the absence of laboratory tests, the angle ϕ may be approximated as follows. For

Dry, loose sand with round grains, uniform gradation, $\phi = 28.5°$ (0.5 rad)
Dry, dense sand with round grains, uniform gradation, $\phi = 35.0°$ (0.61 rad)
Dry, loose sand, angular grains, well graded, $\phi = 34.0°$ (0.59 rad)
Dry, dense sand, angular grains, well graded, $\phi = 46.0°$ (0.80 rad)
Dry, loose silt, $\phi = 27°$ (0.47 rad) to $30°$ (0.52 rad)
Dry, dense silt, $\phi = 30°$ (0.52 rad) to $35°$ (0.61 rad)

When the soil is saturated, the angles of internal friction may be used as stated, considering the *buoyed unit weight* of the soil particles and, in addition, the *hydrostatic pressure of the water*.

FRICTION ON THE FOOTING BASE

The true friction factor which should be used in obtaining the force which resists sliding is that of the shear strength of the soil. In practice, the coefficient used is that of soil on concrete for coarse granular soils, and the *shear strength* or cohesion of the soil for cohesive soils. In the absence of tests, the following values may be used. For

Coarse-grained soil without silt, $f = 0.55$
Coarse-grained soil with silt, $f = 0.45$
Silt, $f = 0.35$

The total frictional force on the base of a footing is

$$F = fN \tag{15.10}$$

where f is the friction coefficient and N the normal force of the footing against the soil.

When the foundation rests on clay the cohesion should be used to determine the force resisting sliding. This cohesion force may be calculated using $\frac{1}{3}$ to $\frac{1}{2}$ of the *unconfined compressive strength* of the soil as the unit cohesion c.

When *pile footings* are used the outer rows should be *battered* (the *batter* is the slope of the pile) to resist the horizontal forces.

For any type of retaining wall, the safety factor against sliding should never be less than 1.5, and should preferably be 2.0 or more.

OVERTURNING

The most hazardous mode of failure of retaining walls is due to overturning because of unbalanced moments. Considering Fig. 15-12, it is apparent that the horizontal force H will tend to overturn the footing about point A. The *overturning moment* will be

$$M_o = Hy_1 \tag{15.11}$$

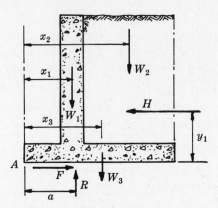

Fig. 15-12

The weight of the stem, the soil and the base will resist the overturning by providing a moment about A equal to

$$M_r = W_1x_1 + W_2x_2 + W_3x_3 \qquad (15.12)$$

which is the *righting moment*. See Fig. 15-12. The safety factor against overturning is

$$\text{SF} = M_r/M_o \qquad (15.13)$$

and should always exceed 1.5, with 2.0 being a desirable value.

The resultant vertical load R should be so located that the stability of the wall is insured. This is usually accomplished by dimensioning the wall so that the resultant force R falls within the *middle third of the base*. However, it is not absolutely necessary that the resultant fall within the middle third of the base, even though this provides the most economical solution. Under certain conditions such as *hydrostatic uplift*, it may not be feasible to have the resultant in the middle third. Also, when several loading conditions must be checked, it may be economical to have the resultant force outside of the middle third for one or more conditions.

When strength design procedures of the 1983 ACI Code are used for proportioning the concrete and steel in the wall and footing, the safety factors are usually of the order 1.8. It is proper that safety factors for sliding and overturning should also be of this order.

When the alternate design method is used for proportioning the concrete and steel, the safety factors are about 2.0 to 2.5. It is proper that the safety factors against sliding and overturning should also be in this range.

SURCHARGE PRESSURES

Loads are often imposed on the soil surface behind a retaining wall. Such forces may be due to loading dock slabs, railroad tracks, roadways, etc. It is common practice to consider such loads as a *surcharge*, and to transform the load into an *equivalent height of soil*.

Fig. 15-13 illustrates a concrete slab supported on the soil. The height of the surcharge is

$$h_s = w_s/w \qquad (15.14)$$

where h_s = equivalent height of soil, ft (m)
　　w_s = pressure of the surcharge, lb/ft^2 (kg/m^2)
　　w = weight of the soil, lb/ft^3 (kg/m^3).

It is seen that the horizontal pressure due to the surcharge is constant throughout the depth of the soil to the base of the footing. The intensity of p_s is obtained from the equation

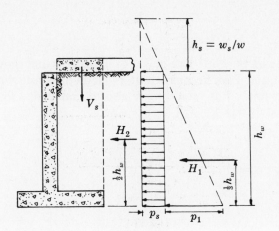

Fig. 15-13

$$p_s = wh_sk_a \tag{15.15}$$

The vertical pressure of the surcharge is equal to w_s.

In the case of a *wheel load*, the weight of the wheel is assumed to be distributed over a stated area. For example, train wheels are usually assumed to act over a width of 14 ft (4.27 m) and a length of 5 ft (1.52 m), considering two wheels (two rails) to act over this area. This is recommended by the Code of the American Railway Engineers' Association (AREA). If the wheels involved are due to highway-type loading, the Code of the American Association of State Highway and Transportation Officials (AASHTO) can be used. If other types of concentrated loads are applied (such as fork lifts), the equipment manufacturer should be consulted.

When a concentrated load or partial uniform load acts at some distance from the rear of the wall, only a portion of the total load affects the wall. A reasonable approach devised by Ketchum is shown in Fig. 15-14.

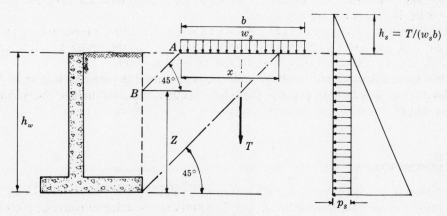

Fig. 15-14

The surcharge pressure p_s may be obtained from

$$p_s = wh_s'k_a \tag{15.16}$$

and h_s' may be determined from

$$h_s' = h_s(x/h) \tag{15.17}$$

using the equivalent height of surcharge in feet (meters) of soil,

$$h_s = w_s/w \tag{15.18}$$

where $\qquad\qquad\qquad\qquad\quad w_s = T/b \tag{15.19}$

Many engineers assume that the effective height of pressure is that shown as dimension Z in Figs. 15-14 and 15-15. Terzaghi and Peck recommend this method, considering that line AB forms an angle of 45° with the horizontal. This procedure is also recommended by the authors.

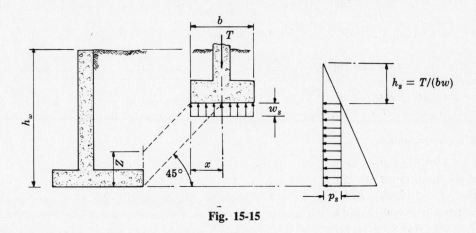

Fig. 15-15

BASE PRESSURE ON FOOTING

The base pressure which will develop on the footing of a retaining wall may be calculated in the same manner as that described for eccentrically loaded footings in Chapter 12. Generally, the pressure diagram will be trapezoidal, as shown in Fig. 15-16. The pressure diagram may be triangular, but will rarely be rectangular due to uniform pressure.

DESIGN PROCEDURE

Dimensioning

In the usual case, the design of retaining walls involves a trial process. Referring to Fig. 15-16, the dimensions may be discussed in general. (See, for example, Fig. 15-1.) Since the vast majority of retaining walls are of inverted-T shape, this discussion will be concentrated in that direction.

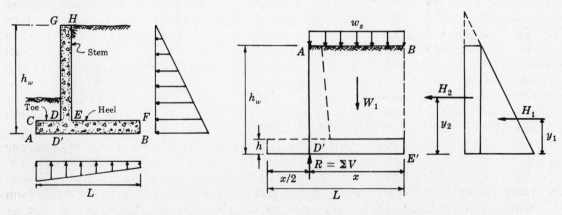

Fig. 15-16   **Fig. 15-17**

The height of the wall will be dictated by the conditions of the problem. The thickness GH of the *stem* is usually at least 8″ (200 mm) at the top, for practical reasons. A larger thickness may be used at the top if the base of the stem is unusually thick or many reinforcing bars are used. The stem thickness at its base, DE, is determined by the forces and moments to be resisted.

A minimum base length is obtained when the resultant vertical load strikes the ground immediately below the front face of the stem, or at D'.

If the *toe* is extended such that CD is exactly $\frac{1}{2}$ of DF, a triangular pressure diagram results if the vertical resultant acts at D'.

If CD is more than $\frac{1}{2}$ of DF, a trapezoidal pressure diagram results when the vertical resultant strikes at D'.

It is important to note that the conditions stated above usually provide a safety factor of about 2 against overturning.

The base thickness h will average from about 7 to 10 percent of the overall height of the wall. A minimum thickness of 10″ to 12″ (250 to 300 mm) is considered to conform to good practice.

Base Length Determination

Professor Ferguson gives an excellent practical method of estimating the base length. Considering Fig. 15-17 and temporarily disregarding the difference in weight of concrete and soil, using W_1 as the weight of a soil block $ABD'E'$, a reasonable estimate of the length X can be obtained. (When surcharge exists, it is considered as added soil.) If $W_1 = wh_w X$, the resultant will pass through D' if $\Sigma M_{D'} = 0$. Then $whX^2/2 = H_1 y_1 + H_2 y_2$, from which

$$X = \sqrt{2(H_1 y_1 + H_2 y_2)/wh_w} \qquad (15.20)$$

The footing base length L may then be *closely estimated* by

$$L = 1.5\sqrt{2(H_1 y_1 + H_2 y_2)/wh_w} \qquad (15.21)$$

Since the concrete density is about 1.5 times that of the earth, a slight additional safety factor is introduced against overturning by disregarding the differences in the weight of soil and concrete.

For the simple special case in which there is no surcharge, that is, in which $H_2 = 0$, $H_1 = k_a wh_w^2/2$ and $y_1 = h_w/3$, the equation simplifies to

$$L = 1.5\sqrt{k_a h_w^2/3} \qquad (15.22)$$

Stem Design

After the initial computations have been completed, an *estimate of the length of the base is known*. The location of the point of intersection of the front face of the wall and the base have also been established.

In order to design the stem, an *estimate of the depth of the base is made*. Considering the base thickness h to be about 7 to 10 percent of the total height, the preliminary wall height h_w can be established. The stem is considered to be a vertical cantilever, springing from the base as shown in Fig. 15-18. The bending moment and shear are obtained per foot of width of wall, and the usual equations for obtaining d, A_s and development length are applied. Working stress design or strength design may be employed for this purpose.

The basic soil pressure, hydrostatic pressure and surcharge pressure due to *permanent* load are considered as *dead load*. Intermittently applied surcharge loads (such as wheel loads or storage loads) are considered to be live load.

In the case of walls of the type shown in Fig. 15-18, the stem reinforcement is required on the *earth side* of the wall. Adequate anchorage of the reinforcement into the base is absolutely necessary.

During construction, a *construction joint* is used at the base of the stem so that the footing and the wall may be constructed separately and not as a unit. The construction joint often takes the form of a *keyway* as shown in Fig. 15-19, but may simply be a roughened surface. The latter is used rather frequently in modern practice because of simplicity. Further, some tests indicate the roughened joint

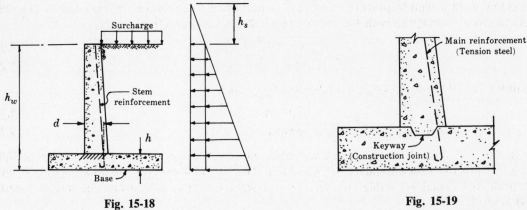

Fig. 15-18   **Fig. 15-19**

to be stronger than the keyway. This construction joint must be capable of transmitting the stem shear into the footing. For this reason the shear force is calculated *at the support*, rather than at a distance d from the support as in other types of beams. It may be necessary to provide dowels for both faces of the wall in order to resist excess shear at the construction joint.

Heel Design

The heel is considered to be a cantilever beam beyond the stem, fixed at the location of the tension steel in the stem. This location is chosen rather than the face of the stem since the tension bars in the stem provide for the transmittal of the reaction forces into the stem. This condition can be approximated fairly accurately by adding 3″ (76 mm) to the projection of the base slab beyond the rear face of the stem. Since the thickness of the stem has been established on the basis of an assumed base thickness, several trial solutions may be necessary in order to make the stem and base calculations compatible.

The moment and shear in the heel are taken at the intersection of the stem tension steel and the assumed top of the heel. Two conditions are considered (see Fig. 15-20):

(1) Moment and shear due to the downward load p_v plus the weight of the footing acting downward, and

(2) Moment and shear due to the net effect of the downward load p_v and the upward loads q_1 and q_2. In this case, the weight of the footing causes downward load and upward pressure which counteract one another.

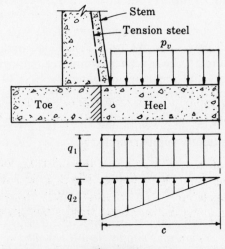

Fig. 15-20

In common engineering practice, Case (1) is permitted to cause 150 percent of the *allowable stresses*. If this is considered with respect to the alternate design method,

$$M_a/S = 1.5f_c = 3f_c/2 \qquad (15.23)$$
$$V_a/bd = 1.5v = 3v/2 \qquad (15.24)$$

If equations (*15.23*) and (*15.24*) are multiplied by $\frac{2}{3}$, then

$$\tfrac{2}{3}M_a/S = f_c \qquad (15.25)$$
$$\tfrac{2}{3}V/bd = v \qquad (15.26)$$

It is practical then, to consider the moment and shear due to Case (2) and compare those quantities to $\frac{2}{3}$ of the values obtained for Case (1). The larger of the two conditions is employed in design using the actual allowable stresses. This procedure is also valid when the strength design method is used.

Toe Design

The toe of the base is considered to be cantilevered from the *edge* of the front face of the stem. The loads employed are those of the base weight acting downward and the base pressures acting upward. The weight of the backfill soil *above the toe* is usually neglected since erosion could cause this counterbalancing load to be diminished.

In cases in which the backfill is *confined* under a slab, these loads should be considered as a second condition in order to determine which case governs the design. Fig. 15-22 shows another condition which could exist. Such construction usually produces complex design calculations which are merely approximate even under the most simple conditions. This situation should be avoided whenever possible.

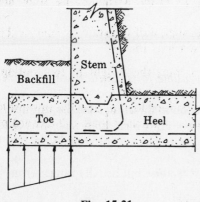

Fig. 15-21

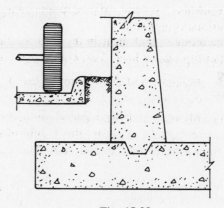

Fig. 15-22

Preliminary Dimensions

For reasonable preliminary estimates, the base length is usually about 40 to 65 percent of the wall height, and the base thickness may be obtained as 7 to 10 percent of the total height.

Shear Keys

When frictional resistance of the footing against the soil is insufficient to provide a satisfactory safety factor against sliding, *shear keys* are employed. The shear key must develop a sufficient passive pressure force to resist the excess lateral force. In order that the passive pressure might develop, the depth of the key and its location along the base must be such that the distance C in Fig. 15-23 will be at least

$$C = a \tan^2 (45° + \phi/2) \qquad (15.27)$$

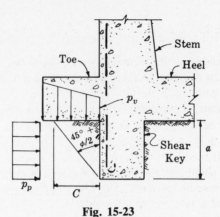

Fig. 15-23

since $(45 + \phi/2)$ is the *shearing angle for passive pressure*. The *passive pressure*, p_p, may be computed as

$$p_p = p_v \tan^2 (45° + \phi/2) \qquad\qquad (15.28)$$

or
$$p_p = p_v(1 + \sin \phi)/(1 - \sin \phi) \qquad\qquad (15.29)$$

which are identical. The multiplier of p_v is the passive pressure coefficient k_p.

Inasmuch as there will be a pressure gradient under the footing in front of the key, the smaller value p_v should be used.

It should be noted that p_v is the pressure beneath the footing, which is pressing downward on the soil as well as upward on the base of the footing.

When possible, the shear key should be located directly beneath the stem in order to provide additional anchorage for the stem reinforcement.

COUNTERFORT RETAINING WALLS

Often it is desirable to use retaining walls which have *counterforts*. Such a wall is illustrated in Fig. 15-24 and is used when the retaining wall must be extremely high.

The structural analysis and design of counterfort walls is quite different than for cantilever walls. The counterforts serve as supports for the wall and footing, both of which are designed as continuous slabs supported at the counterforts. The main reinforcement is placed horizontally in this type of wall.

The spacing of the counterforts depends on a number of factors, but usually ranges from $\frac{1}{3}$ to $\frac{1}{2}$ of the height of the wall.

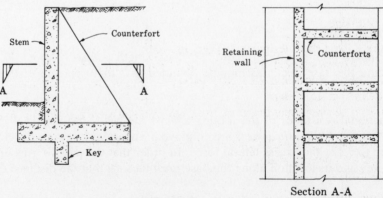

Fig. 15-24

Overturning and sliding are handled in the same manner as for cantilever walls, as is the design of the toe when such is used.

The stem and heel are considered to be one-way slabs, and the 1983 ACI Code coefficients for continuous beams are used in the design. The uniform load per unit width of vertical wall is taken as the average horizontal soil pressure over that strip, as shown in Fig. 15-25.

The counterforts are designed as triangular shaped cantilever beams, fixed at the base slab.

The stem may be used as the flange of a T beam for which the counterfort acts as a web. The maximum depth of the T beam is shown as section A-A' in Fig. 15-26. The design loads on the counterfort are the accumulated reactions from the slab which actually vary linearly. The Code does not specify a critical section for shear for counterforts, because retaining walls are considered by ACI Commitee 318 to be beyond the scope of the Code, but in practice section A'-B' is often used. The shear force should be calculated using the equation

$$V'' = V + (M/d)\tan\beta \qquad (15.30)$$

The effective depth d is obtained using section B'-C', considering d to be the distance from B' to the centroid of the main steel.

Horizontal and vertical bars are provided in both faces of the counterfort to provide for shrinkage. The horizontal bars may be utilized as stirrups and should be formed in the shape of a U in order to hold the main bars in place. It should be noted here that when the counterforts are placed in front of the wall, the term *buttress* is used. The wall is then called a *buttressed wall*. This does not affect the procedure for designing the wall, but the forces on buttresses are reversed from those on counterforts.

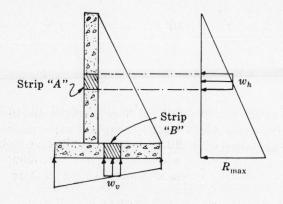

Fig. 15-25

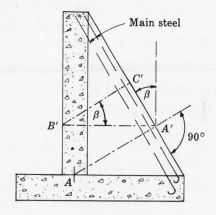

Fig. 15-26

BASEMENT WALLS

The method used for designing basement walls will differ in accord with the method of construction. Four-way support may be obtained from columns or pilasters and the basement and first floor slabs, as shown in Fig. 15-27(a). Two-way support will occur if the slabs alone support the wall, as shown in Fig. 15-27(b).

Axial loads on the wall may be large, in which case *beam-column* action should be considered, as shown in Fig. 15-28.

Chapter 14 of the 1983 ACI Code provides methods for wall design. The Code gives requirements which apply to bearing walls, which includes most basement walls. Those requirements may be listed as follows:

(1) The wall thickness must be at least 1/25 of the unsupported height or width, whichever is the larger.

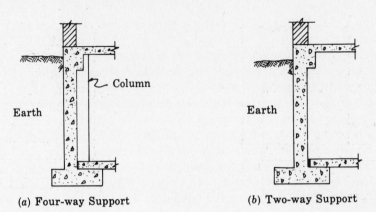

(a) Four-way Support   (b) Two-way Support

Fig. 15-27

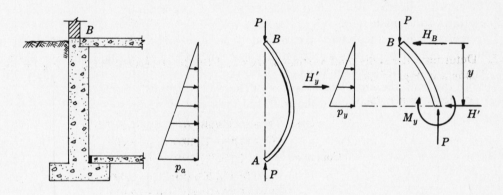

Fig. 15-28

(2) The uppermost 15' (5 m) of wall must be at least 6" (150 mm) thick, and the minimum thickness must be increased 1" (25 mm) for every additional 25' (8 m) or fraction thereof.

(3) The area of horizontal steel must be at least $0.0025bh$ and the area of vertical reinforcement not less than $0.0015bh$.

(4) Walls shall be anchored to the floors, pilasters and intersecting beams using steel area at least equal to No. 3 (10M) bars at 12" (300 mm) oc, for each layer of wall steel.

(5) Exterior walls and basement walls shall not be less than 8" (200 mm) thick.

Solved Problems

15.1. Determine the total horizontal active force and the moment tending to overturn the cantilever wall shown in Fig. 15-29 if $\sin \phi = 0.5$ and the soil weighs 100 lb/ft³. (Use 1 foot of wall width.)

$k_a = (1 - \sin \phi)/(1 + \sin \phi) = (1 - 0.5)/(1 + 0.5) = 0.33$.

$p_{max} = (0.33)(100)(21.67) = 720 \text{ lb/ft}^2$, $H = (720)(21.67)/2 = 780 \text{ lb/ft}$.

Thus $M_o = (7800)(7.22) = 56{,}300 \text{ ft-lb/ft}$ about the toe (point A).

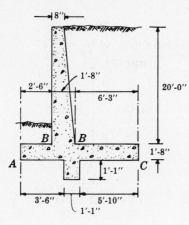

Fig. 15-29

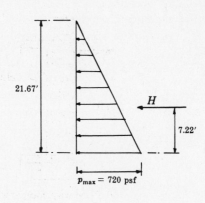

Fig. 15-30

15.2. Determine the individual vertical forces and the total vertical force acting on the structure shown in Fig. 15-29.

Refer to Fig. 15-31.

$$
\begin{array}{lll}
\text{Earth:} & W_1 = (6.25)(20)(100) & = 12{,}500 \text{ lb/ft} \\
& W_2 = (1)(20/2)(100) & = 1{,}000 \\
\text{Concrete:} & W_3 = (1)(20/2)(150) & = 1{,}500 \\
& W_4 = (0.67)(20)(150) & = 2{,}000 \\
& W_5 = (1.67)(10.43)(150) & = 2{,}600 \\
& W_6 = (1.08)(1.08)(150) & = \underline{\quad 175 \quad} \\
& \text{Total Load} & = 19{,}775 \text{ lb/ft}
\end{array}
$$

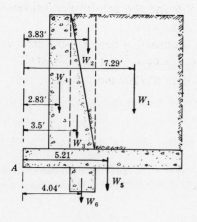

Fig. 15-31

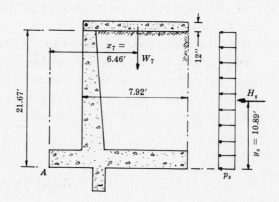

Fig. 15-32

15.3. Determine the righting moment (about A) caused by the vertical loads for the structure of Problems 15.1 and 15.2. The level arm for each force is shown in Fig. 15-31.

The appropriate values are tabulated as follows:

Force, lb/ft	Arm, ft	Moment, ft-lb/ft
12,500	7.29	91,125
1,000	3.83	3,830
1,500	3.50	5,250
2,000	2.83	5,660
2,600	5.21	13,550
175	4.04	710
$\Sigma F = 19{,}775$ lb/ft		$M_r = 120{,}125$ ft-lb/ft

The total vertical load is 19,775 lb/ft and the righting moment is 120,125 ft-lb/ft.

15.4. Determine the safety factor against overturning for Problem 15.3.

From Problem 15.1 and Fig. 15-30, $M_o = 56{,}300$ ft-lb/ft.

From Problem 15.3 and Fig. 15-31, $M_r = 120{,}125$ ft-lb/ft.

Thus $SF = M_r/M_o = 120{,}125/56{,}300 = 2.13$.

15.5. Consider that a 12″ thick slab is placed over the surface of the ground behind the retaining wall in Fig. 15-29. Determine the added horizontal pressure against the wall, the changes in M_o and M_r, and find the new safety factor against overturning. Use results of Problem 15.4 and $k_a = 0.33$ with $w = 100$ lb/ft^3.

Refer to Fig. 15-32.

The slab weight produces a vertical pressure of 150 lb/ft^2. The equivalent height of soil is $h_s = 150/100 = 1.5'$, and the resulting change in horizontal pressure is $p_s = (100)(1.5)(0.33) = 50$ lb/ft^2.

The change in the vertical force is $W_7 = (150)(1)(7.92) = 1188$ lb/ft, and the change in the horizontal force is $H_s = p_s h_w = (50)(21.67) = 1083$ lb/ft. The changes in the moments are

$$\Delta M_o = H_s y_s = (1083)(10.89) = 11{,}800 \text{ ft-lb/ft}$$
$$\Delta M_r = W_7 x_7 = (1188)(6.46) = 7{,}680 \text{ ft-lb/ft}$$

and total moments are

$$M_o = 56{,}300 + 11{,}800 = 68{,}100 \text{ ft-lb/ft}$$
$$M_r = 120{,}125 + 7{,}680 = 127{,}805 \text{ ft-lb/ft}$$

The safety factor against overturning is $SF = M_r/M_o = 127{,}805/68{,}100 = 1.88$.

15.6. Determine the moment at the base of the stem for the wall shown in Fig. 15-29. Use the data given in Problem 15.1 and the results of that problem.

Refer to Fig. 15-29. The pressure is proportional to the distance from the top; thus $p = (0.33)(100)(20) = 660$ lb/ft^2.

The horizontal force on the stem is $H = (660)(20/2) = 6600$ lb/ft, and the moment of that force about the base of the stem is $M = Hy = Hh_w/3 = (6600)(20/3) = 44{,}000$ ft-lb/ft. This moment would be used to design the stem.

15.7. Locate the resultant force R for Problem 15.1 and find the soil pressure at the edges of the toe and heel.

Refer to Fig. 15-33.

The net moment about point A is $M_r - M_o = 120{,}125 - 56{,}300 = 63{,}825$ ft-lb/ft.

The total vertical force R is 19,775 lb, so the location of the resultant with respect to point A is $x_A = (M_r - M_o)/R = 63{,}825/19{,}775 = 3.22'$. Thus the eccentricity from the center of the footing is

$$e = L/2 - x_A = 5.22 - 3.22 = 2.0'$$

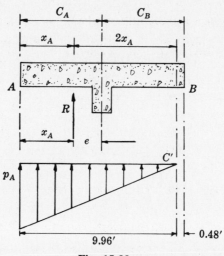

Fig. 15-33

Since e is greater than $L/6$ and therefore outside of the middle third of the base, the pressure diagram does not cover the entire base. The total length of the pressure diagram is $3x_A$ or 9.66′. The effective area of the base will therefore be $A = 9.66$ ft²/ft, and the pressures will be $p_A = 2R/A = (2)(19,775)/9.66 = 4094$ lb/ft² and $p_{C'} = 0$.

15.8. Calculate the shear force and bending moment per foot of width for the retaining wall toe shown in Fig. 15-34 below. Point E represents the face of the stem. The pressures given are *net* values. The difference in density between the concrete and the earth displaced by the base has been deducted from the gross values. (Many designers ignore this quantity, while others ignore the displaced soil and consider the weight of the footing as a downward load.)

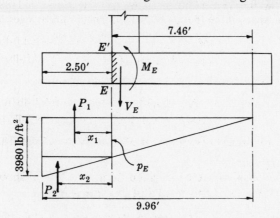

Fig. 15-34

The pressure at A is 3980 lb/ft², so the pressure at E by proportion will be

$$p_E = (3980)(7.46)/9.96 = 2980 \text{ lb/ft}^2$$

The resulting forces are calculated as the volume of the pressure prisms per foot of width; thus

$$P_1 = (2980)(2.5)(1) = 7450 \text{ lb/ft}, \qquad P_2 = (3980 - 2980)(2.5/2)(1) = 1250 \text{ lb/ft}$$

The shear force at E is $V_E = P_1 + P_2 = 7450 + 1250 = 8700$ lb/ft.

The lever arms for forces P_1 and P_2 respectively are $x_1 = \frac{1}{2}(2.5) = 1.25′$, $x_2 = \frac{2}{3}(2.5) = 1.67′$.

The moment at E is $M_E = P_1 x_1 + P_2 x_2 = 11,400$ ft-lb/ft.

The shear force and moment would be used to design the toe structurally using methods explained in previous chapters.

15.9. Determine the preliminary design dimensions for the retaining wall shown in Fig. 15-35(a). The granular backfill weighs 100 lb/ft^3 and has an angle of internal friction $\phi = 30$ degrees. Use $f_c' = 3000$ psi, $f_y = 40{,}000$ psi, and $\rho = 0.014$.

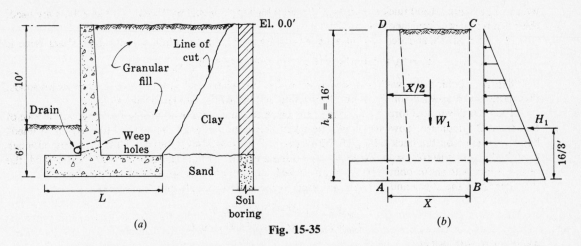

(a)   Fig. 15-35   (b)

Refer to Fig. 15-35(b) and temporarily assume no difference in the weight of soil and concrete. The total weight of block $ABCD$ is $W_1 = hwX - 1600X$, and the lever arm for this force about A is $X/2$.

The active pressure coefficient is $k_a = (1 - \sin \phi)/(1 + \sin \phi) = 0.5/1.5 = \frac{1}{3}$.

The total horizontal force is $H_1 = k_a wh_w^2/2 = (\frac{1}{3})(100)(16)^2/2 = 4260$ lb/ft.

The resultant may be established at point A by using equation (15.22) to determine the base length L. Thus $L = 1.5\sqrt{k_a h_w^2/3} = 8.0'$.

Try $L = 8.0'$ with a toe overhang of $(8/3)'$ or $2'$-$8''$.

Assume the thickness of the footing to be approximately 8% of the total height, or about 1.3'. Thus the cantilever height of the stem will be approximately 14.7', as shown in Fig. 15-36.

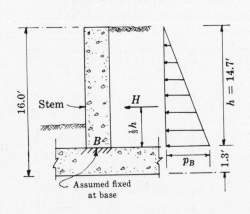

Fig. 15-36

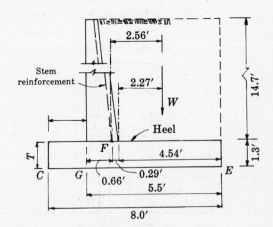

Fig. 15-37

The pressure at the base of the stem is $p_B = k_a hw = (\frac{1}{3})(14.7)(100) = 490$ lb/ft^2.

The horizontal force applied to the stem is $H = p_B h/2 = (490)(14.7/2) = 3600$ lb/ft. Thus the working load shear force at B is 3600 lb/ft. The moment about the base of the cantilever for working loads is $M_B = Hh/3 = (3600)(4.9) = 17{,}650$ ft-lb/ft.

Since the forces are related to dead loads, the ultimate values for use in design are

$$V_u = (1.5)V_B = (1.5)(3600) = 5400 \text{ lb/ft}$$
$$M_u = (1.5)M_B = (1.5)(17{,}650) = 26{,}500 \text{ ft-lb/ft}$$

For $f'_c = 3000$ psi, $f_y = 40,000$ psi, and $\rho = 0.014$, Table 5.2 gives $\phi R_u = 448$ (in psi/ft). The effective depth is therefore

$$d = \sqrt{M_u/\phi R_u b} = \sqrt{(26,500)(12)/[(448)(12)]} = 7.69''$$

which will require a total thickness of 11.5" (to next higher $\frac{1}{2}''$) including 3" cover. If No. 8 bars are used, the effective depth furnished will be 8".

The nominal shear force is $V_n = V_u/\phi = 5400/0.85 = 6353$ lb/ft, while the allowable shear force is

$$V_c = 2\sqrt{f'_c}\, bd = 2\sqrt{3000}\,(12)(8) = 10,615 \text{ lb/ft} > V_n \qquad \text{(O.K.)}$$

The heel behind the wall is designed using $\frac{2}{3}$ of the shear and moment due to downward loads, disregarding the upward soil pressure. The design conditions are shown in Fig. 15-37.

The total downward force is that due to the earth weight on the overhanging heel beyond the face of the stem. The moment, however, is usually taken about the center of the vertical (or slightly inclined) bars in the stem on the back side, at F. The force is $W = (4.54)(14.7)(100) = 6670$ lb/ft, so the ultimate load is $W_u = (1.5)(6670) = 10,000$ lb/ft. The ultimate shear is therefore $V_u = 10,000$ lb/ft, and the ultimate moment about point F is $M_u = (10,000)(2.56) = 25,600$ ft-lb/ft.

Since $\frac{2}{3}$ of the shear and moment are used in the design, the effective values are

$$V_u = (\tfrac{2}{3})(10,000) = 6670 \text{ lb/ft}$$
$$M_u = (\tfrac{2}{3})(25,600) = 17,100 \text{ ft-lb/ft}$$

Again using $\phi R_u = 448$,

$$d = \sqrt{(17,000)(12)/[448(12)]} = 6.18''$$

Now, in terms of allowable shear force, $d = V_c/(2\sqrt{f'_c}\, b)$. In this expression, substitute $V_n = V_u/\phi = 6670/0.85 = 7877$ lb/ft for V_c to obtain

$$d_{min} = 7847/(2\sqrt{3000}\, 12) = 5.97'' < 6.18''$$

Moment governs, and the total thickness must be at least $h = 6.18 + 0.5 + 3.0 = 9.68''$, considering that No. 8 bars will probably be used. This thickness is based on the measurements including $\frac{1}{2}$ bar diameter and 3" clear cover. For practical purposes use $h = 10''$, so the d furnished is 6.5".

The initial assumption for the thickness of the footing was 8% of the base length or 1.3', so the height of stem used was $(1.30 - 0.83) = 0.47'$ smaller than the value provided by the 10" depth.

The slight error involved would be eliminated in a *final review* of the designed section. The review is also made considering the total downward force and the total upward force acting together.

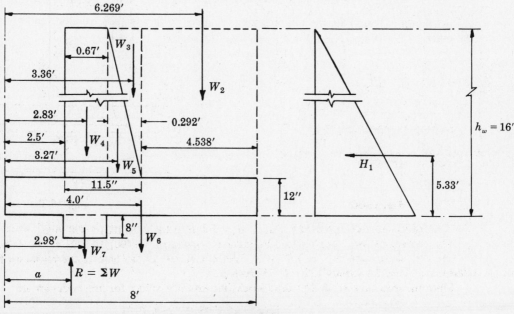

Fig. 15-38

15.10. Determine the safety factor against overturning for the retaining wall shown in Fig. 15-38. Use $k_a = \frac{1}{3}$ and $w = 100$ lb/ft³. Concrete weighs 150 lb/ft³.

Refer to Fig. 15-38 for dimensions and forces.
Vertical forces and righting moments per foot of wall are tabulated as follows:

Calculations	Force, lb/ft	Arm, ft	M_r, ft-lb/ft
$W_2 = (4.538)(100)(15)$ =	6,807	6.269	42,700
$W_3 = (0.292)(100)(15/2)$ =	219	3.360	740
$W_4 = (0.67)(150)(15)$ =	1,508	2.830	4,250
$W_5 = (0.292)(150)(15/2)$ =	329	3.270	1,080
$W_6 = (1)(150)(8)$ =	1,200	4.000	4,800
$W_7 = (0.67)(150)(11.5/12)$ =	97	2.980	290
Totals:	$R = 10,160$ lb/ft		$M_r = 53,860$ ft-lb/ft

The horizontal soil force $H_1 = (\frac{1}{3})(100)(16)^2/2 = 4270$ lb/ft.

The overturning moment $M_o = H_1 h_w/3 = (4270)(5.33) = 22,760$ ft-lb/ft.

For overturning, SF $= M_r/M_o = 53,860/22,760 = 2.37$, which is satisfactory.

15.11. For the retaining wall footing shown in Fig. 15-39, the resisting moment is 52,610 ft-lb/ft and the overturning moment is 23,600 ft-lb/ft. The total vertical load is 9867 lb/ft. Calculate the soil pressures at points C and E.

The location of the resultant force with respect to point C is

$$a = (M_o - M_r)/R = (52,610 - 23,600)/9867 = 2.94'$$

so the eccentricity from the center of the footing is $e = 4.0 - 2.94 = 1.06'$.

The moment of inertia of the footing base per foot width is $I = (1)(8)^3/12 = 42.6$ ft⁴/ft, and the area of the base is $A_F = (8)(1) = 8.0$ ft²/ft.

At any point the pressure is $p = R/A_F \pm Rec/I$ where c is the distance of the point from the center of the footing. Hence

$$p_C = (9867/8) + (9867)(1.06)(4)/42.6 = 2218 \text{ lb/ft}^2, \qquad p_E = 1235 - 983 = 252 \text{ lb/ft}^2$$

As shown in Fig. 15-39, the soil pressure is assumed to vary linearly from one end of the footing to the other.

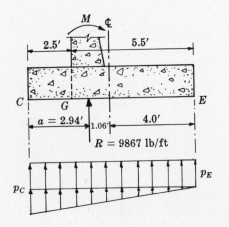

Fig. 15-39

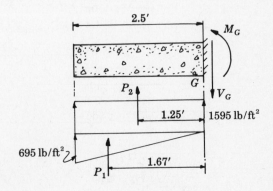

Fig. 15-40

15.12. Determine the moment and shear force for design of the footing toe shown in Fig. 15-40. Consider service loads.

The forces are equal to the volume of the pressure prism per foot; thus

$$P_1 = (695)(2.5/2) = 870 \text{ lb/ft}, \qquad P_2 = (1595)(2.5) = 4000 \text{ lb/ft}$$

The service-load shear force is $V_G = P_1 + P_2 = 4870$ lb/ft.
The service-load moment is $M_G = (870)(1.67) + (4000)(1.25) = 6455$ ft-lb/ft.

If strength design is employed, ultimate shear force and nominal bending moment are required. Because only dead load is involved, the understrength factors (ϕ) are 0.85 for shear and 0.9 for bending. Hence,

$$V_u = 1.4(4870) = 6918 \text{ lb/ft} \qquad V_n = 6918/0.85 = 8138 \text{ lb/ft}$$

and
$$M_u = 1.4(6455) = 9037 \text{ ft-lb/ft}$$

15.13. The basement wall shown in Fig. 15-41 is backfilled with a granular material for which $\phi = 30$ degrees. Determine the design shear and bending moment for the wall if the soil weighs 90 lb/ft^3.

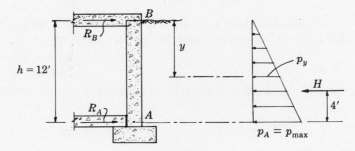

Fig. 15-41

$p_{\max} = wh(1 - \sin \phi)/(1 + \sin \phi) = (90)(12)(1 - \frac{1}{2})/(1 + \frac{1}{2}) = 360$ lb/ft^2.
$R_A = (\frac{2}{3})(12)(360) = 2880$ lb/ft.

Maximum moment occurs at $y = (0.5774)(h) = 6.93'$ from B.

$M_{\max} = (0.1283)Hh = (0.1283)(0.5)(360)(12)(12) = 3326$ ft-lb/ft.

Note. The coefficients for reaction R_A and maximum moment were obtained from Appendix A.

15.14. The counterfort retaining wall shown in Fig. 15-42 is backfilled with a material for which $w = 100$ lb/ft^3 and $\phi = 30$ degrees. Determine the shear and bending moments for designing the wall at its base A at points B and F, considering the wall continuously supported over the counterforts. Assume that restraint develops at A and E.

$$k_a = (1 - \sin \phi)/(1 + \sin \phi) = (1 - 0.5)/(1 + 0.5) = 0.33$$
$$p_{\max} = k_a wh = (0.33)(100)(12) = 400 \text{ lb/ft}^2$$

Consider the load in the lateral direction to be equal to the maximum pressure at the lower end of the wall and constant for the first foot of height; then $w_H = 400$ lb/ft.

Using the 1983 ACI Code coefficients (Chapter 2), the bending moments are approximately

$$M_{B'} = -(\tfrac{1}{10})w_H(L'^2) = -(0.1)(400)(12)^2 = -5750 \text{ ft-lb/ft}$$
$$M_F = (\tfrac{1}{14})w_H(L')^2 = (\tfrac{1}{14})(400)(12)^2 = 4110 \text{ ft-lb/ft}$$

Since the pressure varies linearly from the top of the wall, the uniform load in the transverse direction will also vary in the same way. Thus the moments vary linearly from the top of the wall to the base. Since the pressure is zero at the top, the moments are also zero at the top of the wall.

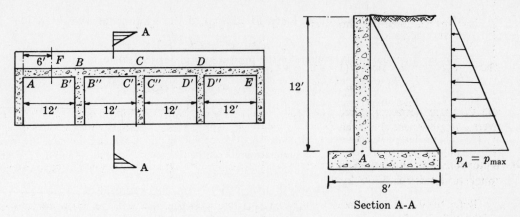

Fig. 15-42

Under strength design (cf. Problem 15.12),

$$M_{uB'} = 1.4(-5750) = -8050 \text{ ft-lb/ft}$$
$$M_{uF} = 1.4(4110) = 5754 \text{ ft-lb/ft}$$

If any type of live load (such as truck or train wheels) were involved as a surcharge on the soil behind the wall, the related forces and moments would have to be calculated separately and factored using the load factor 1.7.

15.15. Determine the horizontal active force and the moment tending to overturn the cantilever wall shown in Fig. 15-43, if $\sin \phi = 0.5$ and the soil density is 1.633 kg/m^3.

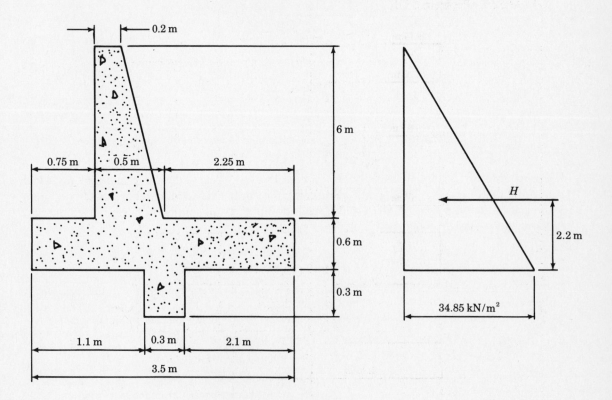

Fig. 15-43

The vertical force (weight) of the soil is $w = mg$, where $g = 9.8$ m/s^2, or

$$w = (1.633)(9.8) = 16 \text{ kN/m}^3$$

Then, since $k_a = (1 - \sin \phi)/(1 + \sin \phi) = (1 - 0.5)/(1 + 0.5) = 0.33$,

$$P_{max} = k_a w h_w = (0.33)(16)(6.6) = 34.85 \text{ kN/m}^2$$

and the total horizontal force due to the soil is $H = p_{max} h_w/2 = (34.85)(6.6)/2 = 115.0$ kN/m. The overturning moment about the toe (point A) is $M_o = H h_w/3 = (115)(6.6)/3.0 = 253.0$ kN·m/m.

15.16. Determine the individual vertical forces acting on the structure shown in Fig. 15-43, if the concrete density is 2.398 kg/m^3.

Refer to Fig. 15-44. Consider 1.0 meter of wall and footing width. The concrete force (weight) is $(2.398)(9.8) = 23.5$ kN/m^3.

Earth:	$W_1 = (2.25)(6)(16)$	$= 216.00$ kN/m
	$W_2 = (0.3)(6)(16)/2$	$= 14.40$
Concrete:	$W_3 = (0.3)(6)(23.5)/2$	$= 21.15$
	$W_4 = (0.2)(6)(23.5)$	$= 28.20$
	$W_5 = (0.6)(3.5)(23.5)$	$= 49.35$
	$W_6 = (0.3)(0.3)(23.5)$	$= \underline{2.12}$
	Total Load	331.22 kN/m

15.17. Determine the righting moment (about A) caused by the vertical loads for the structure of Problems 15.15 and 15.16. The lever arms for the forces are shown in Fig. 15-44. Use the data calculated in Problem 15.16.

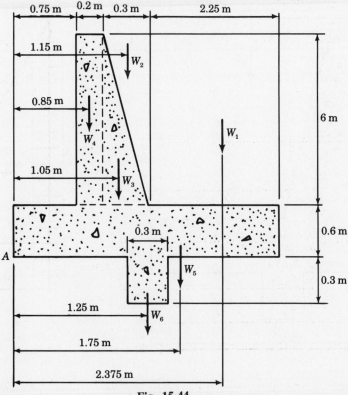

Fig. 15-44

The appropriate values per meter width are:

Force, kN/m	Arm, m	Moment, kN·m/m
216.00	2.375	513.00
14.40	1.15	16.56
21.15	1.05	22.21
28.20	0.85	23.97
49.35	1.75	86.36
2.12	1.25	2.65
$\Sigma F = 331.22$ kN/m		$\Sigma M_r = 664.75$ kN·m/m

15.18. Determine the safety factor against overturning for the structure shown in Fig. 15-44. Use the data obtained in Problems 15.15 through 15.17.

From Problem 15.15, $M_o = 253.0$ kN·m/m; from Problem 15.17, $M_r = 664.75$ kN·m/m. Then, the safety factor against overturning is

$$SF = M_r/M_o = 664.75/253.0 = 2.627 > 2.0 \quad \text{(O.K.)}$$

15.19. Consider that a slab 300 mm thick is constructed over the ground for the structure of Fig. 15-43. Determine the added horizontal and vertical forces, the resulting overturning and resisting moments and the safety factor against overturning. Use data from Problems 15.15 through 15.18. Refer to Fig. 15-45.

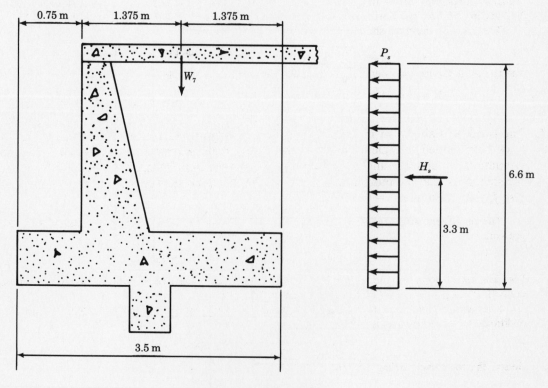

Fig. 15-45

The weight of the slab per meter width is

$$W_7 = (2)(1.375)(0.3)(1)(23.5) = 19.39 \text{ kN/m}$$

and so the vertical pressure beneath the slab is $p_v = 19.39/2.75 = 7.05 \text{ kN/m}^2$.

The equivalent height of surcharge is $(23.5)(0.3)/16 = 0.44$ m, and the resulting horizontal surcharge pressure is

$$p_s = h_s w k_a = (0.44)(16)(0.33) = 2.32 \text{ kN/m}^2$$

The total horizontal force due to the surcharge is $H_s = p_s h_w = (2.32)(6.6) = 15.31 \text{ kN/m}$.

The resulting additional overturning moment is

$$\Delta M_o = H_s h_w/2 = (15.31)(6.6)/2 = 50.52 \text{ kN} \cdot \text{m/m}$$

and the resulting additional righting moment is

$$\Delta M_r = (W_7)(2.125) = (19.39)(2.125) = 41.20 \text{ kN} \cdot \text{m/m}$$

The total moments are

$$M_o = 253.0 + 50.52 = 303.52 \text{ kN} \cdot \text{m/m} \qquad \text{and} \qquad M_r = 664.75 + 41.20 = 705.95 \text{ kN} \cdot \text{m/m}$$

The safety factor against overturning is then $\text{SF} = M_r/M_o = 705.95/303.52 = 2.326$, which exceeds 2.0 and is therefore satisfactory.

15.20. Determine the shear force and bending moment at the base of the stem for the wall of Fig. 15-43 and Problem 15.15. Include the surcharge found in Problem 15.19.

The horizontal pressure at the base of the stem is

$$p_h = wh[(1 - \sin \phi)/(1 + \sin \phi)] = (16)(6)(0.33) = 31.68 \text{ kN/m}^2$$

so that the horizontal force at the base of the wall is $H = p_h h/2 = (31.68)(6)/2 = 95.04 \text{ kN/m}$. The horizontal pressure due to the surcharge is constant and is equal to $p_s = h_s w k_a = (0.44)(16)(0.33) = 2.32 \text{ kN/m}^2$. Hence, the horizontal force due to the surcharge is $H_s = p_s h = (2.32)(6.0) = 13.92 \text{ kN/m}$.

The total service-load shear force at the base of the stem is

$$V = H + H_s = 95.04 + 13.92 = 108.96 \text{ kN/m}$$

and the total service-load bending moment at the base of the stem is

$$M = (Hh/3) + (H_s h/2) = (95.04)(6/3) + (13.92)(6/2) = 232.56 \text{ kN} \cdot \text{m/m}$$

15.21. The concrete slab shown in Fig. 15-45 is used for storage, so that a live load of 8.0 kN/m^2 must be provided for. Determine the new safety factor against overturning, as well as the ultimate shear and bending moment at the base of the stem, using $U = 1.4D + 1.7L$. Check the stem for shear and moment and determine the required steel area at the base of the stem. Use $f_c' = 20$ MPa and $f_y = 400$ MPa.

The equivalent surcharge height for the live load is $h_s = 8.0/16.0 = 0.5$ m. The resulting horizontal pressure is

$$p_s = h_s w k_a = (0.5)(16)(0.33) = 2.64 \text{ kN/m}^2$$

The horizontal force is then $H_s = p_s h = (2.64)(6.6) = 17.42 \text{ kN/m}$, and the additional overturning moment is $\Delta M_o = H_s h/2 = (17.42)(6.6)/2 = 57.49 \text{ kN} \cdot \text{m/m}$.

The vertical load due to the live-load surcharge is $F_v = (8.0)(2)(1.375) = 22.0 \text{ kN/m}$, and the additional righting moment is

$$\Delta M_r = (22.0)(2.125) = 46.75 \text{ kN} \cdot \text{m/m}$$

Hence, the total overturning moment is

$$M_o = 303.52 + 57.49 = 361.01 \text{ kN} \cdot \text{m/m}$$

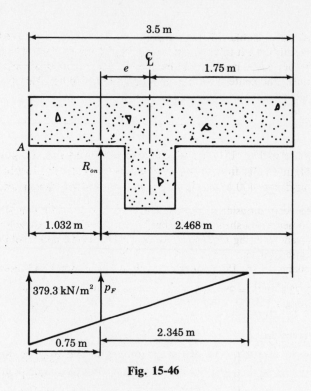

Fig. 15-46

and the total righting moment is

$$M_r = 705.95 + 46.75 = 752.70 \text{ kN} \cdot \text{m/m}$$

The safety factor against overturning is

$$\text{SF} = M_r/M_o = 752.70/361.01 = 2.085 > 2.0 \qquad \text{(O.K.)}$$

The live-load horizontal pressure force at the base of the stem is

$$H_s = p_s h = (2.64)(6) = 15.84 \text{ kN/m}$$

so that live-load shear force is 15.84 kN/m. The live-load bending moment at the base of the stem is

$$M = H_s h/2 = (15.84)(6)/2 = 47.52 \text{ kN} \cdot \text{m/m}$$

The "dead-load" shear force and bending moment at the base of the stem were found in Problem 15.20 to be 108.96 kN/m and 232.56 kN·m/m, respectively. The ultimate design values are therefore

$$V_u = 1.4D + 1.7L = (1.4)(108.96) + (1.7)(15.84) = 179.47 \text{ kN/m}$$
$$M_u = (1.4)(232.56) + (1.7)(47.52) = 406.37 \text{ kN} \cdot \text{m/m}$$

The nominal shear value is $V_n = V_u/\phi = 179.47/0.85 = 211.14$ kN/m, which, together with M_u, is used to design the stem.

For shear, the permissible ultimate shear force on the concrete is $V_c = \sqrt{f_c'}\, bd/6$. Since the wall is designed for 1 m of width, $b = 1000$ mm. The total depth is 500 mm. A cover over the steel of 50 mm is reasonable for No. 20M bars and concrete cast against forms but exposed to earth. The effective depth is then

$$d = 500.0 - 50.0 - (19.5/2) = 440.25 \text{ mm}$$

where 19.5 mm is the diameter of a No. 20M bar. Thus,

$$V_c = [\sqrt{20}\,(1000)(440.25)/6] \times 10^{-3} = 328.16 \text{ kN/m}$$

This is greater than the actual nominal shear force $V_n = 211.14$ kN/m, so the design is satisfactory.

Using $d = 440.25$ mm and

$$\phi R_u = M_u/bd^2 = (406.37 \times 10^6)/[(1000)(440.25)^2] = 2.10 \text{ (in MPa/m)}$$

we may interpolate in Table 5.3 for the required steel ratio: $\rho = 0.0063$. (The minimum steel ratio, $\rho_{min} = 0.0035$, is satisfied.) It follows that $A_s = \rho bd = (0.0063)(1000)(440.25) = 2773.6$ mm^2/m. The area of one No. 20M bar is 300 mm^2, so the number of bars required per meter width will be $2773.6/300 = 9.25$ bars (use 10). The center-to-center spacing of bars will be $1000/10 = 100$ mm.

15.22. For the structure in Fig. 15-43, with the slab shown in Fig. 15-45 and the live-load surcharge given in Problem 15.21, find the soil pressures beneath the footing and design the toe, using $f'_c = 20$ MPa and $f_y = 400$ MPa. Use 1.4 as a dead-load factor and 1.7 as a live-load factor.

It would be very confusing to determine the soil pressures for dead-load forces separately and then to factor the moments and shears thus obtained. It is much easier to factor the "applied forces," such as the overturning and righting moments and the vertical loads, and to obtain the resulting soil pressures from these factored forces.

From Problems 15.16, 15.19, and 15.21, the vertical forces and moments are:

$$R_D = 331.22 + 19.39 = 350.61 \text{ kN/m} \qquad M_o = 303.52 \text{ kN} \cdot \text{m/m} \qquad M_r = 705.95 \text{ kN} \cdot \text{m/m}$$
$$R_L = 22.0 \text{ kN/m} \qquad\qquad\qquad \Delta M_o = 57.49 \text{ kN} \cdot \text{m/m} \qquad \Delta M_r = 46.75 \text{ kN} \cdot \text{m/m}$$

The factored forces are

$$M_{on} = [(1.4)(303.52) + (1.7)(57.49)]/0.9 = 580.73 \text{ kN} \cdot \text{m/m}$$
$$M_{rn} = [(1.4)(705.95) + (1.7)(46.75)]/0.9 = 1186.45 \text{ kN} \cdot \text{m/m}$$
$$R_n = [(1.4)(350.61) + (1.7)(22.0)]/0.9 = 586.95 \text{ kN/m}$$

Note that $\phi = 0.9$ was also used in connection with R_n because this vertical force, along with the moments, is involved in the flexural design of the footing.

With reference to Fig. 15-46, the distance of the resultant force from the toe (point A) is

$$X = (M_{rn} - M_{on})/R_n = (1186.45 - 580.73)/586.95 = 1.032 \text{ m}$$

and the eccentricity $e = 1.75 - 1.032 = 0.718$ m.

The edge of the kern is at a distance $3.5/3 = 1.17$ m; so the resultant force is slightly outside of the kern. The length of footing subjected to pressure is $(3)(1.032) = 3.096$ m. The maximum pressure (ultimate soil pressure) is

$$P_{max} = \frac{2R_n}{A_f} = \frac{2(586.95)}{(3.096)(1.0)} = 379.3 \text{ kN/m}^2$$

and the pressure p_F at the face of the wall behind the toe is

$$p_F = (2.345/3.095)(379.3) = 287.4 \text{ kN/m}^2$$

The pressures are shown in Fig. 15-47 with appropriate distances. We have

$$R_1 = (379.3)(0.75)/2.0 = 142.2 \text{ kN/m} \qquad R_2 = (287.4)(0.75)/2.0 = 107.8 \text{ kN/m}$$

Again, the weight of the soil above the toe has been disregarded because of possible erosion; this is conservative.

The shear force at the face of the stem is

$$V_u = 142.2 + 107.8 = 250 \text{ kN/m} \qquad \text{or} \qquad V_n = 250/0.85 = 294.1 \text{ kN/m}$$

and the bending moment is

$$M_u = (142.2)(0.5) + (107.8)(0.25) = 98.1 \text{ kN} \cdot \text{m/m}$$

For shear, the allowable force on the concrete is $V_c = \sqrt{f'_c}\, bd/6$, with minimum effective depth, $d = d_{min}$, corresponding to $V_c = V_n$; thus,

$$d_{min} = \frac{6V_n}{\sqrt{f'_c}\, b} = \frac{6(294.1 \times 10^3)}{\sqrt{20}\,(1000)} = 395 \text{ mm}$$

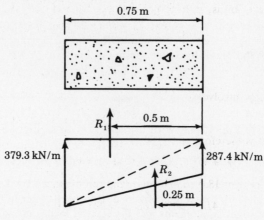

Fig. 15-47

Now, the assumed total depth was 0.6 m, or 600 mm, for the footing. The clear cover over the reinforcement must be 70 mm. Assuming No. 20M bars with a diameter of 19.5 mm, the effective depth furnished is

$$d = 600.0 - 70.0 - 19.5/2.0 = 520.25 \text{ mm}$$

which is greater than the value 395 mm required for shear.

Corresponding to $d = 520.25$ mm,

$$\phi R_u = \frac{M_u}{bd^2} = \frac{98.1 \times 10^6}{(1000)(520.25)^2} = 0.3624$$

But, in Table 5.3, this answers to a ρ-value smaller than $\rho_{min} = 0.0035$. Hence, minimum reinforcement will be used: $\rho = 0.0035$, $\phi R_u = 1.208$, and

$$d = \sqrt{\frac{M_u}{(\phi R_u)b}} = \sqrt{\frac{98.1 \times 10^6}{(1.208)(1000)}} = 285.00 \text{ mm}$$

Because this is smaller than the required minimum effective depth for shear (395 mm), use the latter as the final effective depth. Required clear cover is 70 mm, so that with No. 20M bars ($d_b = 19.5$ mm), the total depth is

$$h = 395 + 70.0 + (19.5/2) = 475 \text{ mm} < 600 \text{ mm assumed} \qquad \text{(O.K.)}$$

The steel area required is $A_s = \rho bd = (0.0035)(1000.0)(395) = 1383$ mm^2/m, and since each No. 20M bar has an area of 300 mm^2, the number required per meter is $1383/300 = 4.61$, or 5. The spacing between bars is then $1000/5 = 200$ mm.

Section 7.6.5 of the 1983 ACI Code states that primary flexural reinforcement in slabs shall not exceed 3 times the slab thickness nor 450 mm. This condition is satisfied in the above design.

Nothing in the 1983 ACI Code directly specifies minimum reinforcement in the perpendicular direction for wall footings. However, many designers interpret the "shrinkage reinforcement" requirements of Section 9.12.2 of the 1983 ACI Code to apply to footings. Despite some arguments to the contrary, it is recommended that the minimum transverse steel should be $0.0020bh$ when Grade 300 deformed bars are used and $0.0018bh$ when Grade 400 bars are used. In this case,

$$A_s = (0.0018)(1000)(475) = 855 \text{ mm}^2/\text{m}$$

using No. 20M bars, and the number required per meter is $855/300 = 2.85$. Using 3 bars per meter, the spacing is $1000/3 = 333$ mm. This satisfies Section 7.12.3 of the 1983 ACI Code which states that the maximum spacing of shrinkage (and temperature) reinforcement is $5h$ ($5 \times 475 = 2375$ mm), but not more than 500 mm.

15.23. For Problem 15.22 (and related previous problems) design the heel of the footing. Again use $f_c' = 20$ MPa and $f_y = 400$ MPa, and refer to Fig. 15-47.

The weight of the soil is, from Problem 15.16, $W_1 = 216$ kN/m. The weight of the concrete slab is

$$W_9 = (2.25)(0.3)(1.0)(23.5) = 15.86 \text{ kN/m}$$

The surcharge force is

$$W_s = (8.0)(2.25)(1.0) = 18.00 \text{ kN/m}$$

and the footing weight involved is

$$W_{10} = (0.6)(2.25)(23.5) = 31.73 \text{ kN/m}$$

The total dead load is therefore

$$W_1 + W_9 + W_{10} = 216.00 + 15.86 + 31.73 = 263.59 \text{ kN/m}$$

Since the live load is $W_s = 18.00$ kN/m, the factored load is

$$F = 1.4D + 1.7L = (1.4)(263.59) + (1.7)(18) = 399.63 \text{ kN/m}$$

The cover over the vertical steel in the stem is 50 mm and the diameter of a No. 20M bar is 19.5 mm; so the distance from the center of the bar to the face of the concrete is $50 + 19.5/2 = 59.75$ mm (use 60 mm).

The lever arm of the force F from the center of the reinforcing bars in the stem is

$$x = 2.25 + 0.06 = 2.31 \text{ m}$$

so that the moment is $Fx = (399.63)(2.31) = 923.15$ kN·m/m. The design is made for $\frac{2}{3}$ of the shear and moments, so the nominal design shear force and ultimate moment are

$$V_n = (\tfrac{2}{3})(399.63)/0.85 = 313.4 \text{ kN/m} \qquad M_u = (\tfrac{2}{3})(923.15) = 615.4 \text{ kN·m/m}$$

Assuming No. 20M bars with 70-mm cover, the effective depth is $d = 520.25$ mm (cf. Problem 15.22), and the allowable shear force on the concrete is

$$V_c = \tfrac{1}{6}\sqrt{f_c'}\, bd = \tfrac{1}{6}\sqrt{20}\,(1000)(520.25)/1000 = 387.8 \text{ kN/m}$$

which exceeds 313.4 kN/m   (O.K.).

From $M_u = \phi R_u bd^2$,

$$\phi R_u = \frac{M_u}{bd^2} = \frac{615.4 \times 10^6}{(1000)(520.25)^2} = 2.273$$

and so, by interpolation in Table 5.3, $\rho = 0.0098$. Consequently,

$$A_s = \rho bd = (0.0098)(1000)(520.25) = 5098 \text{ mm}^2$$

As each No. 20M bar has area 300 mm^2, the number of bars required per meter of width is $5098/300 = 16.99$, or 17 bars. Their center-to-center spacing is $1000/17 = 58.8$ mm.

Supplementary Problems

15.24. Solve Problem 15.1 if $\phi = 28$ degrees and the soil weighs 90 lb/ft^3.   *Ans.* $M_o = 55{,}000$ ft-lb/ft

15.25. Solve Problem 15.3 if the soil weighs 90 lb/ft^3 and the stem has a constant thickness of 15". The length of toe is unchanged.   *Ans.* $M_r = 111{,}300$ ft-lb/ft

15.26. Determine the safety factor against overturning for Problem 15.24.   *Ans.*   SF = 2.02

15.27. Solve Problem 15.5 if a live load of 250 lb/ft^2 is placed over the slab.   *Ans.*   SF = 1.60

15.28. Determine the design moment for the base of the stem of the wall in Fig. 15-29 for $\phi = 35$ degrees and $w = 110$ lb/ft^3.   *Ans.* $M = 39{,}700$ ft-lb/ft

15.29. Solve Problem 15.8 if the pressure at A is 2600 lb/ft^2.   *Ans.* $M_E = 7460$ ft-lb/ft

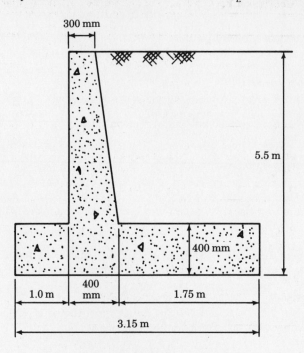

300 mm

5.5 m

400 mm

1.0 m   400 mm   1.75 m

3.15 m

Fig. 15-48

15.30. Investigate completely the retaining wall shown in Fig. 15-48. Check wall stability and check shear in the stem, toe, and heel. Determine required steel areas in the stem, toe, and heel. Consider No. 25M bars, $f'_c = 20$ MPa, $f_y = 400$ MPa. Use approximate $\rho = 0.01$ with strength design. Soil weight is 16 kN/m^3 and concrete weight is 24 kN/m^3. Use a 20-kPa surcharge as live load, with $k_a = 0.3$ and sliding friction coefficient $f = 0.5$.
Partial Ans.   (a) stem $A_s = 3050$ mm^2; (b) stem $V_n = 165$ kN/m; (c) toe $V_n = 225$ kN/m; (d) toe $A_s = 1100$ mm^2

Appendix

To convert from	to	multiply by
Length		
inch (in)	millimeter (mm)	25.4
inch (in)	meter (m)	0.0254
foot (ft)	meter (m)	0.3048
yard (yd)	meter (m)	0.9144
Area		
square foot (sq. ft or ft^2)	square meter (m^2)	0.09290
square inch (sq. in or in^2)	square millimeter (mm^2)	645.2
square yard (sq. yd)	square meter (m^2)	0.8361
Volume		
cubic inch (cu. in or in^3)	cubic millimeter (mm^3)	16387.1
cubic foot (cu. ft or ft^3)	cubic meter (m^3)	0.023832
cubic yard (cu. yd)	cubic meter (m^3)	0.7646
gallon (gal) Canadian	liter (L)	4.546
gallon (gal) Canadian	cubic meter (m^3)	0.004546
gallon (gal) U.S.	liter (L)	3.785
Force, Weight		
kip	newton (N)	4448.2
kip	kilonewton (kN)	4.4482
pound (lb)	newton (N)	4.4482
Force per Unit Length		
pound per foot (lb/ft)	newton per meter (N/m)	14.5939
kip per foot (kip/ft)	kilonewton per meter (kN/m)	4.5939
Force per Unit Area		
pound per square foot (lb/ft^2)	newton per square meter (N/m^2)	47.8803
Stress, Modulus of Elasticity, Pressure		
*pound per square inch (psi)	megapascal (MPa)	0.006895
pound per square foot (lb/ft^2)	pascal (Pa)	47.8803
kip per square inch (kip/in^2)	megapascal (MPa)	6.895
Moment		
foot kip (ft-kip)	kilonewton meter (kN·m)	1.3558
foot pound (ft-lb)	newton meter (kN·m)	1.3558
Mass, Density		
pound (lb)	kilogram (kg)	0.4536
pound per foot (lb/ft)	kilogram per meter (kg/m)	1.488
pound per square foot (lb/ft^2)	kilogram per square meter (kg/m^2)	4.882
pound per cubic foot (lb/ft^3)	kilogram per cubic meter (kg/m^3)	16.02

* The formula $Q = 1.0\sqrt{f'_c}$, in which both f'_c and Q are in psi, is equivalent to the formula

$$Q = 0.083036\sqrt{f'_c}$$

with f'_c and Q now in MPa.

A-2. Properties of Reinforcing Bars

ASTM STANDARD REINFORCING BARS

BAR SIZE DESIGNATION	WEIGHT POUNDS PER FOOT	NOMINAL DIMENSIONS – ROUND SECTIONS		
		DIAMETER INCHES	CROSS SECTIONAL AREA · SQ. INCHES	PERIMETER INCHES
#3	.376	.375	.11	1.178
#4	.668	.500	.20	1.571
#5	1.043	.625	.31	1.963
#6	1.502	.750	.44	2.356
#7	2.044	.875	.60	2.749
#8	2.670	1.000	.79	3.142
#9	3.400	1.128	1.00	3.544
#10	4.303	1.270	1.27	3.990
#11	5.313	1.410	1.56	4.430
#14	7.65	1.693	2.25	5.32
#18	13.60	2.257	4.00	7.09

METRIC REINFORCING BARS

BAR DESIG-NATION	MASS WEIGHT kg/m	NOMINAL DIMENSIONS		
		DIAMETER mm	CROSS SECTIONAL AREA mm²	PERIMETER mm
10M	0.785	11.3	100	35.5
15M	1.570	16.0	200	50.1
20M	2.355	19.5	300	61.3
25M	3.925	25.2	500	79.2
30M	5.495	29.9	700	93.9
35M	7.850	35.7	1000	112.2
45M	11.775	43.7	1500	137.3
55M	19.625	56.4	2500	177.2

The nominal dimensions of a deformed bar are equivalent to those of a plain round bar having the same mass per metre as the deformed bar.

A-3. Densities of Floors, Ceilings, Roofs, and Walls

Material	Density, lb/ft^2	Material	Density, lb/ft^2
CEILINGS		PARTITIONS	
Channel suspended system	1	Clay Tile	
Lathing and plastering	See Partitions	3 in.	17
Acoustical fiber tile	1	4 in.	18
		6 in.	28
FLOORS		8 in.	34
Concrete-Reinforced 1 in.		10 in.	40
Stone	$12\frac{1}{2}$	Gypsum Block	
Slag	$11\frac{1}{2}$	2 in.	$9\frac{1}{2}$
Lightweight	6 to 10	3 in.	$10\frac{1}{2}$
Concrete-Plain 1 in.		4 in.	$12\frac{1}{2}$
Stone	12	5 in.	14
Slag	11	6 in.	$18\frac{1}{2}$
Lightweight	3 to 9	Wood Studs 2 × 4	
Fills-1 in.		12–16 in. o.c.	2
Gypsum	6	Steel partitions	4
Sand	8	Plaster 1 in.	
Cinders	4	Cement	10
Finishes		Gypsum	5
Terrazzo 1 in.	13	Lathing	
Ceramic or Quarry Tile $\frac{3}{4}$ in.	10	Metal	$\frac{1}{2}$
Linoleum $\frac{1}{4}$ in.	1	Gypsum Board $\frac{1}{2}$ in.	2
Mastic $\frac{3}{4}$ in.	9		
Hardwood $\frac{7}{8}$ in.	4	WALLS	
Softwood $\frac{3}{4}$ in.	$2\frac{1}{2}$	Brick	
		4 in.	40
ROOFS		8 in.	80
Copper or tin	1	12 in.	120
3-ply ready roofing	1	Hollow Concrete Block	
3-ply felt and gravel	$5\frac{1}{2}$	(Heavy Aggregate)	
5-ply felt and gravel	6	4 in.	30
Shingles		6 in.	43
Wood	2	8 in.	55
Asphalt	3	$12\frac{1}{2}$ in.	80
Clay tile	9 to 14	Hollow Concrete Block	
Slate $\frac{1}{4}$	10	(Light Aggregate)	
Sheathing		4 in.	21
Wood $\frac{3}{4}$ in.	3	6 in.	30
Gypsum 1 in.	4	8 in.	38
Insulation-1 in.		12 in.	55
Loose	$\frac{1}{2}$	Clay tile	
Poured in place	2	(Load Bearing)	
Rigid	$1\frac{1}{2}$	4 in.	25
		6 in.	30
		8 in.	33
		12 in.	45
		Stone 4 in.	55
		Glass Block 4 in.	19
		Windows, Glass, Frame & Sash	8
		Structural Glass 1 in.	15
		Corrugated Cement	
		Asbestos $\frac{1}{4}$ in.	3

A-3. (*cont.*)

Material	Density, kg/m²
FLOORINGS	
normal-density concrete topping, per 10 mm of thickness	24
semi-low-density concrete (1900 kg/m³) topping, per 10 mm	19
low-density concrete topping, per 10 mm	15
22 mm hardwood floor on sleepers, clipped to concrete without fill	24
40 mm terrazzo floor finish directly on slab	97
40 mm terrazzo floor finish on 25 mm mortar bed	152
25 mm terrazzo finish on 50 mm concrete bed	182
20 mm ceramic or quarry tile on 12 mm mortar bed	82
20 mm ceramic or quarry tile on 25 mm mortar bed	108
8 mm linoleum or asphalt tile directly on concrete	6
8 mm linoleum or asphalt tile on 25 mm mortar bed	60
20 mm mastic floor	46
hardwood flooring, 22 mm thick	19
subflooring (softwood), 20 mm thick	13
asphaltic concrete, 40 mm thick	92
CEILINGS	
12.7 mm gypsum board	10
15.9 mm gypsum board	12
19.0 mm gypsum board directly on concrete	24
20.0 mm plaster directly on concrete	26
20.0 mm plaster on metal lath furring	41
suspended ceilings, *add*	10
acoustical tile	5
acoustical tile on wood furring strips	15
ROOFS	
five-ply felt and gravel (or slag)	32
three-ply felt and gravel (or slag)	27
five-ply felt composition roof, no gravel	20
three-ply felt composition roof, no gravel	15
asphalt strip shingles	15
slate, 8 mm thick	58
gypsum, per 10 mm of thickness	8
insulating concrete, per 10 mm	6

A-3. *(cont.)*

| Material | Density, kg/m^2 | | |
	Unplastered	One side plastered	Both sides plastered
WALLS			
100 mm brick wall	190	214	238
200 mm brick wall	384	408	432
300 mm brick wall	570	594	618
100 mm hollow normal-density concrete block	140	164	188
150 mm hollow normal-density concrete block	170	194	218
200 mm hollow normal-density concrete block	215	239	263
250 mm hollow normal-density concrete block	255	279	303
300 mm hollow normal-density concrete block	300	324	348
100 mm hollow low-density block or tile	110	134	158
150 mm hollow low-density block or tile	130	154	178
200 mm hollow low-density block or tile	165	189	213
250 mm hollow low-density block or tile	195	219	243
300 mm hollow low-density block or tile	230	254	278
100 mm brick, 100 mm hollow normal-density block backing	330	354	378
100 mm brick, 200 mm hollow normal-density block backing	405	429	453
100 mm brick, 300 mm hollow normal-density block backing	490	514	538
100 mm brick, 100 mm hollow low-density block backing	300	324	348
100 mm brick, 200 mm hollow low-density block backing	355	379	403
100 mm brick, 300 mm hollow low-density block backing	420	444	468
Windows, glass, frame and sash			39
100 mm stone			264
Steel or wood studs, lath, 20 mm plaster			88
Steel or wood studs, lath, 15.9 mm gypsum board each side			29
Steel or wood studs, 2 layers 12.7 mm gypsum board each side			45

A-4.   Minimum-Design Uniformly Distributed Live Loads, Impact Included*

Occupancy or Use	Loads, lb/ft^2 (kN/m^2)
Auditoriums	
Fixed seats	60 (2.87)
Movable seats	100 (4.79)
Garages, for passenger cars	50 (2.39)
Hospitals	
Operating rooms, laboratories and service areas	60 (2.87)
Patients' rooms, wards and personnel areas	40 (1.92)
Libraries	
Reading rooms	60 (2.87)
Stack areas—books and shelving 65 lb/ft^3 (1040 kg/m^3)	150 (7.18)
Lobbies, first floor	100 (4.79)
Manufacturing	125–250 (6–12)
Office buildings	
Corridors above first floor	80 (3.83)
Files	125 (6.00)
Offices	50 (2.39)
Residential	
Apartments and hotel guest rooms	40 (1.92)
Attics, uninhabitable	20 (0.96)
Corridors (multifamily and hotels)	80 (3.83)
One- and two-family	40 (1.92)
Retail stores	
Basement and first floor	100 (4.79)
Upper floors	75 (3.59)
Schools	
Classrooms	40 (1.92)
Corridors	80 (3.83)
Toilet areas	40 (1.92)

* See local building code for permitted reductions for large loaded areas.

A-5. Pressures, lb/ft^2 (kPa), for Winds with 50-Year Recurrence Interval

Height Zone, ft (m) above curb	EXPOSURES								
	110-mph (180-km/h) basic wind speed[a] (Coastal areas, N.W. and S.E. United States)			90-mph (145-km/h) basic wind speed[a] (Northern and central United States)			80-mph (130-km/h) basic wind speed[a] (Other parts of the United States)		
	A^b	B^c	C^d	A^b	B^c	C^d	A^b	B^c	C^d
0–50 (0–15)	20 (0.96)	40 (1.92)	65 (3.11)	15 (0.72)	25 (1.20)	40 (1.92)	15 (0.72)	20 (0.96)	35 (1.68)
50–100 (15–30)	30 (1.44)	50 (2.39)	75 (3.59)	20 (0.96)	35 (1.68)	50 (2.39)	15 (0.72)	25 (1.20)	40 (1.92)
100–300 (30–90)	40 (1.92)	65 (3.11)	85 (4.07)	25 (1.20)	45 (2.15)	60 (2.87)	20 (0.96)	35 (1.68)	45 (2.15)
300–600 (90–180)	65 (3.11)	85 (4.07)	105 (5.03)	40 (1.92)	55 (2.63)	70 (3.35)	35 (1.68)	45 (2.15)	55 (2.63)
over 600 (180)	85 (4.07)	100 (4.79)	120 (5.75)	60 (2.87)	70 (3.35)	80 (3.84)	45 (2.15)	55 (2.63)	65 (3.11)

[a] At 30-ft (9-m) height above ground surface.
[b] Centers of large cities and rough, hilly terrain.
[c] Suburban regions, wooded areas and rolling ground.
[d] Flat, open country of coast, and grassland.

A-6. Roof Design Loads, lb/ft^2 (kN/m^2), for Snow Depth with 50-Year Recurrence Interval

Regions (U.S.A.) (other than mountainous)[a]	Roof angle with horizontal, deg (rad)			
	0–30 (0–0.5)	40 (0.7)	50 (0.9)	60 (1.0)
Southern states	5 (0.24)	5 (0.24)	5 (0.24)	0
Central and northwestern states	10 (0.48)	10 (0.48)	5 (0.24)	5 (0.24)
Middle Atlantic states	30 (1.44)	25 (1.20)	20 (0.96)	10 (0.48)
Northern states	50 (2.39)	40 (1.92)	30 (1.44)	15 (0.72)

[a] For mountainous regions, snow loads should be based on analysis of local climate and topography.

A-7. Geometrical Properties of Sections

Dash-and-dot lines are drawn through centers of gravity

A = area of section I = moment of inertia r = radius of gyration

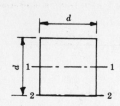

$$A = d^2$$
$$I_1 = \frac{d^4}{12}$$
$$I_2 = \frac{d^4}{3}$$
$$r_1 = 0.2887d$$
$$r_2 = 0.5774d$$

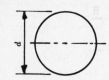

$$A = \frac{\pi d^2}{4} = 0.7854d^2$$
$$I = \frac{\pi d^4}{64} = 0.0491d^4$$
$$r = \frac{d}{4}$$

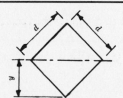

$$A = d^2$$
$$y = 0.7071d$$
$$I = \frac{d^4}{12}$$
$$r = 0.2887d$$

$$A = 0.8660d^2$$
$$I = 0.060d^4$$
$$r = 0.264d$$

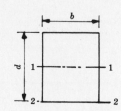

$$A = bd$$
$$I_1 = \frac{bd^3}{12}$$
$$I_2 = \frac{bd^3}{3}$$
$$r_1 = 0.2887d$$
$$r_2 = 0.5774d$$

$$A = 0.8284d^2$$
$$I = 0.055d^4$$
$$r = 0.257d$$

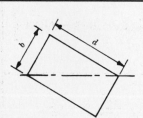

$$A = bd$$
$$y = \frac{bd}{\sqrt{b^2 + d^2}}$$
$$I = \frac{b^3d^3}{6(b^2 + d^2)}$$
$$r = \frac{bd}{\sqrt{6(b^2 + d^2)}}$$

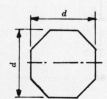

$$A = bd - ac$$
$$I = \frac{bd^3 - ac^3}{12}$$
$$r = \sqrt{\frac{bd^3 - ac^3}{12(bd - ac)}}$$

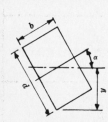

$$A = bd$$
$$y = \frac{b \sin \alpha + d \cos \alpha}{2}$$
$$I = \frac{bd(b^2 \sin^2 \alpha + d^2 \cos^2 \alpha)}{12}$$
$$r = \sqrt{\frac{b^2 \sin^2 \alpha + d^2 \cos^2 \alpha}{12}}$$

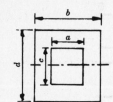

$$A = \pi(d^2 - d_1^2)/4$$
$$= 0.7854(d^2 - d_1^2)$$
$$I = \pi(d^4 - d_1^4)/64$$
$$= 0.0491(d^4 - d_1^4)$$
$$r = \frac{1}{4}\sqrt{d^2 + d_1^2}$$

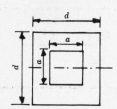

$$A = d^2 - a^2$$
$$I = \frac{d^4 - a^4}{12}$$
$$r = \sqrt{\frac{d^2 + a^2}{12}}$$

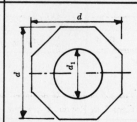

$$A = 0.8284d^2 - 0.7854d_1^2$$
$$= 0.7854(1.055d^2 - d_1^2)$$
$$I = 0.055d^4 - 0.0491d_1^4$$
$$= 0.0491(1.12d^4 - d_1^4)$$
$$r = 0.257d - 0.25d_1$$
$$= 0.25(1.028d - d_1)$$

A-8. Fixed-End Moments*

Loading	Moment at A	Moment at B
IF $a = b = \ell/2$ M_A Max. $a = \ell/3$	$M_A = \dfrac{Pab^2}{\ell^2}$ $M_A = \dfrac{P\ell}{8}$ $M_A = \dfrac{4P\ell}{27}$	$M_B = \dfrac{Pba^2}{\ell^2}$ $M_B = \dfrac{P\ell}{8}$
	$M_A = \dfrac{Pa}{\ell}$	$M_B = \dfrac{Pa}{\ell}$
n Loads equally spaced $a = \ell/(x+1)$	$M_A = \dfrac{Pa}{12} n(n+2)$	$M_B = \dfrac{Pa}{12} n(n+2)$
	$M_A = \dfrac{w}{12\ell^2}\left[(\ell-a)^3(\ell+3a) - b^3(4\ell-3b)\right]$	$M_B = \dfrac{w}{12\ell^2}\left[(\ell-b)^3(\ell+3b) - a^3(4\ell-3a)\right]$
	$M_A = \dfrac{wa^2}{12}\left(6 - 8\dfrac{a}{\ell} + 3\dfrac{a^2}{\ell^2}\right)$	$M_B = \dfrac{wa^3}{12\ell}\left(4 - 3\dfrac{a}{\ell}\right)$
	$M_A = \dfrac{w\ell^2}{12}$	$M_B = \dfrac{w\ell^2}{12}$
	$M_A = \dfrac{w_0\ell^2}{30}$	$M_B = \dfrac{w_0\ell^2}{20}$
IF $a = \ell/2$	$M_A = M_0\left(-1 + 4\dfrac{a}{\ell} - 3\dfrac{a^2}{\ell^2}\right)$ $M_A = \dfrac{M_0}{4}$	$M_B = M_0\dfrac{a}{\ell}\left(2 - 3\dfrac{a}{\ell}\right)$ $M_B = \dfrac{M_0}{4}$

* Reproduced with permission of Canadian Portland Cement Association.

A-9. Reactions, Moments, and Displacements in Beams Supported at Both Ends (Reproduced by permission of United States Steel Corporation)

UNIFORM LOAD PARTIALLY DISTRIBUTED

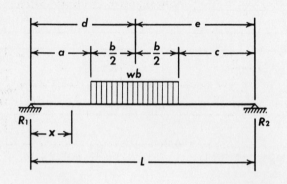

UNIFORMLY DISTRIBUTED LOAD

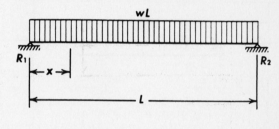

$$R_1 = \frac{wbe}{L}$$

$$R_2 = \frac{wbd}{L}$$

at $x = a + \dfrac{R_1}{w}$

$$M_{\text{max. positive}} = R_1 \left(a + \frac{R_1}{2w} \right) = \frac{w}{2L^2} (2abeL + b^2e^2)$$

when $0 < x < a$

$$M_x = R_1 x - \frac{wbex}{L}$$

$$\triangle_x = \frac{wbex}{24EIL} (b^2 + 4e^2 - 4L^2 + 4x^2)$$

when $a < x < (a+b)$

$$M_x = R_1 x - \frac{w}{2} (x-a)^2$$

$$\triangle_x = \frac{w}{24EIL} [bex(b^2 + 4e^2 - 4L^2 + 4x^2) - L(x-a)^4]$$

$$R_1 = R_2 = \frac{wL}{2}$$

at $x = \dfrac{L}{2}$

$$M_{\text{max. positive}} = \frac{wL^2}{8}$$

$$\triangle_{\text{max.}} = -\frac{5wL^4}{384EI}$$

when $0 < x < L$

$$M_x = \frac{wx}{2} (L - x)$$

$$\triangle_x = \frac{wx}{24EI} (2Lx^2 - L^3 - x^3)$$

A-9. *(cont.)*

CONCENTRATED LOAD AT ANY POINT	CONCENTRATED LOAD AT CENTER

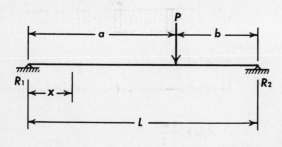

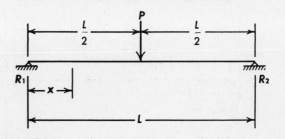

$$R_1 = \frac{Pb}{L}$$

$$R_2 = \frac{Pa}{L}$$

at $x = a$

$$M_{\text{max. positive}} = \frac{Pab}{L}$$

$$\triangle_a = -\frac{Pa^2b^2}{3EIL}$$

at $x = \sqrt{\dfrac{L^2 - b^2}{3}}$ when $a > b$

$$\triangle_{\text{max.}} = -\frac{Pb}{3EIL}\left(\frac{L^2 - b^2}{3}\right)^{\frac{3}{2}}$$

when $0 < x < a$

$$M_x = \frac{Pbx}{L}$$

$$\triangle_x = \frac{Pbx}{6EIL}(b^2 - L^2 + x^2)$$

$$R_1 = R_2 = \frac{P}{2}$$

at $x = \dfrac{L}{2}$

$$M_{\text{max. positive}} = \frac{PL}{4}$$

$$\triangle_{\text{max.}} = -\frac{PL^3}{48EI}$$

when $0 < x\dfrac{L}{2}$

$$M_x = \frac{Px}{2}$$

$$\triangle_x = \frac{Px}{48EI}(4x^2 - 3L^2)$$

A-9. *(cont.)*

**UNIFORMLY VARYING LOAD
PARTIALLY DISTRIBUTED**

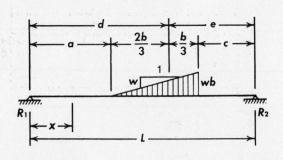

$$R_1 = \frac{wb^2e}{2L}$$

$$R_2 = \frac{wb^2}{2} - R_1 = \frac{wb^2d}{2L}$$

at $x = a + b\sqrt{\dfrac{e}{L}}$

$$M_{\text{max. pos.}} = \frac{wb^2e}{6L}\left(3a + 2b\sqrt{\frac{e}{L}}\right)$$

when $0 < x < a$

$$M_x = R_1 x = \frac{wb^2ex}{2L}$$

$$\triangle_x = \frac{wb^2x}{3240EIL}[45b^2c + 17b^3 - 270de(L+e) + 270ex^2]$$

when $a < x < (a+b)$

$$M_x = \frac{w}{6L}[3b^2ex - L(x-a)^3]$$

$$\triangle_x = \frac{w}{3240EIL}\left\{b^2x[45b^2c + 17b^3 - 270de(L+e) + 270ex^2]\right.$$
$$\left. - 27L(x-a)^5\right\}$$

when $(a+b) < x < L$

$$M_x = \frac{wb^2d}{2L}(L-x)$$

$$\triangle_x = \frac{wb^2}{3240EI}\left\{\frac{x}{L}[17b^3 - 45b^2(L-c) - 270e(L^2-e^2)\right.$$
$$\left. + 270ex^2] + b^2(45a + 28b) - 270(x-d)^3\right\}$$

UNIFORMLY VARYING LOAD

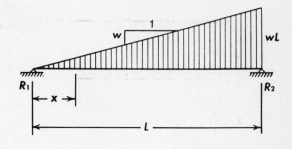

$$R_1 = \frac{wL^2}{6}$$

$$R_2 = \frac{wL^2}{3}$$

at $x = \dfrac{L}{\sqrt{3}} = 0.5774L$

$$M_{\text{max. positive}} = \frac{wL^3}{9\sqrt{3}} = 0.06415wL^3$$

at $x = L\sqrt{1 - \sqrt{\dfrac{8}{15}}} = 0.5193L$

$$\triangle_{\text{max.}} = -0.006522\frac{wL^5}{EI}$$

when $0 < x < L$

$$M_x = \frac{wx}{6}(L^2 - x^2)$$

$$\triangle_x = -\frac{wx}{360EI}(7L^4 - 10L^2x^2 + 3x^4)$$

A-9. *(cont.)*

MOMENT APPLIED AT ONE END   **MOMENT APPLIED AT ANY POINT**

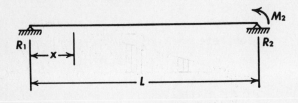   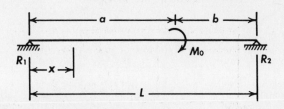

$R_1 = \dfrac{M_2}{L}$ $R_1 = -\dfrac{M_0}{L}$

$R_2 = -R_1 = -\dfrac{M_2}{L}$ $R_2 = \dfrac{M_0}{L}$

$M_2 = M_{\text{max. positive}}$ at $x = a_{\text{(left)}}$

at $x = \dfrac{L}{\sqrt{3}} = 0.5774L$ $\quad M_{\text{max. negative}} = R_1 a$

                                                 at $x = a_{\text{(right)}}$

$\triangle_{\text{max.}} = -\dfrac{M_2 L^2}{9\sqrt{3}EI} = -0.06415\dfrac{M_2 L^2}{EI}$ $\quad M_{\text{max. positive}} = R_1 a + M_0$

when $0 < x < L$ at $x = \sqrt{\dfrac{L^2 - 3b^2}{3}}$ if $a > 0.4226L$

$\quad M_x = \dfrac{M_2 x}{L}$ $\quad \triangle_{\text{max. positive}} = \dfrac{M_0}{3EIL}\left(\dfrac{L^2 - 3b^2}{3}\right)^{\frac{3}{2}}$

$\quad \triangle_x = \dfrac{M_2 x}{6EIL}(x^2 - L^2)$ at $x = L - \sqrt{\dfrac{L^2 - 3a^2}{3}}$ if $a < 0.5774L$

                                                 $\quad \triangle_{\text{max. negative}} = -\dfrac{M_0}{3EIL}\left(\dfrac{L^2 - 3a^2}{3}\right)^{\frac{3}{2}}$

                                                 when $0 < x < a$

                                                 $\quad M_x = R_1 x$

                                                 $\quad \triangle_x = -\dfrac{M_0 x}{6EIL}(3b^2 - L^2 + x^2)$

                                                 when $a < x < L$

                                                 $\quad M_x = R_1 x + M_0$

                                                 $\quad \triangle_x = \dfrac{M_0(L - x)}{6EIL}(3a^2 - 2Lx + x^2)$

A-10. **Section Properties for Shear Stress Computations (Courtesy of Portland Cement Association, Skokie, Illinois)**

INTERIOR COLUMN

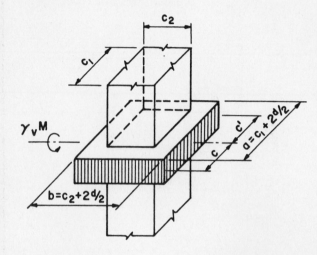

Concrete area of critical section:

$$A_c = 2(a + b)d$$

Modulus of critical section:

$$\frac{J}{c} = \frac{J}{c'} = [ad(a + 3b) + d^3]/3$$

where

$$c = c' = a/2$$

CORNER COLUMN

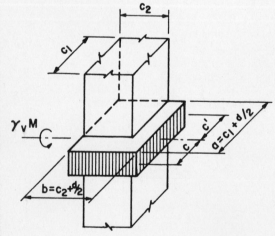

Concrete area of critical section:

$$A_c = (a + b)d$$

Modulus of critical section:

$$\frac{J}{c} = [ad(a + 4b) + d^3(a + b)/a]/6$$

$$\frac{J}{c'} = [a^2d(a + 4b) + d^3(a + b)]/6(a + 2b)$$

where

$$c = a^2/2(a + b)$$
$$c' = a(a + 2b)/2(a + b)$$

A-10. *(cont.)*

EDGE COLUMN (Bending Parallel to edge)

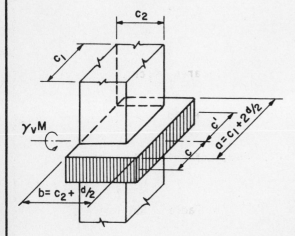

Concrete area of critical section:

$$A_c = (a + 2b)d$$

Modulus of critical section:

$$\frac{J}{c} = \frac{J}{c'} = [ad(a + 6b) + d^3]/6$$

where

$$c = c' = a/2$$

EDGE COLUMN (Bending perpendicular to edge)

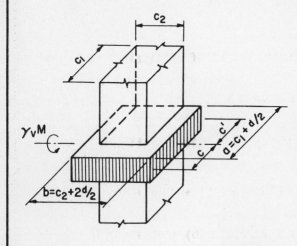

Concrete area of critical section:

$$A_c = (2a + b)d$$

Modulus of critical section:

$$\frac{J}{c} = [2ad(a + 2b) + d^3(2a + b)/a]/6$$

$$\frac{J}{c'} = [2a^2d(a + 2b) + d^3(2a + b)]/6(a + b)$$

where

$$c = a^2/2(a + b)$$
$$c' = a(a + b)/(2a + b)$$

Index

Catalog

If you are interested in a list of SCHAUM'S
OUTLINE SERIES send your name
and address, requesting your free catalog, to:

SCHAUM'S OUTLINE SERIES, Dept. C
McGRAW-HILL BOOK COMPANY
1221 Avenue of Americas
New York, N.Y. 10020